Instruments, Travel and Science

Nowadays it is a matter of mere common sense that science is to a large degree an instrumental activity, producing numbers, measurements and graphs by means of sophisticated devices. We take it for granted that scientific data are valid inasmuch as they can travel beyond the space of a laboratory, be reproduced elsewhere or collated with other data, produced in similar conditions. This book investigates the historical development of the underlying relationship between instruments, travel and natural knowledge that gave rise to modern science. The contributors trace the displacement of instruments across the globe, the spread of practices of precision, and the circulation and appropriation of skills and knowledge.

Through comparative and contextual approaches, the volume confronts the tension between the local and the global, examining the process of the universalisation of science. Bringing together case studies ranging from the seventeenth to the twentieth centuries, contributors discuss French, German and British initiatives, as well as the knowledge and techniques of travellers in countries such as India, Africa, Southeast Asia and the Americas.

Marie-Noëlle Bourguet is Professor of History at the University of Paris 7 – Denis Diderot.
Christian Licoppe is Head of the social science research laboratory at France Télécom R&D.
H. Otto Sibum is Director of an Independent Research Group 'Experimental History of Science' at the Max-Planck-Institute for the History of Science, Berlin.

Routledge Studies in the History of Science, Technology and Medicine

Edited by John Krige, Georgia Institute of Technology, USA

Routledge Studies in the History of Science, Technology and Medicine aims to stimulate research in the field, concentrating on the twentieth century. It seeks to contribute to our understanding of science, technology and medicine as they are embedded in society, exploring the links between the subjects on the one hand and the cultural, economic, political and institutional contexts of their genesis and development on the other. Within this framework, and while not favouring any particular methodological approach, the series welcomes studies which examine relations between science technology, medicine and society in new ways, e.g. the social construction of technologies, large technical systems.

1 **Technological Change**
Methods and themes in the history of technology
Edited by Robert Fox

2 **Technology Transfer out of Germany after 1945**
Edited by Matthias Judt and Burghard Ciesla

3 **Entomology, Ecology and Agriculture**
The making of scientific careers in North America, 1885–1985
Paolo Palladino

4 **The Historiography of Contemporary Science and Technology**
Edited by Thomas Söderquist

5 **Science and Spectacle**
The work of Jodrell bank in post-war British culture
Jon Agar

6 **Molecularizing Biology and Medicine**
New practices and alliances 1910s–1970s
Edited by Soraya de Chadarevian and Harmke Kamminga

7 **Cold War, Hot Science**
Applied research in Britain's defence laboratories 1945–1990
Edited by Robert Bud and Philip Gammett

8 **Planning Armageddon**
Britain, the United States and the Command of Western Nuclear Forces 1945–1964
Stephen Twigge and Len Scott

9 **Cultures of Control**
Edited by Miriam R. Levin

10 **Science, Cold War and the American State**
Lloyd V. Berkner and the balance of professional ideals
Alan A. Needell

11 **Reconsidering Sputnik**
Forty years since the soviet satellite
Edited by Roger D. Launius

12 **Crossing Boundaries, Building Bridges**
Comparing the history of women engineers 1870s–1990s
Edited by Annie Canel, Ruth Oldenziel and Karin Zachmann

13 **Changing Images in Mathematics**
From the French Revolution to the new millennium
Edited by Umberto Bottazzini and Amy Dahan Dalmedico

14 **Heredity and Infection**
The history of disease transmission
Edited by Jean-Paul Gaudillière and Ilana Löwy

15 **The Analogue Alternative**
The electric analogue computer in Britain and the USA 1930–1975
James S. Small

16 **Instruments, Travel and Science**
Itineraries of precision from the seventeenth to the twentieth century
Edited by Marie-Noëlle Bourguet, Christian Licoppe and H. Otto Sibum

17 **Birth of a Scourge**
The anti-cancer war in France: 1890–1940
Patrice Pinell

Instruments, Travel and Science

Itineraries of precision from the seventeenth to the twentieth century

Edited by Marie-Noëlle Bourguet, Christian Licoppe and H. Otto Sibum

ROUTLEDGE
Taylor & Francis Group
London and New York

First published 2002
by Routledge
2 Park Square, Milton Park, Abingdon, Oxfordshire OX14 4RN

Simultaneously published in the USA and Canada
by Routledge

711 Third Avenue, New York, NY 10017

First issued in paperback 2014

Routledge is an imprint of the Taylor & Francis Group, an informa business

Transferred to Digital Printing 2008

Typeset in 10/12pt Baskerville by Graphicraft Limited, Hong Kong

British Library Cataloguing in Publication Data
A catalogue record for this book is available from the British Library

Library of Congress Cataloging in Publication Data
A catalog record for this book has been requested

ISBN 978-0-415-27295-7 (hbk)
ISBN 978-1-138-00854-0 (pbk)

Contents

List of figures

Notes on contributors

Jim Bennett is Director of the Museum of the History of Science, University of Oxford. He works on the history of scientific instruments, practical mathematics, astronomy and museums. Recent publications include (with Scott Mandelbrote) *The Garden, the Ark, the Tower, the Temple: Biblical Metaphors of Knowledge in Early Modern Europe* and a chapter on 'Shopping for Instruments in Paris and London' in P. Smith and P. Findlen, *Merchants and Marvels: Commerce, Science, and Art in Early Modern Europe.*

Christophe Bonneuil is Researcher in the History of Science at the Centre Alexandre Koyré d'Histoire des Sciences et des Techniques, Paris. His Ph.D. thesis (1997) was entitled 'Mettre en ordre et discipliner les tropiques: les sciences du végétal dans l'empire français', and since 1990 he has written several articles on 'Science, Nature and Colonialism'.

Marie-Noëlle Bourguet is Professor of History at the University of Paris 7 – Denis Diderot. She is the author of *Déchiffrer la France. La statistique départementale à l'époque napoléonienne* (Paris: Éd. des archives contemporaines, 1988) and co-author of *L'invention scientifique de la Méditerranée (Égypte, Morée, Algérie)* (Paris, 1998). Her current work is focused on the history of naturalist travels, sea voyages and scientific expeditions in the early modern period (seventeenth to nineteenth century).

Christian Licoppe is Head of the social science research laboratory at France Telecom R&D, in Paris. He is the author of *La formation de la pratique scientifique. Le discours de l'expérience en France et en Angleterre, 1630–1820* (Paris, 1996). He is currently working on the history and sociology of the uses of information technologies.

Giuliano Pancaldi is Professor of the History of Science at the University of Bologna. He is the author of *Darwin in Italy* (Indiana University Press, 1991), and *Enlightenment and the Battery* (Princeton University Press, 2002).

Kapil Raj is Maître de conférences (Associate Professor) at the École des Hautes Études en Sciences Sociales in Paris, France and is a member of the Centre Alexandre Koyré d'Histoire des Sciences et des Techniques, Paris. He has published extensively on the reception and practice of modern science in India. His current research is focused on the construction of the field sciences (mainly natural history and geography) through intercultural encounter in the colonial context. He is preparing a book on the history of the Survey of India, 1760–1885.

Simon Schaffer is Reader in History and Philosophy of Science at the University of Cambridge. He is the co-author of *Leviathan and the Air Pump: Hobbes, Boyle and the Experimental Life* (Princeton, 1985), and co-editor of *The Sciences in Enlightened Europe* (Chicago, 1999). He has published on the social history of natural philosophy, of astronomy, and of Victorian physics.

H. Otto Sibum is Director of an Independent Research Group 'Experimental History of Science' at the Max-Planck-Institute für Wissenchaftsgeschichte, Berlin. His research covers the history of the physical sciences (seventeenth to twentieth century). Recent publications include: 'Les gestes de la mesure. Joule, les pratiques de la brasserie et la science', *Annales. Histoire, Science Sociale* (1998); 'The Golden Number of the Century. The History of a Scientific Fact', *Quaderni Storici* (2001). Current project: science and non-literary knowledge traditions around 1800.

Richard Staley is Assistant Professor in the History of Science Department at the University of Wisconsin – Madison. He teaches history of modern physics, and his research has explored physics circa 1900 from a number of perspectives. Recent publications include papers on the meteorological background to C.T.R. Wilson's early cloud chamber experiments, and on scientists' use of research histories in the early development of relativity.

David Turnbull is Senior Lecturer in Science Studies at Deakin University, Australia and has recently published *Masons, Tricksters and Cartographers: Comparative Studies in the Sociology of Scientific and Indigenous Knowledge* (Amsterdam, 2000). He is currently working on knowledge and space in differing traditions.

Acknowledgements

The making of this book owed much to travel, thanks to the Max Planck Institute for the History of Science's generous support and persistent encouragement that historians and researchers from various fields and countries meet and exchange their ideas. Several workshops sponsored by the Institute in Berlin and by the Centre de recherche en histoire des sciences et des techniques at La Villette (Paris) enabled the contributors to explore the issues and give shape to the overall argument of the book. Among the scholars who accompanied this project, Lorraine Daston deserves our thanks for her unfailing support and acute comments, and we are grateful to the many members and visitors at the MPI in Berlin, especially at the Experimental History of Science Group, who offered wise remarks as well as linguistic help at several stages of our writing. The task of assembling and editing the book was made possible by the dedication and good humour of Bettina Schütz, through countless electronic exchanges. Finally, our gratitude goes to our fellow authors for their scholarship and for the patience they showed over the time it has taken to bring our project to completion, and to John Krige and Iain Hunt for having given our adventure its editorial shelter at Routledge.

Acknowledgements

1 Introduction

Marie-Noëlle Bourguet, Christian Licoppe and H. Otto Sibum

The engraved frontispiece of Francis Bacon's *Instauratio Magna* (1620) shows two ships, one sailing off the pillars of Hercules at Gibraltar in a boundless ocean, the other, in the foreground, preparing to return from its voyage of exploration freighted with wealth. At a time of western European political and economic expansion, the ships evoke travel, commerce and conquest. Yet, as the biblical text inscribed at the bottom makes clear, the engraving also symbolises the adventures and possibilities of the sciences once freed from the bonds of tradition. ('Multi pertransibunt et augebitur scientia': Many shall go to and fro and knowledge shall be increased [Daniel 12: 4].) However, such a great renewal of the sciences was difficult to achieve, as Francis Bacon tells his readers:

> [Men] are like fatal pillars of Hercules to the sciences; for they are not stirred by the desire or the hope of going further . . . This should not be taken to imply that nothing at all has been achieved in so many centuries, with so much effort . . . but before the ocean could be crossed and the territories of the new world revealed, it was necessary to have a knowledge of the nautical compass as a more reliable and certain guide.

By linking science and voyages in a common endeavour, by making instrumentation their new guide and the ship an emblem for the process of the sciences, the new natural philosophers regarded the relationship between travel, instruments and natural knowledge as crucial for the emergence of the modern sciences and their future development.[1]

When Bacon laid out his programme for a new science, rare things, local singularities and strange phenomena were privileged objects of concern in his culture for learned and curious travellers. Wonders, rarities and curiosities were to be witnessed, gathered, then displayed in cabinets as objects of admiration. Except for the representation of a higher world of regularity ruled by the deity and represented by mathematics and other sciences, explorers offered an image of nature as heterogeneous, singular and variable. Furthermore, the lower world of ordinary experience pertained to the realm of the arts, in which craftsmen used their skills to give it a limited order, despite its obdurate resistance. But from the mid-seventeenth century onwards, European natural philosophers

Figure 1.1 The frontispiece of Francis Bacon's Instauratio Magna of 1620. Gravure by Simon van de Passe. Copyright © The Trustees of the National Library of Scotland, Edinburgh.

increasingly insisted that natural knowledge lay as much in craftsmen's hands as in philosophers' minds. By employing experiments, instruments, machines and material devices, they demonstrated hitherto unrecognised regularities in the mundane world and constructed orderly representations of it.[2] Travellers were instructed to carry instruments on their journeys, such as microscopes, barometers or thermometers, and to use them for making observations and measurements. Meteorological data were symptomatic, for they would allow comparisons across the world: 'There are now small portable thermometers that are quite convenient and proper to that purpose', suggested Charles César Baudelot de Dairval in his instructions to travellers, published in 1688. A new regime of knowledge, grounded on empirical trials and the use of instruments, was thus pursued as a challenge to previous observational practices. This elaboration of natural knowledge flourished throughout the eighteenth century, in natural philosophers' laboratories as well as in travellers' fieldwork on sea voyages, topographical surveys or mountain explorations. In 1799, Alexander von Humboldt could declare to Joseph Banks: 'I cannot stand staying longer in Europe. I want to set off, for I have gathered all the instruments I need for a long voyage.' Data were then about to become the stuff of the sciences and instrumental procedures the path towards scientific achievement.[3]

We are now accustomed to conceive of science as an instrumental activity, producing numbers, measurements and graphs by means of sophisticated devices, whether immured within the space of a laboratory or taken on field expeditions. Scientific data are judged valid to the extent that they can travel beyond the limits of the laboratory or the exotic site, be reproduced elsewhere, and collated with other precise and calibrated data produced by other instruments in similar indoor or outdoor conditions. So, looking back at Bacon's image, we might wonder about the long-term historical process that gave rise to this instrumental culture and made it so pervasive in nineteenth- and twentieth-century sciences as now to seem a matter of mere common sense. Both geographical distance and cultural diversity came to be regarded as obstacles to scientific practices when they would not allow for meaningful comparisons. Instruments, measures and data were meant to travel and provide templates for standardisation and accountability to varied experiences and encounters made in far-flung sites. But how was that ability historically achieved? Was the diversity of local knowledge traditions indeed progressively replaced by this instrumental culture, or did it rather increase diversity as much as achieve commensurability? What role did instruments play in establishing the regularity of the new natural order? These are among the issues this book investigates. It traces the displacement of instruments across the globe, the spread of practices of precision and the circulation and appropriation of skills and knowledge. It thus throws new light on an essential question in the history and philosophy of science: the conditions for the emergence of a universalist conception of nature and of a science that describes it.[4]

This book advances a historiographical approach commonly labelled 'contextual'. Over the last two decades, historians of science engaged in writing contextually have argued against a view of the development of knowledge, and particularly of scientific knowledge, as a unilinear process, and against the notion that the universality of science progressively imposes itself by the sheer force of the uniformity of the laws of nature. Conversely, these historians have described through multiple case studies how the universal dimension of science was, in fact, predicated upon the context of its making and deeply grounded in locality. Adopting this contextual approach to the sciences allows us now to take a further step of confronting anew the tension between the local and the global and the process of the universalisation of science. We will scrutinise in vivid detail the historical process of knowledge's decontextualisation or, to maintain the imagery of travel, its delocalisation and displacement. This process made it possible for science to be uprooted from its local embeddedness and develop as a global and universal undertaking.

Because we take displacement and travel – of things, instruments, data, and people – as our starting point, and because this very process is so important over broad spans of time and space, we have the opportunity to set aside both the usual historical division between early modern and modern periods, and the common focus on western European cultures. This book brings together case studies ranging from the seventeenth to twentieth centuries, and puts forward a comparative perspective. The authors discuss French, German and British initiatives, and move beyond the European sphere to explore the relationship between these enterprises and the knowledge and techniques of travellers in other cultural spaces such as India, Africa, Southeast Asia and the Americas. The rich and complex historical itineraries, which such a combination delineates, testify, we hope, to the fruitfulness of that choice.

Science as travel

In October 1761, John Harrison's timekeeper was embarked aboard the *Deptford* at Portsmouth docks, secured by four locks whose keys were held by different persons. The journey was planned to test the accuracy of his watch on a voyage to Jamaica, so as to measure the longitude at sea by comparing standard time carried by the watch with local time determined from the position of the sun or stars. This travelling experiment involved a complex regime of actions: local time at Portsmouth was measured by a member of the Royal Naval Academy, ships' officers were appointed witnesses of Harrison's son's attempts to take the time, and other competent persons were engaged either to determine local time at arrival or to determine longitude by the eclipses of Jupiter's satellites.[5] This journey, one among many longitude trials, is emblematic of the scientific endeavour and the meaning of travel and displacement within this process. Before some agreement can be made about trials or observations that have taken place at a distance, before various data can be compared, something – whether it be a person, a material device, a textual inscription, a number, or all of them together – has to be put in motion and circulate. There is no science which does not involve some form of displacement.

A variety of patterns of travel, including the stories told in this book, illustrate this point. Sometimes a traveller is on the move in person, with the explicit brief of bringing back data and constructing, by means of tools and calibrated instruments, a common ground of knowledge about different and distant settings or experiences. Such is a naturalist like Ramond de Carbonnières, exploring the Pyrenees at the end of the eighteenth century and taking his barometer to the top of the Pic du Midi, or like Alexander von Humboldt making series of observations and measurements in the Cordillera. Such is an engineer-geographer engaged in mapping Languedoc for the French monarchy or an Indian pundit surveying Tibetan territory for the British imperial administration.[6] There are also situations in which an instrument is destined for calibration. In the late seventeenth century, the hydrostatic balance which Robert Boyle designed was soon employed to assay the density of gold and deal with African traders, as well as to unmask English forgers; the watch built by Harrison in the mid-eighteenth century was to be used to keep track of time and measure the longitude at sea; in the second half of the nineteenth century professional botanists tried to set precise standards for the delimitation and naming of plant species, while physicists and engineers worked hard with their ingeniously mobile instruments and machines to make such parameters as the mechanical equivalent of heat into the 'golden number' of the age, valid everywhere.[7]

There are situations in which an experiment has to be reproduced in a distant place and time, in a way that has to be negotiated, so as to create a degree of consensus. As the case of the voltaic battery shows, the replication of a trial in different settings may sometimes seem easy, as long as rather little agreement is needed about what is actually going on. After Alessandro Volta's first public announcement in Como in 1800, the pile was soon being built and put to work throughout Europe. But was this pile an object for curiosity and entertainment, a chemical device, or an electrical machine? The quality of a place matters much to the status of the fact derived from the experimental work that has been performed there. Some mundane, if puzzling, phenomena such as the higher rate at which warm rather than cold water freezes, the so-called Mpemba effect, can be observed quite freely in disparate places without apparently ever quite achieving the status of a reputable scientific fact.[8] There are, finally, situations where scientists themselves go on journeys, such as the American physicists Henry Rowland and Abraham Michelson, who performed grand tours of European laboratories in the late nineteenth century in order to familiarise themselves with the state-of-the-art in a scientific field, to learn from the field's masters as well as to instruct them.[9] Distinctions between these various situations and types of displacement are of course not clear-cut. The cases which follow weave together most of these features in locally distinctive combinations. Yet in all cases something is set in motion and displaced, some kind of travel is involved and empirical trials are performed in different settings. Each enterprise is launched in the hope of bridging the gap between distant experiences and of telling something about nature, a nature whose uniformity and regularity are at once asserted and scrutinised by the instruments.

As these examples make clear, the process of travel, as we understand it, involves much more than simply moving something from one place to another so that, in the end, it remains the same even if found elsewhere. There is no displacement, whether it be of people, instruments, objects, or data, without some sort of work, material as well as social, which enables their extraction from a given place and context, and insertion in a new place and new set of relations.[10] Think, for instance, of the uprooting of a plant, however delicately, from its nurturing soil and its transportation, either as a living specimen to be acclimatised in a botanical garden or as a dried sample to take its place in a herbal; of the transport of a sophisticated astronomical instrument to the top of a mountain; of the installation of a chronometer on board a ship; or, simply, of scientists departing for a long expedition or to stay in a foreign laboratory. As soon as Michelson, an *habitué* of such laboratories, left Potsdam to try his optical experiment in other places, he was to experience all the idiosyncrasies of the sites he visited as forms of resistance which, in turn, helped him to reshape his own experimental apparatus and progressively make it into a robust measuring and travelling device.[11] Any itinerary is paved with unexpected situations and encounters, which induce necessary reshaping or adjustment of objects as well as of persons.

Islands of knowledge: a rationale for travel

When the English met the Akan on the coast of Guinea, they encountered a people whose gold traders maintained their own complex system of values. When Australian Aboriginals construct maps of their land, their practices and products do not match modern scientific ways of map-making. When Rowland went to Britain, in order to meet James Clerk Maxwell and to discuss his own discovery in relation to Maxwell's new theory of electromagnetism, he was surprised to discover there a local culture of experimental practices and ways of conceptualisation hardly recognised elsewhere in Europe. In Germany the theory of electromagnetism was in turn completely different. These examples, drawn from varying places and times, all tell a similar story: knowledge in the process of its making, is, first of all, situated in a particular site and embedded in a specific context. What makes a context specific is the way it is embodied.

The locally grounded quality of knowledge is first experienced by scientists today, as it was by seventeenth-century naturalists and experimental philosophers, through their embodied engagements with material objects, instruments and apparatus in practical places. By emphasising the embodiment of knowledge, we wish to distance ourselves firmly from any form of an idealism in which knowledge, particularly scientific knowledge, develops continuously and imprints itself on the minds of those who are exposed to it by the sheer force of the truths it claims to unravel. But we also wish to avoid the pitfalls of material determinism. The use of things in practice is neither random nor determined, nor is the knowledge that comes from it. Following a tradition that conceives cognition as an embodied activity, we therefore consider bodies, devices and instruments as bristling with enabling aids and means which guide the way we grasp them without predetermining our gestures. They are embedded in procedures, routines and knowledge, which guide without fully prescribing the way we use them. We deliberately break away from a static concept of bodiless intellectual knowledge towards a conception of knowledge that involves a dynamic process of emergence and development. From the perspective of the embodied character of knowledge, its place, especially the actions performed there, is inseparably linked with the knowledge created.[12]

Instead of providing a general discussion of the historical embodiment of natural knowledge, we underline here its specific implications for our understanding of instruments, especially the integration of instrumentation with human performance and the establishment through such performances of trustworthy accounts of remote phenomena. Instruments are not to be solely identified with wood, metal, or glass machines. A human body whose walking pace and perceptual skills have been trained and disciplined is also functioning as an instrument. For his first attempt to climb Mont Blanc, Horace-Bénédict de Saussure minutely calibrated his pace to keep track of the distance covered. To map the distantly inaccessible reaches of Tibet, the British had to train pundits to walk in a regular gait that could be measured and recorded. Furthermore, the distinction between corporeal and material, machine-like, devices is not relevant here. It would reassert an artificial dualism of things and actions, of instruments and gestures. These distinctions need to be explored through their history, not taken for granted. Instruments' meanings are constituted by human performances; instruments and body techniques work together. Regimes of mutual trust and sociability help secure the complex entanglement of performance and instrument. In controversies, however, the bodily performance is often analytically split from the instrument, then called into question. There were good reasons to insist, though, in certain situations, on the integration of savants' tools and bodies. Experimental natural philosophers sought to establish themselves as a special entity, combining their own bodies with their experimental tools to make themselves a unique organ for the experience of nature. The tense distinction between bodies and instruments was a constant and fluctuating problem, still present in the late eighteenth and early nineteenth centuries. The romantic attitude of Goethe or Ramond, which proclaimed the authenticity of gazing at Nature through one's own senses, was to be contrasted with the figure of the enlightened or Victorian naturalist-traveller, overburdened with meteorological, botanical and geological instruments and tools. A protagonist of self-experimentation in fields such as chemistry and galvanism, Humboldt himself dreamt of a harmonious integration of his senses and his instruments, his perceptions and his measures, in a continuum of experience.[13] Moreover, this tension culminated in the nineteenth-century enthusiasm for self-registering instruments as the ultimate means of achieving objectivity.[14]

The relation between the place of knowledge and the complex of instruments and human performances matters for the politics of testimony too. Any new knowledge originates in a specific place and is linked to particular experiences, then has to be shared and communicated in order to be accepted as a proper fact on which a new science could be grounded. Hence arises the puzzle of reliable testimony of remote events and their performative production. In systems of embodied knowledge trust relationships must necessarily bridge across temporal and spatial gaps. John Locke's graphic parable of the King of Siam's incredulity when told of icy rivers by the Dutch ambassador neatly encapsulates these troubles of travellers' tales.[15] It has thus been elaborated extensively by successive philosophers and in the recent history and sociology of the sciences.

Information about the exotic or unfamiliar might indeed be obtained by immediate visitation and personal engagement in the investigation of nature. It might, on the other hand, be obtained indirectly, if direct access to the relevant place were impossible and reliance on correspondence or travel stories thus necessary. The relevance of accounts has to be judged, adjusted and negotiated in order to make sense of knowledge claims about any phenomena at such distant settings.

Instruments of precision

This book illustrates the ways instruments of precision have become a privileged means of bridging the gap between distant heterogeneous places. We need a workable account of the meaning of 'instruments of precision'.[16] Precision – a word which came to be systematically linked to scientific practice in the second half of the eighteenth century, after some sporadic occurrences in the earlier period – is importantly involved in the commensurability of work at separated sites. Precision devices allow for the travel of data by imposing forms of equivalence and modes of comparison between them. The precondition is the instruments' successful displacement in unusual, foreign or strange places: whereas a traveller is expected to be modified by his encounters and surroundings, instruments are not; they are supposed to go on recording faithfully the properties of the environment they are embedded in. The repertoires of these instruments may be diverse. Instruments considered in this book were, among other tasks, used to refine the currency or position boats, describe the weather or map peaks, classify exotic plants or manage industrial production. Between the seventeenth and nineteenth centuries the legitimacy of creating comparative repertoires for these various domains was constructed or grounded through the use of such instruments of precision: a hydrostatic balance to measure gold density; a barometer plus a hypsometric formula to determine mountain heights; an optical apparatus for spectroscopic standards. Besides the common range of measurement apparatus, the notion of instruments of precision has to be extended beyond the strictly quantitative. For example, plant taxonomy and the practices that go with it can induce forms of qualitative precision which are very much to the point in this analysis. In the great herbaria of colonial botany, the numbering and labelling of specimens involved a host of disciplinary instrumental practices on whose integrity the good order of knowledge relied. By investigating these devices of precision, this book seeks to account for the social, cultural and material history of the investments which have been necessary for the apparent decontextualisation of local knowledges into global sciences.

The following chapters treat two kinds of investments employing precision instruments: mapping and experiment. There is an intimate relation between these two enterprises. Both depend on the ordering and co-ordination of places of performance and dialogue, of accounting and recounting. For many centuries the building of a map was based on the compilation and exegesis of ancient geographic descriptions and travel accounts. However, since the seventeenth century, the emphasis of geographic practices shifted gradually from the exegetic tradition to the building of maps based on astronomical observations and triangulation measures. Whereas previous maps relied on textual knowledge stored in libraries, their authority would now rest on a collection of figures and tables, produced by astronomers, topographers and their instruments. The eighteenth century is emblematic of this change. In France regulated map-making then became an especially important enterprise in forging a putatively unified and centralised state. Reliance on travellers' tales and local reports set up an important tension in the methods to be used for charting such positions. Entrepreneurial geographers relied on sifting narratives and printed charts, not direct surveys, then used their maps to display administratively important networks, such as places for the collection of the salt tax, postal routes, or parliamentary administration. In contrast, astronomer-geometers such as the Cassini clan were rather more impressed by the possibilities offered by reliable and precision instrumentation in liberating surveyors from a supposedly humiliating dependence on local informants and story-tellers. It was imagined that the construction of a geometrical grid would provide a sure and powerful means of defining position and territory throughout the kingdom. The apparent substitution of travellers' tales by instrumental data marks an important cultural shift in the investments which underlay the very possibility of delocalising or decontextualising knowledge, going from an emphasis on the exegesis and comparison of textual accounts towards a commitment to the use, calibration and comparability of geodetic and physical instruments. For part of his life Humboldt shifted between a personal engagement in scientific travel, whether in America, Asia or Italy, and the constitution of a network of travellers and local observers, who sent him samples and data from all over the world. He believed in the construction of a grid of local facts which he could turn into a global map of nature. Local set-ups, whether in colonial Indian surveys or African botanical gardens, simultaneously involved the intimate negotiation of situated facts about territorial positions or plant species, and the apparently general mapping of vast geographical or botanical spaces. Thus, the local is dependent also on the global, and the one makes sense with respect to the representation of the other.

The second major set of investments in decontextualising and contextualising knowledge which characterises this period concerns practices of experimentation, in the accounts of which the need to assert repeatability in varied locations is both pervasive and prominent. If, for example, it proves impossible to gain direct access to an appropriate site, the empirical facts can be established at a distance by replication of the experiments performed at that site. But getting experimental set-ups to make sense with one another across space and time is a complex accomplishment. The same can be said of using a given instrumental set-up away from its usual context of use. The instrument must first be disentangled from the initial contexts of its production or prior uses in order to be embedded into a new setting, while yet retaining enough of its identity to allow for the comparison of its data in both situations. In particular, to enforce modes of commensurability between distant places, objects or phenomena, instruments have first to be calibrated and made comparable under a common standard.

This involves a range of disciplinary practices. For instance, in order to compare thermometric measurements made in different places and gather them in meaningful tables, the eighteenth-century savant René Antoine Ferchault de Réaumur formulated a set of rules which aimed at disciplining the making of the instruments and the building of comparable devices. Meanwhile, he also calibrated the thermometer scale to discipline the use and reading of its graduations. Aware that his programme demanded that the instrument be usable by anybody and everybody, not only by its maker or owner, who could be expected to be attuned to their device's every whim, Réaumur was also anxious to train and discipline the users themselves. He therefore sent instructions to his correspondents, describing the place, time and conditions in which the observations had to be made. He reckoned these rules, both social and material, would ensure that anybody could understand the 'language' of a thermometer built according to his standards. The instrument's quantitative language could thus become the basis for a shared experience of heat and, through a collective and cumulative process, lead to a spatial mapping of the climates throughout the world. To play their part in this endeavour, his thermometers had first to be disentangled from their context of production or from that of personal and singular use, and in principle become commodities that could travel everywhere and be read by anybody.[17]

The longitude debates around Harrison's watch bring out a similar point about disentanglement from initial exigencies. In its arguments against Harrison's claims, the Board of Longitude relied heavily on the fact that, unlike Réaumur's thermometers, Harrison's timekeeper lacked a method which would allow it to become a replicable commodity. In the view of the Board's members, the watch could indeed keep the time perfectly, but as long as it was bound to the unique craftsmanship of its maker it could not be deemed a discovery of a method for the finding of longitude. Only if the watch mechanism could be explicated and produced according to a general principle, and so be separated from Harrison's personal and local skills, could he be construed as a discoverer of an accurate and certain method and win the prize. The chapters in this book which treat late nineteenth-century physics illustrate how dependent the expanding physics community of the industrial epoch was on such commonly agreed-upon measurement standards. At that epoch, Michelson's interference apparatus could only shift from being an almost solitary experimental set-up to become a genuine measuring instrument if its measurements could be replicated to determine spectroscopic standards. His American colleagues hoped that if this could indeed be achieved, then costly, remote and unreliable expeditions to measure the transit of Venus would no longer be necessary. Precision travel on a global scale would be displaced in the regime of astronomical metrology by more secure, laboratory-controlled, experimentation. Similarly, the point of Henry Rowland's ingenious new techniques for checking the errors of his superb optical gratings by watching for spectral 'ghosts' was precisely to make these commodities reliable enough to leave his Baltimore laboratory, then travel back to Europe and in turn generate a standardised repertoire for world-wide spectroscopy.

Instruments and experimental apparatus retain their meaning if, and only if, they can be displaced or replicated. However constructed and negotiated, this quality is an absolute necessity, intrinsic to their very identity as instruments. It can indeed be achieved in quite different ways, but in all cases what is at stake is the construction of a shared experience, a common knowledge, forms of intersubjectivity, trust and consensus. This book therefore concerns itself with the long and complex history of this issue of precision and commensurability, approached through the social and moral as well as material aspects of co-ordination. It is a history in which some kind of globally objective regularity was achieved by using instruments that carried this universalist ethos. It is a history of these instruments' displacements and travels, of the investments that allowed these journeys to provide data that made sense within a global knowledge.

The traveller's encounter: reshaping the social and natural order

We have referred in passing to practices of co-ordination, to negotiation and agreement, to shared experiences. Yet we do not want our description to convey an image of the sciences as a neatly ordered world, nor see knowledge as an unproblematically universal consensus based on communities of practices and shared meanings, nor erase the social and material work that was required in the very making of those agreements. We want to look, rather, at how the travel of precision instruments accompanied a redefinition of categories and representations and the imposition of some form of order on both nature and society. From that perspective, instruments of precision and quantitative practices are to be seen, following Michel Foucault's expression,[18] as 'technologies of power' which simultaneously redefine the perception of the self, the representation of Nature and the principles underlying the social order. This is what makes the itineraries of precision, as we try to retrace them, so valuable to the historian as well as to the historian of science. 'No traveller ever comes back the same.' In that sense, Rowland's journey round European laboratories was as formative as the Grand Tour performed by young British aristocrats in the seventeenth and eighteenth centuries. Indeed, the common saying also applies to travellers using precision instruments. All the travellers evoked in this book encountered new worlds of experience through the graduations of their instruments. We want to show what role they play, as 'travellers of precision', in the global and complex process of contextualisation and decontextualisation of knowledge.

This type of travel was not without some kind of risk and danger. Once again, the late eighteenth and early nineteenth centuries were a crucial period, raising the doubts and worries inherent in the very use of precision instruments in the process of decontextualisation and, more generally, in the whole endeavour of a precise and quantifying science. Investigators, particularly Goethe, then expressed their fear of a loss of authenticity in the aesthetic and emotional experience of Nature, if instruments were to be used in a systematic approach towards the world. However passionate about measurements and calculations, Humboldt himself sometimes feared that a quantified science would drive humanity away from a holistic experience of Nature and cause an impoverishment in one's sense of self.[19] While the process of decontextualisation of knowledge through precision instruments was questioned, the very experience of travel equally threatened the traveller's persona and with it the process of contextualisation. La Condamine told how, during a local festival, the population of Tarqui in the Cordillera performed as French academicians by mimicking their gestures, a telescope constantly turned towards the sky. Torn from the world where they had been produced to be brought to work in another distant one, on such occasions the precision instruments that the travellers carried with them were rather like disengaged objects around which the respective identities and worlds of experience of the participants were at once displayed and reshaped. Of the many ways in which this could happen, a singularly striking one is the case of the pundits of the nineteenth-century Indian survey, whose prayer wheels had been customised with rolls of paper that allowed them to record and measure their daily walks. This transformation of a religious object into a measuring device meant the pundits, being given the multiple guises of pilgrims, spies and topographers, were at risk to perform in inappropriate roles.

Besides affecting the traveller's persona, travel also impinges on the categories of the natural and the social orders. Cassini's project of a map of France, for instance, aimed both at producing an objective and somewhat definitive map and at collecting all the proper topographical names which were to be placed on the map, as well as their correct spelling. Such information could only be obtained in the field from local informants. Mapping and naming were therefore co-produced in fieldwork, which involved a dialogue between the bearers of local customary topographical knowledge and the engineers in charge of global systematic mapping. This process could not work without much mutual involvement; its outcome depended on the course of the dialogue between local informants and the astronomers. For instance, the names which were eventually incorporated into the map may well not have been those most commonly used by local people, especially if reluctant to answer the engineers. However, because of their association with the royal map project, these very names were to gain an unexpected legitimacy, leading to a reshaping of the spatial representation of the province.

The exploration of mountains offers a similar case. Designations and categories defining places or groups were profoundly remodelled when instruments began to be carried to the summit. For instance, once a method for measuring the height of mountains had been agreed upon at the end of the eighteenth century, a mountain's precise height, just as much as its name, became part of its identification. Such was the case with Mont Blanc, which schoolchildren are taught to associate with the figure of 4,807 metres. The categories of social order could be similarly affected and reshaped through experimental trials and encounters in the field. Thus, Saussure's barometric measurements on top of Mont Blanc and the relationship he established between height, the beating of the human pulse and altitude sickness allowed him to rank the members of his expedition according to the height at which each of them began to feel nauseous. Fortunately (and somewhat suspiciously), it so happened that Saussure himself, a gentleman and Genevan natural philosopher, was able to climb without discomfort to a higher altitude than his own servant. Traditional hierarchies were reasserted on this occasion, but on others they could be endangered by new criteria for evaluating people. Measures of human capacity could help subvert the normal topographies of social or cultural identity. Observing that all persons sickened equally above a certain altitude ('we all complained in a similar manner'), Humboldt did not hesitate to fill his expedition's mountaineering team with a mixture of Indians, Creoles and Europeans. The social organisation corresponding to barometric measurements could sometimes transgress and reshape cultural and anthropological boundaries.[20]

In the case of Boyle's hydrostatic balance, we can see how experimental trials and encounters at the periphery, at the crossroads between different cultures and worlds of experience and practice, became milestones in the construction of new hierarchies of colonial power. It was agreed that this balance was able to measure and so compare the density of gold and, moreover, that this density provided a reliable scale to assess the quality of a given gold currency, according to which the value and circulation of all commodities could be regulated. This agreement did not impinge solely on the natural world. It could also shape trade relations with the Akan people whom the English traders encountered in the gulf of Guinea. At the far reaches of his metrological network, Boyle counterposed his instruments to the Akan fetishes. From the seventeenth-century European standpoint, this encounter would oppose, on the one hand, the figure of the trader as an experimenter, whose honesty and trustworthiness lay in plying the trade faithfully under Boyle's experimental discipline, and, on the other hand, the Akan, who would remain cunning and shifty figures, attached to their fetishes and the layers of superstition that allegedly surrounded them. At the other extremity of Boyle's network, such encounters could also occur in London. There too the balance imposed a drastic revision of the social status of some – namely, the forgers. When the Royal Mint and the English judges joined forces to revalue gold currency, those whose practice now appeared as a form of forgery were condemned to die on the gallows. In their efforts to establish commensurability early modern naturalists tried to define reasonably reliable measures while simultaneously mapping distinctly unreliable cultures.

Compare the social and natural milieux of late nineteenth-century physicists such as Michelson and Rowland, who encountered new worlds of experience through precision instruments. They could already draw on well-developed shared standards. But they had to know precisely the margins of error embodied in their instruments and accompanying performances to judge their experience on a case-by-case basis, either as evidence for a new natural effect or else as an error. Links between individual workers such as Michelson and more widely extended networks of instrument-makers and manufacturers were decisive resources in these improvised judgements. So while he initially gave specific details and provenance for his apparatus, later Michelson tried to extend the validity of his results by referring merely to general classes of instruments, distancing his interferometer from some of its local contexts. Bent on constructing a universal and fully decontextualised system of units, Michelson's correspondent Henry Rowland emphasised that the regime of absolute standards employed in his experiments would lead to new and quite unexpected effects. He thus experienced the characteristic dilemma of the modern scientific world as a system of rigidly determined measures and yet of fluid and flexible experimental opportunism. The archaeology of scientific practices such as these demonstrates the dialectic of generalised rules and local improvisations in shaping the natural and the social orders. Local, situated and embodied practices on the one hand and global, universal knowledge on the other are always reshaped, rewoven and redefined with respect to one another. The relation between the local and the global, far from being unidirectional, instead provides the impetus for a dynamic and open-ended process.

Itineraries of precision

The itineraries of precision displayed in this book, which infringe the conventional divisions of early modern and modern periods and of European and extra-European geographies, raise the question of the links between precision and politics. For it is not purely by chance that the book's chronology matches that of the imperial age. It opens with a chapter about the English gold trade in the seventeenth and early eighteenth centuries and continues during the nineteenth and twentieth centuries to follow the development of science in the industrial world, as well as to the limits of colonialism. Nor is it by accident that the disciplines treated in its chapters range so widely between hydrostatics, cartography, botany, navigation, galvanism, light, thermodynamics and electrodynamics. Clearly, the historical interactions between our object of investigation – the links between the sciences and travel – and the economic and political evolution undergone by Western countries and their empires since the seventeenth century are many and various.

The development of European state bureaucracies, capitalist economies and colonial empires needed reliable information and implied a process of administrative rationalisation and a growing globalisation of exchange. These systems found resources in the instrumental and quantifying procedures of precision which allowed for the delocalisation and travel of distant data or experiences. So travelling scientists were often mandated by the state or imperial institutions and had to integrate institutional demands within their own concerns. In England, Newton's metrological activities at the Mint mobilised the interests of the Crown to promote regulation of the gold trade. The Board of Longitude became involved in assessing the achievement of Harrison's watch because finding the longitude at sea was crucial for British naval expansion. In France, at the same period, the astronomer Cassini and his collaborators were advancing astronomical and geodetic mapping of the whole kingdom, the programme of which went back to the alliance of the absolutist regime and academic science at the time of Louis XIV. Because of the overwhelming importance of economic botany in natural resource imperialism, the search for an accurate definition of species by the professionals of European gardens was part of their concern with controlling the work of collectors and field naturalists in the colonies. The mapping of Tibet by the pundits, after their discipline and training by the Scottish military engineer Thomas George Montgomerie, blended geography with the specific character of British colonial rule in nineteenth-century India and answered the exigencies of the Great Game in high Asia. In the projects of late nineteenth-century physicists, precision measurement of the key variables of industrial technology was intimately linked with the emergence of international commercial networks and mass production.

From a broader perspective, in consonance with recent arguments in the historiography of the sciences,[21] this set of cases reveals the long-term, pervasive and intertwined development of an ethos of quantification and precision and of the modern economic and political world. Many of the historical examples presented here illustrate how tightly the structure of centre and periphery which characterises metrological networks may be coupled with the hierarchy and centralisation of social and political systems. The mapping of France by C.F. Cassini de Thuzy is a case in point. The astronomer himself stood at the centre with his Board and the Académie royale des sciences. At the periphery developed a vertical hierarchy, in which the engineers and observers in charge of the triangulation were ranked according to their expertise and equipped with smaller instruments and paid lower wages, as their role in the measurement process was deemed more base.[22] The story of the eighteenth-century map of Languedoc lets us push this issue one step further: the failure of a regional project, funded by local institutions and led by gentlemanly academicians using their leisure time to survey the province and produce a map, suggests that the process of the decontextualisation of knowledge founded on precise measurement matched the royal administration's ambition for hierarchy and centralisation. What is true in this respect of the eighteenth-century French monarchy is true also of India under the Raj or of the vertically integrated firms of late nineteenth-century large-scale industry. Even in the field or in the laboratory, the travelling scientists were part and parcel of this centralising process and had to face the distrust of the local people, to overcome their resistance. Consider, for instance, the difficulties that all eighteenth-century French astronomers and surveyors had to face when they encountered the local peasants and inhabitants during their measuring campaigns, whether it be Cassini in the Auvergne, Bouguer and La Condamine in the Cordillera, or Delambre and Mechain in France and Spain during the measure of the meridian at the time of the Revolution.

But this view, which stresses the links between journeys of precision, the development of centralised projects and the emergence of the bureaucratic ethos, is still too unidirectional. The travel of precision practices may also involve the shaping of political concerns or interests to practitioners' own ends. This is evident, for example, in Réaumur's almost single-handed attempt of the 1730s to initiate a meteorological survey. By distributing his newly invented thermometers to travellers and colonial administrators of his acquaintance, the academician indeed bound his endeavour to the hierarchical and colonial order of the Old Regime. Yet the travel of his instruments implied in turn the retrieval and processing of the data which had been collected. While compiling his correspondents' reports in comparative tables about the degrees of heat in various distant places, within the range of opportunities that the centralised system of the monarchy offered, Réaumur began to push for his own metrological network with himself at the centre.[23]

The travel of instruments does not necessarily entail vertically graded and centralised forms of organisation. On the contrary, it may give way to other types of structures, combining many centres and places of measurement and calculation into a more widespread, horizontally distributed complex. We must speak of different cultures of experiment in early nineteenth-century London and Paris to make sense of the very different meanings the voltaic battery carried there, just as we must note the highly uneven degree to which late nineteenth-century local field collectors followed the instructions of metropolitan botanists, or the strikingly heterogeneous field of standards and conventions in which optical physicists then plied their trade. The disciplinary co-ordination of widely distributed scientific work was never as unquestionably robust nor as steeply graded as a simple model of central political hierarchy might imply. Neither the travels of French provincial naturalists like Ramond de Carbonnières or Giraud-Soulavie, nor those of Alexander von Humboldt, were directly funded by any state institution. Humboldt's attempt at a comprehensive approach to nature by processing multiple series of measurements relied instead on a distributed network of observers, compilers and travellers in the field with whom he corresponded, as well as sending his own data to other savants. He aimed at a global description or, ideally, an all-encompassing map which could be shared by a wide and learned audience. As Humboldt understood it, the true achievement of science would be a grid of data, disembodied from their local materiality and processed in linear and graphic representations, sundered from the historical vicissitudes of their production. Thus he reflected, in terms simultaneously made familiar by Constantin Volney's *Ruines* or Percy Shelley's *Ozymandias*, when he arrived at Yarouqui in the Cordillera. Humboldt there contemplated the ruins of the pyramids which the French academicians Bouguer and La Condamine had built, some sixty years earlier, to testify to the glory of the King of France and their own achievement in measuring the shape of the earth. In Humboldt's view, neither the materiality of an inscription nor the authority of a king, which would both crumble away, were the right grounds on which to found the scientific fame of these two travellers. Exactitude was an immaterial and disembodied achievement. The elusive order of the world, he reckoned, could only be expressed through series of measures and tables of figures.[24]

In 1883 William Thomson (Lord Kelvin) spelt out at its clearest the meaning of this kind of measurement for the creation of scientific knowledge:

> I often say that when you can measure what you are speaking about, and express it in numbers, you know something about it; but when you cannot

> measure it, when you cannot express it in numbers, your knowledge is of a meagre and unsatisfactory kind: it may be the beginning of knowledge, but you have scarcely advanced to the stage of science, whatever the matter may be.

In order to illustrate the philosophy of measurement, the Scottish natural philosopher, engineer and keen yachtsman imagined an ideal traveller on a scientific tour through the universe, cut off from all connection with the earth, applying canny ingenuity and a knowledge of absolute standards to 'make measurements which shall be definitely comparable with those which we now actually make, in our terrestrial workshops and laboratories.'[25] Bacon's Atlantic vessels had by now given way to Glaswegian space travel. In Thomson's utopia every traveller in possession of Nature's true representatives, absolute standards and precision instruments, could in principle rebuild his culture anywhere. The point of his traveller's tale was that full integration in a universally distributed system of disciplinary order would therefore grant every traveller the possibility of complete autonomy. The itineraries of precision mapped in this book reveal the complex dialectic in play in the search for self-reliant independence through reliance on the manifold networks of science, technique and power. Our travellers' encounters disclose in many different ways the seemingly endless struggle between locally contextual entanglements and globally decontextualised extensions. Local contexts never vanish; they may indeed generate resistances, but they remain the source of change in the sciences. This dynamic drives the kind of history of science we propose. The histories recounted in this book matter because every meeting involves a kind of reunion. During each of their journeys, travellers' successive encounters draw their meaning and power from the context of all those previous rendez-vous.

Notes

1 A.D. Burnett, 'The engraved title-page of Bacon's *Instauratio Magna*: an icon and paradigm of science and its wider implication', *The Durham Thomas Harriot Seminar, Occasional Paper* no. 27, Durham, 1998. F. Verulam, 'The Great Renewal. Preface on the state of sciences, that is neither prosperous nor far advanced; and that a quite different way must be opened up for the human intellect than men have known in the past, and new aids devised, so that the mind may exercise its right over nature', in L. Jardine and M. Silverthorne (eds) *Francis Bacon. The New Organon*, Cambridge Texts in the History of Philosophy, Cambridge: Cambridge University Press, 2000, pp. 6–10. For a discussion of the role of the instrument, i.e. the magnetic compass, and its meaning for science, see D. Warner, 'Terrestrial magnetism: for the glory of God and the benefit of mankind', in A. van Helden and T.L. Hankins (eds) 'Instruments', *Osiris*, vol. 9, 1994, 67–84. On the ship as an instrument, see also Richard Sorrensen, 'The ship as a scientific instrument in the eighteenth century', *Osiris*, vol. 11, 1996, 221–36.

2 L. Daston and K. Park, *Wonders and the Order of Nature, 1150–1750*, New York: Zone Books, 1998; P. Findlen, *Possessing Nature: Museums and Collecting in Early Modern Italy*, Berkeley: University of California Press, 1994; J.V. Field, *Renaissance and Revolution: Humanists, Scholars, Craftsmen and Natural Philosophers in Early Modern Europe*, Cambridge

and New York: Cambridge University Press, 1993; H. Vérin, *La Gloire des ingénieurs: L'Intelligence technique du XVIe au XVIIIe siècle*, Paris: Albin Michel, 1993; Thomas L. Hankins and Robert J. Silverman, *Instruments and the Imagination*, Princeton: Princeton University Press, 1995; C. Licoppe, *La formation de la pratique scientifique. Le discours de l'expérience en France et en Angleterre (1630–1820)*, Paris: La Découverte, 1996.

3 Ch. C. Baudelot de Dairval, *Mémoire de quelques observations générales que l'on peut faire pour ne pas voyager inutilement*, Bruxelles: J. Leonard, 1688, p. 6; A. von Humboldt to J. Banks, 15 August 1798, in I. Jahn and F.G. Lange (eds) *Die Jugendbriefe Alexander von Humboldts, 1787–1789*, Berlin: Akademie-Verlag, 1973, p. 637. Within recent scholarship the multiple meanings of the term 'instrument' in the natural philosophy of this period of cultural change have been regarded as an opportunity of investigation. See introductions to A. van Helden and T.L. Hankins, 'Instruments', op. cit., and to T.L. Hankins and R.J. Silverman, *Instruments and the Imagination*, op. cit. For the most extensive exploration of the status of instruments in the seventeenth century, see S. Shapin and S. Schaffer, *Leviathan and the Air-Pump: Hobbes, Boyle, and the Experimental Life*, Princeton: Princeton University Press, 1985. Hankins and Silverman date the introduction of precision measuring instruments in experimental physics in the second half of the eighteenth century. See also M.-N. Bourguet and C. Licoppe, 'Voyages, mesures et instruments: une nouvelle expérience du monde au siècle des lumières', *Annales. Histoire, Sciences Sociales* 52 (5), 1997, pp. 1115–51.

4 The relationship between instruments, precision measurement and the rise of quantifying science has been discussed in several publications with the aim to regard instruments, practices of precision and accuracy as having lives of their own, which are to be looked at from a historical perspective. See especially M.N. Wise (ed.) *The Values of Precision*, Princeton: Princeton University Press, 1994; Peter Galison, 'History, philosophy, and the central metaphor', *Science in Context*, vol. 2 (1988), pp. 197–212, and idem, *Image and Logic: A Material Culture of Microphysics*, Chicago: University of Chicago Press, 1997; T. Frängsmyr, J.L. Heilbron and R.E. Rider (eds) *The Quantifying Spirit in the 18th Century*, Berkeley: University of California Press, 1990; Bourguet and Licoppe, 'Voyages, mesures et instruments', op. cit.; H.O. Sibum, 'Les gestes de la mesure. Joule, les pratiques de la brasserie et la science', *Annales. Histoire, Sciences Sociales*, 53 (4–5), 1998, pp. 745–74.

5 Bennett in this volume.

6 Bourguet, Licoppe, Raj in this volume.

7 Schaffer, Bennett, Bonneuil, Sibum in this volume.

8 Pancaldi, Turnbull in this volume.

9 Staley, Sibum in this volume.

10 A. Appadurai (ed.) *The Social Life of Things. Commodities in Cultural Perspective*, Cambridge: Cambridge University Press, 1986.

11 Staley in this volume.

12 On the issue of historical embodiment of knowledge, see H.O. Sibum, 'Les gestes de la mesure', op. cit., and idem, 'Experimental history of science', in S. Lindqvist (ed.) *Museums of Modern Science*, Nobel Symposium 112, Canton, Mass.: Science History Publications, 2000, pp. 77–86. On the anthropological tradition: M. Mauss, 'Les techniques du corps', *Journal de psychologie normale et pathologique*, vol. 32, 1935, reprinted in *Sociologie et anthropologie*, Paris: PUF, 1985.

13 S.F. Cannon, 'Humboldtian science', in *Science in Culture: The Early Victorian Period*, New York: Science History Publications, 1978, pp. 73–110; M.-N. Bourguet, 'La république des instruments. Voyage, mesure et science de la nature chez Alexandre de Humboldt', in M.C. Hoock-Demarle, É. François and M. Werner (eds) *Marianne–Germania. Deutsch-französischer Kulturtransfer im europäischen Kontext*, Leipzig: Leipziger Universitätsverlag, 1998, pp. 405–36; M.-N. Bourguet and C. Licoppe, 'Voyages, mesures et instruments', op. cit.

14 L. Daston and P. Galison, 'The image of objectivity', *Representations*, 40 (1992), 81–128; R.M. Brain, 'The graphic method: inscription, visualization, and measurement in 19th-century science and culture', Phil. Dissertation, University of California, Los Angeles, 1996; S. de Charadevian, 'Graphical method and discipline: self-recording instruments in 19th-century physiology', *Studies in History and Philosophy of Science* 24, 1993, 267–91.

15 The tale reports an encounter between the King of Siam and the Dutch Ambassador who claims that in Holland water could become so hard as to bear the weight of an elephant. As Steven Shapin expresses it: 'Whenever, and for whatever reasons, those who judge observation-claims cannot be at the place and time where the phenomena are on display, then judgement has to be made "at a distance". The trust relationship is, in that sense, inscribed in space' (S. Shapin, *A Social History of Truth. Civility and Science in Seventeenth-Century England*, Chicago: The University of Chicago Press, 1994, p. 245); see also Turnbull's retelling of this tale in this volume.

16 In the current scholarly literature, the term 'instrument' is used rather broadly and still remains ambiguous: 'Sometimes ambiguity is a virtue, and until we have a better understanding of the role of instruments in natural science, we are better off leaving to the term "scientific instrument" its traditional vagueness' (A. van Helden and T.L. Hankins, 'Instruments', op. cit., p. 5). In a similar way the historical investigation of the culture of precision has to be further developed. Contributors to the most recent book on the subject, *The Values of Precision*, edited by M.N. Wise (supra, note 4), also conclude that the changing meanings of 'precision' still require further investigation. This volume is an attempt to study the role instruments of precision played in the decontextualisation of local knowledge with the aim to establish a universally accepted system of natural knowledge.

17 Licoppe in this volume.

18 Following M. Foucault, *Discipline and Punish: The Birth of the Prison* (translated from the French by A. Sheridan), New York: Pantheon, 1977.

19 On Goethe's criticism of mathematical physics, and his fear that 'number and proportion, in their nakedness, destroy all form, and banish the spirit that informs real perception', see 'Tibia und fibula' (1824) in *Goethes Werke, II. Abt. Naturwissenschaftliche Schriften*, vol. 8, Weimar: H. Böhlau, 1893, p. 219 (quoted in G.A. Wells, *Goethe and the development of science, 1750–1900*, series Science in History 5, Alphen aan den Rijn, Sijthoff & Noordhoff, 1978, p. 100). A. von Humboldt, *Cosmos. Essai d'une description physique du monde*, Paris: Gide, 1847, 1, pp. 19–20.

20 A. von Humboldt, *Reise auf dem Rio Magdalena durch die Anden und Mexico* (ed. M. Fagt), Berlin: Akademie-Verlag, vol. 1 (1986), p. 183.

21 M.N. Wise (ed.) *The Values of Precision*, op. cit.

22 As a comparison, see L. Daston, 'Enlightenment calculations', *Critical Inquiry*, 21 (Autumn 1994), 182–202.

23 See B. Latour, *Science in Action: How to Follow Scientists and Engineers through Society*, Cambridge, Mass.: Harvard University Press, 1987, pp. 215ff.

24 A. von Humboldt, *Reise*, op. cit., pp. 186–9.

25 Sir W. Thomson, 'Electrical units of measurement. (A lecture delivered at the Institution of Civil Engineers on May 3, 1883 . . .)', in Idem, *Popular Lectures and Addresses*, London: Macmillan and Co., 1889, pp. 73–136, p. 73.

2 Golden means

Assay instruments and the geography of precision in the Guinea trade

Simon Schaffer

Since Christ said to the Thief, that he should be with him that day in Paradise, there hath been more search after the place of Paradise than before, not for the pleasures of that place, but for its Neighbourhood, wherein 'tis said There was Gold which was good.

John Pettus, *Fleta Minor, containing Essays on Metallick Words* (1683)[1]

Cast it in the Southern Seas, Put on what Spectacles You please, Your Guinea's but a Guinea still.

Jonathan Swift, *The South Sea* (1720)

Early modern European natural philosophers held that the constancy of divine creation allowed the global translation of their instruments and theories. Then the very success of these translations was supposed to demonstrate the wise order of God's world. This interdependence of natural theology and global mobility was secured by institutions for the accumulation of goods and knowledge, and the assay of instruments and techniques. Such institutions were elements in large-scale commercial and political empires. Extended systems of travel and commerce in material artefacts became the precondition of the apparently world-wide grasp of the new sciences. Fragile relations between curious travellers, traders and investigators relied on these institutions' work.[2] Recent studies of early modern Iberian empires, the Dutch VOC, the Society of Jesus, or the Linnaean naturalists of the Enlightenment have shown the role played by calibrated instrumentation and technique, and accumulated goods and specimens, in keeping these networks working.[3]

English naturalists' capacity to act globally may seem an odd case in this impressive list, because of the traditional image of England's weak polity. Extending attention from the naturalists' world to the fiscal system of cash and goods helps correct this prejudice. After 1688 the kingdom raised huge sums to pay for almost continuous warfare, with an excise network large by contemporary European standards and staffed by an unusually extensive cadre of officers expert in measuring, calculating and accounting. John Brewer shows that 'its key technology was not derived from the arts of war but from the counting-house – slips of paper and slide rules'.[4] Metrological work through the state apparatus accompanied tighter definitions of commodities. Contests around the manufacture and trade in such vital goods as tobacco, sugar or spirits saw struggles between the Customs officers and customary practice.[5] Control of commodity samples in the assayers' rooms was difficult to apply beyond their confines. The chasm between projectors' closets and the created order is the point of monetary satires of the period, like those of Jonathan Swift and his allies.[6] To make calculators' techniques work elsewhere, it was necessary to exercise control over comparatively wide spaces. Sites of metrological work, such as mines, markets, mints and, especially, the slave plantations, were increasingly subjected to strict discipline, often with state force.

Historians have noted resistance to centralised metrology within metropolitan societies. They have also challenged a Eurocentric picture of imperial expansion, whether in the Atlantic, Indian Ocean or the South Pacific. The indispensable role of expert indigenous traders and navigators, and the catastrophes which ensued when these relations collapsed, has been well documented, especially in the long history of slavery, conquest and commerce.[7] Fernand Braudel stressed the dynamism of West African states in the period of Portuguese, Dutch and English activity in Guinea, and pointed especially to these states' possession of 'standards of recognised value', strong political systems and economic vigour.[8] But those zones which seemed to Europeans to lie beyond the grasp of their own disciplinary geography were often judged as irretrievably errant. In the English worldview, this might mean Ireland or West Africa, but it could also mean London's shadier quarters. Contemporary London writers well knew the intimate relation between the troubles of metropolitan and global commerce and natural history. In the 1720s Swift used the exotic travels of Lemuel Gulliver to represent the crises of metropolitan values and the imposition of false coin on his Irish compatriots; Daniel Defoe told stories about stock-jobbers and pirates alongside fantasies of urban crime and West African gold traders.[9] There was a close link between the construction of means of measuring commodities within reliable margins of error and the definition of those marginal sites judged ineluctably erroneous.

One eminent eighteenth-century English politician defined dirt as 'matter out of place'. The aim here is to explore some relations between impurity and geography in the classical age. This was a moment especially concerned with the purification of commodities and the production of counterfeits. By tracing the making, refining and distribution of goods we can illuminate the way in which some places were judged sources of purity, others seen as sites of pollution.[10] This chapter examines some cases where naturalists tried to define reasonably reliable measures and to map distinctly unreliable cultures. The commodity *par excellence* is gold; the instruments are those used to assay precious metal and make it into good coin. The institutions which counted include the systems of coastal trade in West Africa and the Royal Mint. The chapter tracks the metrological concerns of Robert Boyle and Isaac Newton, protagonists of the gold system. The interaction between natural philosophical instruments and the techniques of coinage helps define the geography of reliability and error. The journey starts with Boyle's laboratory trials, trade commissions and useful gossip, reaches the beaches, forts and forests of the Gold Coast of Guinea, then returns to Newtonian London and the metrological regime at the Mint. On the way, both contrasts and symmetries emerge in London and the Gold Coast, especially in the assay of fiscal commodities and the political theology of natural order. A particular concern here is evidence of individual variation in natural and social kinds, the 'species scepticism' adopted by Boyle and his allies in their more nominalist moods. Institutions for accumulation and assaying did vouchsafe the world order, but their work also showed that the world was full of singular individuals hard to dragoon into universal classes. The puzzle of reconciling intrinsic variation and global classification was apparent in natural history and in fiscal systems too. Empires strike back. So, especially around 1700, did the vast and complex networks of informal and formal monetary exchange. Historians write of 'metallurgical creativity', in which alongside and against official minting there were powerful and widely acceptable systems of autonomous coining.[11] The systems of instrumentation and coinage show that any hegemony of early modern metrology always faced the resistance of both metropolitan and exotic cultures.

In 1690, at the end of his life, the Christian virtuoso Robert Boyle published a metrological text, *Medicina hydrostatica*, on the values of items in standard London pharmacopoeias, and a long advertisement for a range of quantitative techniques to reform medical and chemical assays. Boyle, here as elsewhere, was cautious in distinguishing between the workable reliability of human measures (which he accepted) and an underlying mathematical structure of creation (which he tended to deny). Steven Shapin has well described Boyle's claim that the properties of individual members of a putative natural kind might vary intrinsically, and so undercut mathematical dogmatists' assumption that natural kinds were truly uniform.[12] 'There will scarcely be found so great an Uniformity in Qualities and particularly in Specifick Weight among Bodies of the same Kind or Denomination as there is generally presum'd to be.'[13] Boyle held that natural kinds did not exist in absolutely uniform individuals, so that imposed measures were at best a means of provisional, doubtless artificial, classification.

In the same year, 1690, Boyle's friend and executor John Locke published an *Essay concerning Human Understanding*, in which species scepticism was given a formidable defence. One natural kind of peculiar concern in Locke's text was gold, defined through the assay procedures to which it was subject: 'yellowness, great weight, ductility, fusibility, and solubility in *aqua regia* all united in an unknown substratum'.[14] Locke's best example was the 'convenience that made men express several parcels of yellow matter coming from Guinea and Peru under the same name'. It was not 'the true and precise nature of things' but 'the convenience of language and quick despatch' which prompted convenient names for 'true gold, perfect metal'.[15] Gold was of salient importance for Boyle too, and not only because of his long-term interest in philosophical alchemy, a recipe for which he passed to Locke at his death in 1691 and which Locke then discussed with Newton.[16] Alchemical projection certainly demanded expertise in the assay of true gold and its counterfeits.[17] Some argued that since commercial laws specified gold by its outward qualities, anything that shared these qualities would

Figure 2.1 Early modern European assay procedures for gold: in front, hydrometric weighing of gold-bearing silver ore; at rear, an assay furnace and a large flask for parting gold from impurities with acids. From John Pettus, *Fleta Minor* (London, 1683), p. 153, figure 19.

therefore be legal gold. Transmutation would allegedly succeed if such surface properties could be reproduced.[18] To check these enterprises, most mines and mints employed skilled assayers. Their precision methods normally involved recovering putatively pure gold from ore by cupellation (heating with excess lead to absorb the dross), then washing with strong acid to extract silver mixed in the gold sample (see Figure 2.1). Because of his interest in assays, Boyle added to the *Medicina hydrostatica* a long appendix on 'a hydrostatical way of estimating ores'.

He here amplified the results of a project initiated thirty years earlier. In 1669 and 1675 he told the Royal Society of a glass instrument made of a bulb and a graduated stem (later baptised a 'hydrometer'), which could be used to estimate the densities of various liquids, and, counterbalanced against a piece of metal, could be used to estimate the density of the metallic sample[19] (see Figure 2.2).

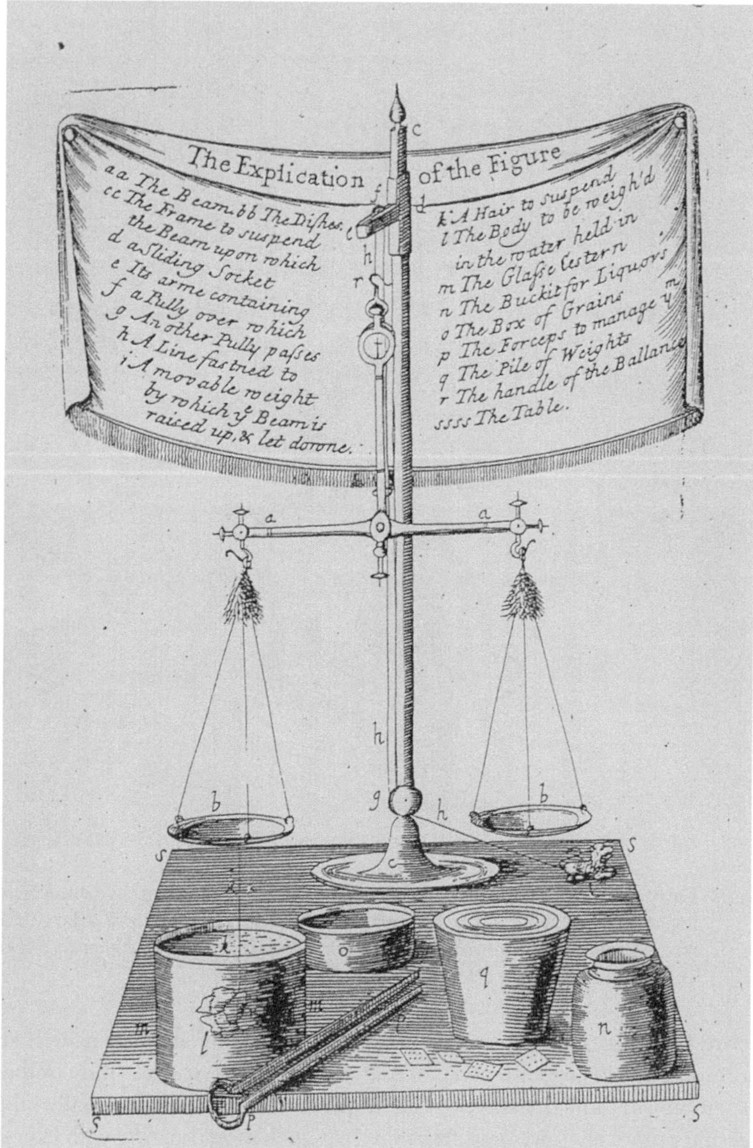

Figure 2.2 Robert Boyle's hydrostatic assay instrument perfected in 1690. From Robert Boyle, *Medicina hydrostatica* (London, 1690), frontispiece.

Pieces of gold, the heaviest metal, would always be of lesser volume than any other samples of the same weight. Hydrometry was not news. From 1681 to 1683 the Royal Society sponsored a new translation of Lazarus Ercker's definitive work on German assays, which long encouraged refinement of the density method 'based on true natural causes'.[20] What Boyle proffered, rather, was a reliable device to render hydrometry applicable more globally.

Boyle's enemies were, for the moment, the coiners who counterfeited gold and the clippers who pared it away. He claimed hydrometry was better than touchstones or acids. Clipped coins would be very much lighter than forged ones, since 'men will scarce venture their lives to steal but three or four grains from a true Guiny, and much less from a false one'. There were many cautions, some due to the intrinsic variability of natural kinds, some to the vagaries of art. 'I have found that Guinys are not all precisely of the same weight nor all waters neither,' Boyle remarked. 'If a falsifier of money have the skill, by washing or otherwise, to take off much of the quantity or substance of the Guiny without altering or impairing the figure or the stamp, the piece of Coin will not be able to depress our Instrument to the usual mark, and thereby make it to be judg'd counterfeit, when 'tis indeed but too light.' So the judgement of forgery was never better than 'probable'.[21]

The authority and use of Boyle's new assay instrument relied on reasonable social judgements about human interest and gullibility. Indeed, Boyle often used the assay of gold coin as an emblem of right judgement:

> I am wont to judge of opinions as of coins: I consider much less, in any one that I am to receive, whose inscription it bears, than what metal it is made of. It is indifferent enough to me whether it was stamped many years or ages since, or came but yesterday from the Mint. Nor do I regard through how many, or how few, hands it has passed for current, provided I know by the touchstone or any sure trial purposely made, whether or no it be genuine, and does or does not deserve to have been current . . . if I find it counterfeit, neither the prince's image or inscription, nor its date (how ancient soever) nor the multitudes of hands through which it has passed unsuspected will engage me to receive it.[22]

The trope of words as coins was a commonplace. Thomas Hobbes used the image tellingly in *Leviathan* to separate right reason from mere authority.[23] Boyle's extended metaphor suggested the hazy basis of value. How should they behave who lacked access to his esoteric metrological technique and were thus conned into accepting coins (or facts) at face value? Boyle so closely studied this matter that he publicly hinted at 'another possible way of counterfeiting Guinys', but prudently forbore to 'teach bad men a skill that probably they will not otherwise acquire'.[24] The west country divine John Beale, a staunch admirer of Boyle's new metrology, confirmed that such 'bad men' were endemically at work. 'Few of us are not sometimes cheated with counterfeit Money, and I think there is scarce a House-keeper which is not abused by false Mettals', he told the Royal

Society. The king's hydrographer should be commissioned to market Boyle's hydrostatic devices.[25] Beyond the known and knowable realm of good order and careful valuation, both epistemic and social principles subverted standards and made measures artificial and provisional. Natural kinds did not necessarily appear uniformly in all instances. Metropolitan coiners and clippers were constantly at work manipulating and changing the measure of exchange.

Boyle extended his analysis of these errant margins of measurement in his 1690 essay on the hydrostatic balance. He again insisted on species scepticism, advising that readers should use his measures with caution, 'neither Nature nor Art being wont to give all the Productions that bear the same name a Mathematical Preciseness either in Gravity or other Qualities', and again commended his hydrostatic balance as the most probable means of distinguishing true gold.[26] His concerns with the social subversion of value broadened to imperial trade. Boyle was instrumental in the new regime's 1689 repeal of the law against alchemical gold-making. He reckoned the repeal would promote mineral exploration. All gold obtained by processes now allowed under law was to be sent to the Mint.[27] Boyle's contacts with Irish projectors and 'a considerable Number of Navigators and other great travellers and with divers Persons that had settled themselves in the Indies' all encouraged great expectations. He had served on the Council for Foreign Plantations, conversed with the English governor of the new colony in Jamaica to confirm the possibility of mines of true gold, witnessed the presentation to Charles II of samples from such diggings, and gossiped with the governor of the English base on the Gold Coast at Cape Coast Castle.[28] Trustworthy informants' status and work helped reinforce a boundary between reasonable certainty and dubious error. The chief trouble which Boyle here identified lay in Guinea, the very source of the gold from which coins took their name.

Gold pieces were first struck from African gold in 1663, with the arms of the Africa Company on their reverse.[29] London street slang baptised them 'gory', neatly combining the name of one of the principal trading stations in Guinea (Gorée) with a reference to bloody laws which sustained their value.[30] More than half a million guineas were made in London from Guinea gold before 1713.[31] Just as in the counterfeit world of London so in West Africa, Boyle reported, traders were in constant danger of being duped. Boyle made notes on the reports of such traders as Richard Jobson, who travelled for a London company in Guinea. He conversed with others such as Jean Barbot, a Huguenot refugee in London from Catholic France who had much experience in the West African gold trade in the early 1680s.[32] Armed with a hydrometer, Boyle claimed, a trader

> may oftentimes prevent that chief Fraud of the Negroes, whereof several Traders to the Golden Coast are not a little apprehensive . . . For they complain, that though the Blacks be otherwise, for the most part, but a dull sort of People; yet they have often made a shift to cheat the Traders, by clandestinely mixing, with the right Sand-Gold, Filings of Copper, or rather of Brass, whose Colour does so resemble that of Gold, that the Fraud is not easily discerned.[33]

Guinea gold dust, Boyle reckoned, should be turned into a standard, then used to calibrate the value of the Guinea trade helped by hydrostatics.

The formation of the Africa Company and determination to engross Dutch power in the Guinea trade in gold and slaves were enterprises of a newly expansionist and ambitiously imperial state. Natural philosophers like Boyle and Locke recognised their connection with the African trade. In 1664 the Royal Society launched a project to draw up inquiries world-wide after hearing their treasurer's 'queries for Guinea', with questions about the gold supply there.[34] Boyle contributed several schedules to this project, including a long set on mines and ores.[35] When a Lutheran pastor from Hamburg who served a Gold Coast trading company in the 1660s reported on the gold trade, his countryman Henry Oldenburg, the Royal Society's secretary, at once adapted this report in queries sent from London to Cape Coast Castle. He wanted to know where Guinea gold came from and how it was assayed.[36] In his *History of the Royal Society* (1667), Thomas Sprat described the new Company of Royal Adventurers, which imported West African gold to make English guineas, as 'the twin-sister of the Royal Society'.[37] There was a good scriptural precedent for the gold trade in Solomon's commerce with Ophir, a fabled source of precious metal whose exact location preoccupied English travel writers. Jobson argued that Ophir was in Guinea. One of Boyle's closest alchemical confidants claimed Ophir lacked gold mines, but that Solomon had his secret laboratories there. Publicists for the new Africa Company reminded Londoners of the Solomonic gold to be found in Guinea.[38] John Ogilby, who orchestrated Charles II's coronation, also presented him with a luxurious account of the Guinea trade. Sprat added that the English monarch's support for both the Africa Company and the Royal Society was comparable with 'the wisest of ancient Kings: who at the same time sent to Ophir for Gold and compos'd a Natural History'.[39]

State-sponsored and commercial natural history mattered to Locke too. He was secretary to the Carolina proprietors in 1668, a member of the Royal African Company founded in 1672, and secretary of the Council of Trade and Plantations the following year. This gave Locke's philosophical views on the intrinsic variation between Guinea and Peruvian gold a commercial point, and involved him in trade in metal and slaves.[40] In the Council for Foreign Plantations, Boyle and his allies among the American merchants sought to regulate trade and religion in the colonies. Even before his own work in Ireland and for the plantations, Boyle contrasted the bearers of proper knowledge and 'those Indian Caciques who in an envy'd affluence of all the World calls goode, are covetous of nothing but counterfeit beads, Glasses, painted whistles, and such other Childish gugaws, that we that know their value, count but trifles'.[41]

The Glorious Revolution of 1688 saw new economic and military attempts in Ireland and overseas, spawning an appalling currency crisis as gold bullion flowed out of Britain and clipping and coining became endemic. Those 'that know value' grew rich. Boyle's *Medicina hydrostatica* was published at the same time as a petition to Parliament from London goldsmiths to stop bullion export and to police minting. Locke produced an ultimately successful series of pamphlets on

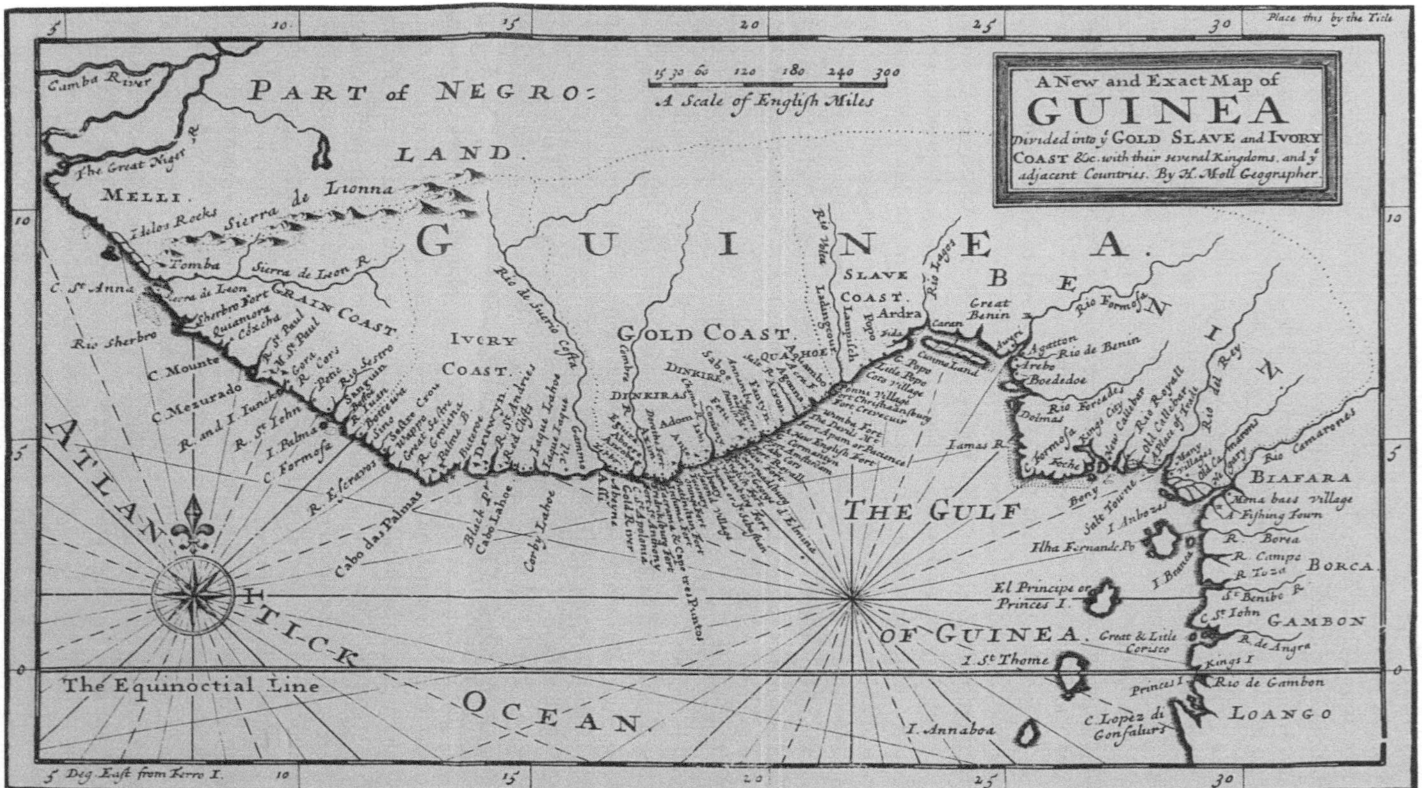

Figure 2.3 The Gold Coast mapped by Hermann Moll in 1705. From Willem Bosman, *A New and Accurate Description of the Coast of Guinea* (London, 1705), frontispiece.

recoinage, the so-called 'social crime' which saw all British silver coin re-issued with a metal content equivalent to its elevated face value.[42] By this act, the government created a situation in which silver could command more payment in gold overseas than in England, thus promoting a huge drain of silver coin out of the country and the effective establishment of gold coin as the standard of the nation's currency. The same year, 1695, Locke was appointed to the powerful position of Commissioner of Trade where he further developed a strikingly mercantilist economic policy towards Ireland and the American colonies.[43] A 1695 Act against Coining also set up a money police of the places in London where 'minting' independently went on, such as Wapping and Whitefriars, soon turned into sanctuaries of criminal resistance beyond the easy reach of state power.[44] High profits from coining and public tolerance made this policing weak. The monetary disciplines of the 1690s showed the importance of the relationship between the metropolitan establishment of regulated values and the marginalisation of errant groups.

Boyle's comments on the errors of coiners, caciques and crafty Guineans illuminate the relation between metropolis and margin. Damned as beyond the moral and the economic pale, these others were often understood as shady masters of the relations of exchange by whom honest traders and naturalists could easily be duped. The hydrometer thus became a moral emblem and an assay tool to aid the fiscal order's relationship with what lay just beyond its boundaries. The fragility of the relation was especially acute in the Guinea trade. During the seventeenth century Guinea, and especially the eponymous Gold Coast which stretched for 250 miles between Axim and the mouth of the Volta river, was a principal source of the precious metal and a scene of competitive commerce between Danes, Portuguese, French, Dutch and English traders, all in complex negotiation with the Akan peoples of the region (see Figure 2.3). It was well known to Europeans that the Akan did not control the origin of all this gold. Long-range internal trade networks, beyond European reach, were crucial: 'they know which goods to buy on the coast for the interior markets, they are at all the distant places where gold occurs'.[45] Gold, in the form of dust or small nuggets, was the local currency. What the Europeans sought as a valuable commodity was instead in Guinea a continuous medium of local exchange. All Akan traders carried individual sets of weights to measure out appropriate amounts of gold dust in each transaction. Akan gold-weights were elaborate series of metal figures, often representing proverbial scenes of animals and humans, alongside scales, spoons, sieves and touchstones. Each weight user would check their weights against others involved in every transaction. Gold was also valued by the Akan for its political use, in making symbols of office and offering the measure of authority. Akan goldsmith guilds worked the metal, made the weighing instruments, and asserted their rights through descent from the ancestor who first descended from the skies to teach gold-working.[46] These weights, whose measures derived from a mixture of Islamic, Portuguese and troy values, thus became the focus of European traders' intense anxiety and of continuous native recalcitrance and ingenuity.[47]

The indigenous middlemen who dominated such dealings were the masters of skilled manual art, weight scales and their handling. A French traveller reported in 1669 that such native traders 'always have gold: if the merchandise has a low price they take gold with the finger, increasing or decreasing the amount as they deem necessary; if the price is high they weigh the gold'.[48] The Guinea trade required a confessedly tenuous system of metrology and credit. One expert Akan broker for the Royal African Company demanded in 1683 that the English give him 'an account of company scales and weights, for he finds himself a loser by taking money with his smaller scales and weights';[49] Henry Oldenburg's German informant reported that some Europeans tried to ban hand-held gold scales, instead hanging them from a large cross-beam. 'In one pan the sub-factor was to place the weight, in the other the black trader was to lay the gold'[50] (see Figure 2.4). This scheme soon failed. Credit and commodities flowed from the

Figure 2.4 Gold-weighing in Guinea as imagined by the Dutch in 1704. From Willem Bosman, *Nauwkeurige Beschryving van de Guinese Goud-Kust* (Utrecht, 1704), frontispiece.

English bases and their ships through the middlemen into the interior, and then back along the gold routes.[51] Akan traders were sometimes damned as incompetent by Europeans unable to command these routes. Pieter de Marees, author of the most influential Dutch account of the Gold Coast, claimed 'they are not very good at counting, especially in arithmetic . . . they become confused and cannot get out of their muddle; instead they sit stammering or mumbling, so that they forget their figure and have to start anew each time'. But he also described the work of intermediary interpreters and brokers, counselled against the practice of credit and an overtrusting treatment of native weights, then pointed out that 'for us it is difficult to weigh with such Scales: one has to be very experienced to know how to use them. If one weighs small quantities of Gold with them, one has to watch very carefully . . . But among themselves they know how to weigh so accurately that there is never a mistake.'[52] Simultaneous images of credulity and cunning were characteristic of the Gold Coast reports. The security and profitability of dealing there relied on definitions of value which graphically highlighted the relative power of metropolitan and marginal measures.

In this gold network relations between Europeans and Africans were mediated by another powerful term, 'fetish'. Initially conceived with medieval church doctrines of witchcraft and the real potency of charms and icons, thus used by the Portuguese who reached Guinea in the later fifteenth century, the term then acquired a novel sense as part of an attribution by Dutch and other commentators to the Africans of a deluded belief in 'quasi-personal powers and material objects capable of being influenced through acts of worship and through manipulations of material substances'.[53] It is plausibly argued that this shift in the sense of the fetish accompanied two decisive developments in European notions of materiality. Matter is impersonal, so superstition is the personification of impersonal natural forces, and 'the truth of material objects' has its reality proved by their 'silent' (and doubtless commercial) 'translatability across alien cultures'.[54] These developments in doctrines and practices of material and marketable value characterise many of the anecdotes to which such authors as Boyle appealed: the deluded sanctification of European technical artefacts; the identification of truly valuable commodities whose current owners did not recognise this value (it was often alleged that Africans worshipped trifles and underestimated the value of goods); and the ever-present danger to Europeans offered by really valueless objects treated with reverence by the indigenous peoples (the admixture or adulteration of true gold with baser metals, the social power of the fetish).

This fetishism was of especial interest to Boyle's hydrometry. Because Guinea gold was shipped as 'sand or gravel', Boyle emphasised that 'in making Estimates of the Genuineness and the degrees of Purity of these native Fragments of Gold, our Hydrostatical way of exploring may be of no small use'.[55] Boyle saw danger in taking gold dust mixed with brass filings. But such mixtures were themselves called fetishes, and the resultant adulterate named 'fetish-gold'.[56] Boyle's London informant Barbot reported that a French officer had recently brought home such dross, gold adulterated with copper filings, even though all knew that 'such filings will shew twice as large as the same weight of gold, being so much the more ponderous'. This was the basis of Boyle's hydrometric strategy against fetish gold.[57] Willem Bosman, Chief Merchant in the 1690s of the Dutch West Indies Company, explained in his definitive *Description of the Coast of Guinea* that

> Fetiches are a sort of Artificial Gold composed of several Ingredients . . . this Artificial Gold is frequently mixed with . . . silver and copper, and consequently less worth, and yet we are pestered with it on all parts of the Coast, and if we refuse to receive it, some Negroes are so unreasonable that they will undeniably take back all their pure Gold: so that we are obliged sometimes to suffer them to shuffle in some of it.[58]

According to Bosman, all such things 'made in honour of their false gods' were thus named. He drew the conclusion that in view of 'their ridiculous ceremonies' surely Roman Catholics would have the best chance of converting such people to Christianity. The Huguenot Barbot made the point explicit. He had experienced Louis XIV's military campaigns against Protestants in France, 'cruel monsters' whom Barbot judged fiercer than any fetishes found in Guinea.[59] Irrational superstition, whether European or Akan, sustained a challenge to the values of these traders.[60]

Many traders combined stories that made Africans seem foolish (so to encourage their readers to commerce) with accounts of indigenous artfulness (thus making themselves the indispensable mediators between European and African cultures) (see Figure 2.5). Dutch merchants retold old if dubious reports of a 'silent trade' in the forested interior, hundreds of miles from the more familiar coastal strip where Europeans worked. There in the forest the Akan 'do their trade in the open, in a forest which is made into a market'.[61] Robert Boyle's source Richard Jobson recounted in some detail how this up-country silent gold trade worked. Different groups would leave gold samples on the ground, then leave, when those wishing to buy the gold would arrive and place their trade goods on the ground. Thus commerce would allegedly take place without language, solely by signs and tacit agreement.[62] Europeans were fascinated by this silent trade because it so well symbolised the puzzles of the gold economy and its meaning in African culture, especially since European traders had never gained access either across the Sahara or from the coast into the gold-rich interior. William Smith, despatched from London to Guinea by the Royal African Company, reported that though no African could follow gold veins they banned European miners from them and had remarkable ability in locating gold dust.[63] Bosman, Smith's source, made a similar point at greater length. After long passages on African irrationality, he told his European readers that 'the Negroe women know the exact value of [fetish-gold] so well at sight that they are never mistaken' and that 'the Negroes are very subtle Artists in the sophisticating of gold'.[64]

This meant traders needed their own skills. Bosman derided the technique of dissolving samples in *aqua fortis*, which Boyle had commended, since it was too precise, and would show the gold sample false even if it had a small impurity.

Figure 2.5 European and African traders in Guinea in the 1660s. From Olfert Dapper, *Naukeurige Beschrijvinge der Afrikaensche Gewesten* (Amsterdam, 1668), p. 354. This work was published by John Ogilby in England in 1670.

'I can assure you that the present times will not admit of such useless niceness.'[65] Alongside the tense relationship between the skills of Africans and their unreason, murderously institutionalised by the rapidly expanding slave trade which by 1700 had begun swiftly to displace the gold business, there was a similar relation between the allegedly general authority of European measures and their extreme local fragility.[66] Europeans themselves must become skilful judges, more devoted to local custom than absolute standards. Bosman equally lauded those standards by which he judged all African cultures. Fetish-gold, it was said, 'must be judged of by the touchstone and the skill of the buyer you employ'.[67] De Marees described how Akan traders, equipped with remarkable series of their own goldweights, just the objects so often derided as fetishes, would

> weigh against the Factor's [weights], and if [the gold] does not reach the full weight they take their [own] weight together with the Scales, weigh it themselves and check whether it agrees with their weight or not . . . They also inspect our weights and yardsticks or measuring-tapes and make markings on them; and when they come back, they inspect these measures and weights again to see whether they have not been changed for heavier weights and longer measures with which one would cheat them.[68]

Here metrological technique made it possible, but hard, to maintain the unequal exchange Europeans wanted. 'They are more cautious about Gold', de Marees said, 'than people would ever think of being in the Netherlands.' In the complex set-up of trader, middleman, interpreter and gold-bringer, errors and cheats were of course endemic.[69] Balance-pans were fixed, or blown into. Europeans often kept heavier weights to buy gold and lighter weights to pay it away again, just as did Akan traders themselves.

Reporters often described the sacred and the base status of instruments in the Guinea trade. Smith gave an anecdote of the Gambia River where his theodolite had been treated as a bewitched fetish by the natives.[70] Such European devices – surveying instruments such as theodolites, assay devices such as hydrometers – themselves became fetishes, merely material objects invested by the English traders with the admittedly fragile power to adjudicate between truth and falsehood. The charge that Africans were easily duped into seeing mundane objects as sacred could be easily laid against Europeans. De Marees was astonished that the Akan 'could not understand' that their goods 'came from God, saying it was not God who gave them the Gold, but rather the earth'.[71] English writers alleged that other cultures 'make this metal their God' yet freely confessed that 'all of us seem this day to be guilty of this Metallick idolatry'.[72] Europeans were the idolaters and their commodity fetishism was obvious. De Marees might fill his pages with accounts of Akan fetishism, and report that 'they are very keen on their Gold', but the Akan traders 'know very well there is no gold in Holland, that it is for its sake that we come here, and that so much diligence is applied to get it, and therefore they say that Gold is our God'.[73]

A troubling symmetry seemed in play in the gold trade. Fetishistic faith in the spiritual power of mere matter was apparent on all sides, and alleged by all participants.[74] Error was noticeable because commensurability seemed temporarily established. There were for example many forms in which 'gold' was obtained: as fetishes, mixed with many other metals, as dust or plate, as ornaments, through different weighing systems. 'No doubt you have been under the impression that this metal has everywhere the same value and purity, but this is an incorrect view you must reject', explained Barbot, Boyle's French informant. 'The gold varies, Sir, not only inasmuch as it is varyingly adulterated by the natives, but also because it naturally differs in the form which it leaves the mines.'[75] 'Species scepticism' about gold was easily warranted. The obviously arbitrary stipulation that there was a pure commodity form of gold, from which all others variously departed, was not the cause but the result of this complex set of techniques and interactions. The difference between pure gold and its adulterates was made into the difference between European reason and indigenous errors and crimes. This distinction – partly reinforced by the hydrometric project – then helped establish the contrast between slavers and slaves, between 'honest traders' and the sly or superstitious. The new metrologies of English natural philosophy – institutionalised by the Newtonians in London – were closely linked with the measures required by the gold and slave trades.[76]

The distinctions viciously constructed in Guinea played their role back in the metropolis. There was a significant coincidence between the English currency stabilisation, a trebling of excise revenue from 1688 to 1714, and the transatlantic trade in gold and slaves. This was a crucial element in the formation of the English maritime zone of commerce and information.[77] The gold coins that flowed into Britain to pay for this trade imbalance became a major part of British currency liquidity. In the Anglo-Dutch capital of London, the new institutions of the 1690s financial regime tried to separate a knowable administered world of measure from the marginal errors beyond the walls of the Bank of England and the Royal Mint.[78] The Mint, where Guinea gold was turned into golden guineas, was a notorious sanctuary where the writ of the police scarcely ran, yet it was the chief metrological site of the Kingdom. Adam Smith famously described its function during these eighteenth-century crises:

> the operations of the Mint were somewhat like the web of Penelope; the work that was done in the day was undone in the night. The Mint was employed not so much in making daily additions to the coin as in replacing the very best part of it which was daily melted down.[79]

Within the Tower of London, the Royal Mint was a self-contained community, with its own pub, two dozen moneyers and an establishment, including its own priest and porters, of about thirty (see Figure 2.6). It was run by a Comptroller as business manager, a Warden, the monarch's representative and chief legal officer, and a Master, who contracted for coin production. During the Restoration the Master was Henry Slingsby, FRS, and Secretary of the Council for Foreign Plantations.[80] Boyle consulted him about the hydrometer project, telling the Royal Society that their fellow-member's 'exactness and diligence allows us to expect that no injury that care and skill can prevent shall be done to the Coin'.[81] Their hydrometric interest included the source of the Mint's income, a customs duty on imported liquor that required good density measures for a range of excised liquids. Slingsby was also a patron of the Huguenot engineer Pierre Blondeau, who during the 1660s mechanised Mint production. Metal bars would henceforth be rolled thin with horse-powered iron turning cylinders. Skilled and strenuous manpower was used to punch out discs, weigh, soften and (a secret process, this) mill the new coin.[82] Because this produced such fine machined coin, and the extant bad coin was not called in, Mint currency was hoarded. In the wartime economy of the later 1680s and 1690s bad money dominated circulation and clipping and forging became endemic.[83]

'Great contentions do daily arise among the King's Subjects in Fairs, Markets, Shops,' reported the Treasury secretary in the 1690s. 'Persons before they conclude in any Bargains are necessitated first to settle the Price of the Value of the very Money they are to Receive for their Goods; and if it be in Guineas at a High Rate or in Bad Moneys they set the Price of their Goods accordingly.' It was claimed in 1696 that 'in most places the people has got such a way of taking money now as was never in use before: that is, they take all by weight. Everyone

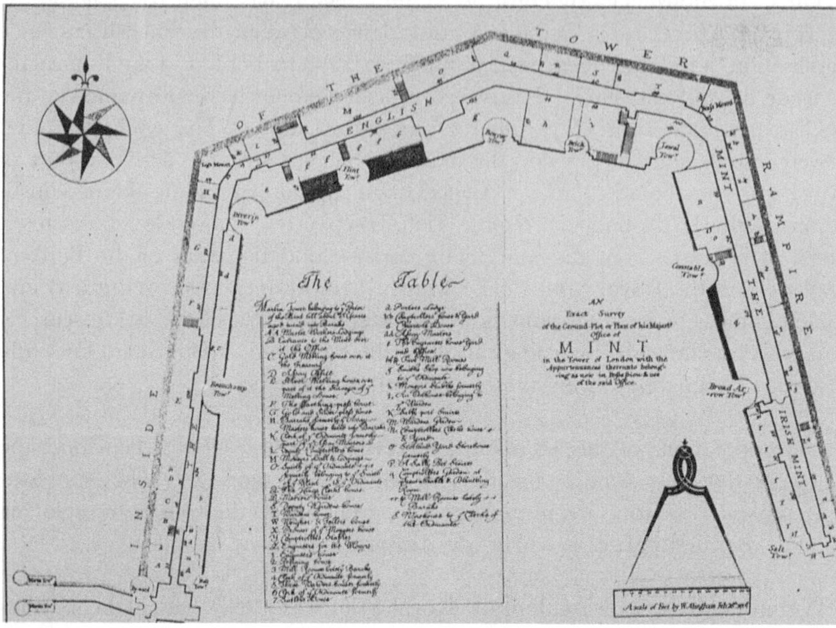

Figure 2.6 The Royal Mint in the Tower of London in 1701. From William Allingham's plan reprinted in *Twenty First Annual Report of the Royal Mint* (London, 1890).

carries a pair of scales in his pocket . . . if it want but two or three grains they refuse it.'[84] Londoners were behaving towards each other, it was feared, like Guinea traders. There were many ways in which Guinea custom was visible in the destabilised metropolis. One Whig journalist penned a 'vision or allegory' of Public Credit, 'seated on a Throne of Gold' like some Guinea prince. Under the threats of anarchy, tyranny and atheism, however, 'the great heaps of Gold on either side the Throne now appeared to be only heaps of Paper'. Only the Protestant succession and the triumph of the Glorious Revolution could make 'the heaps of Paper' change back into 'pyramids of Guineas'.[85] The metropolitan coinage crisis was but one aspect, though fundamental, of the shaky settlement of the new Anglo-Dutch regime after 1688. Expansion of uncensored print after the lapse of state licensing and the plethora of paper money schemes coinciding with the foundation of the Bank of England gave the English the sense that their world was based on nothing but feebly manipulable paper signs.[86] The recoinage run by Locke and Newton was immediately interpreted in terms which bound together the political theology of the Kingdom, the moral condition of mankind, and the instrumentation of forgery and coining.[87] A sermon on the recoinage preached in London in 1697 argued that

> our spiritual state, like that of the coin, is impaired. Our divisions have been to us what the shears and such like instruments have been to the money . . . When

> the coin was debased, they corrupted it with baser metals and placed a counterfeit stamp upon it. This was our ruin at first. By the Fall, the image of God was lost and defaced, and a contrary image stamped upon the soul. There must be a restoring of the King's image again, the debased coin must be broken with the hammer, melted with the fire, and made susceptible of a new stamp.[88]

Such linkages between the economies of signs and of commodities had a political and theological implication. Jacobite nonjurors accused their enemies of belief in a convenient fiction – that the former king had been legitimately replaced by William of Orange through God's will. Conversely, William's supporters accused the nonjurors of idolatry, a Popish faith in outdated signs and fetishes.[89] The London of the 1690s was obsessed with counter-accusations of idolatry and fetishism and the vulnerable basis of value and credit. Counterfeiting was high treason against the monarch and his regime. This was when Boyle issued his *Medicina hydrostatica*, and Locke and his colleagues, such as Isaac Newton, angrily debated the means of restoring the coin's value by a 'return to nature'. Newton wrote in favour of devaluation, against Locke's opinion. His recipe was characteristically administrative, calling for 'quick dispatch in the Mint' and 'a Commission of fit Persons' set 'over the Trade of the City with Power to set Prices upon Wares' to counter inflation. Once the recoinage project began, in 1696, Newton was made Warden by his patron the Whig grandee Charles Montagu, then became Master from 1700.[90]

Newtonian projects were not, of course, limited to fiscal standardisation. Between 1709 and 1713, ably assisted by the young Cambridge mathematician Roger Cotes, Newton prepared a revised edition of his *Principia mathematica*. Some 'hypotheses' which had prefaced its third book in 1687 now became reworked as '*regulae philosophandi*'. The second rule stated that 'the causes assigned to natural effects of the same kind must be, so far as possible, the same. Examples are . . . the falling of stones in Europe or America.'[91] The *Principia*, in this sense, was a handbook for travellers. Newton described the celebrated marvels of tidal ebb and flow in the East Indies, the Straits of Magellan and the Pacific. Keen to show the universal grip of his gravitational model of lunar pull, Newton here faced characteristic troubles of trust in travellers' tales.[92] Against Leibnizian rivals, Newton and Cotes now sought massively to reinforce the apparent precision of their measures.[93] They discussed whether to omit or include tide data from variably reliable mariners using assumptions about such parameters as the earth's density.[94] The link between trust in persons and in creation's constancy was even clearer in their work on the length of isochronic pendulums in Europe, America and in Africa too. In the 1680s Newton had hoped that 'the excess of gravity in these northern places over the gravity at the Equator' would be 'determined exactly by experiments conducted with greater diligence'.[95] Cotes now 'considered how to make that Scholium appear to the best advantage as to the numbers'.[96] They would make a table of the variations in the length of a seconds pendulum at different points on earth, visibly accurate over very small length differences of fractions of an inch. Cotes held 'that the generality of Your Readers must be gratified wth such trifles, upon which they commonly lay ye greatest stress'.[97] To reach such exactness they had to judge the uniformity of the kind of matter of which the earth was made and the uniformity of the standards met by (mainly French) travellers.

Jean Richer's celebrated ten months' work at Cayenne in 1672, when the shortening of the seconds pendulum was first detected, was taken as the standard, even though the French astronomers themselves long doubted whether Richer's data were reliable, and instead supposed seconds pendulums must be the same length everywhere.[98] But Richer's measures let Newton project from local manipulations of a pendulum clock to the shape of the planet: 'this diligence and caution seem to have been lacking in other observers'.[99] One puzzle was the limit which Newton should allow for the difference between pendulum lengths in France and at the Equator. Some numbers fell outside his bound of two and a quarter lines (one line = one-twelfth of an inch). Thus the astronomer Claude-Antoine Couplet's measures during a voyage from France via Portugal to Guiana from 1697–8 were dismissed: 'he is less trustworthy because of the crudity of his observations', as was the Minim mathematician Louis Feuillée in Martinique in 1704.[100] Gérard-Paul des Haye's values from Gorée in Guinea in 1682 and at Cayenne in 1700 were also treated this way.[101] Newton and Cotes agreed that 'the differences between the measurements' of the different Frenchmen 'are nearly imperceptible' – they amounted to fractions of a line – 'and could arise from imperceptible errors in the observations'.[102] A decade later, for the final edition of the book, Newton expanded this useful appeal to imperceptible but nevertheless certain local variability. 'This discrepancy could have arisen partly from errors in observations, partly from the dissimilitude of the internal parts of the earth and from the height of mountains, and partly from the differences in heat of the air.'[103] Incorrigible variations of humans and of Nature were at last used to explain away variations in measures. Then these measures were used to justify a magisterial projection of Newtonian uniformity, to be assayed in its turn by French, Spanish and Swedish surveyors of the earth's figure in the Andes and the Arctic during the 1730s.[104]

Like Boyle's provisional assays of gold, so Newton's judicious estimates of pendulums were projects in which global uniformities were constructed through precision measures and moral judgements. Boyle's hydrostatic balance was an instrument in the gold trade, where assaying was a prerequisite of exchange and true gold was a candidate for the universal monetary standard. The seconds pendulum had been proposed as the basis of a natural standard of length capable of sustaining a newly universal metrology. Newton's concerns with the gold standards were immediate. In the year after 1709, more than 20,000 guineas of African gold were made at his Mint. While revising the *Principia* during spring 1712, Newton was also engaged in tortuous negotiations with the Treasurer about new coinages for Scotland and Ireland, the assay of gold and copper, and the prevention of forgery. His office weighed 30,000 gold pieces to establish their weight: 'its [*sic*] convenient that the coins should bear the same proportion to one another in both kingdoms [Britain and Ireland] for preventing all fraudulent practices'.[105] Making an Atlantic space of reliable pendulum measures was like making an Atlantic space of reliable coin.[106] At the Mint, measured concern with the causes of difference between subtly different values turned into political and economic issues of moment for the kingdom.

Newton's administrative metrology at the Mint accompanied his equally fierce attack on idolatry and his insistence on divine power.[107] A former MP in the Convention Parliament that legitimated William's regime, and an expert in monarchical law, Newton often linked right government with God's dominion. In early 1712, during his work with Cotes on the new *Principia* edition, using his massive research on the scriptural and prophetic texts concerning God's rule, Newton now publicly argued that God was 'Lord of all' (*universorum dominus*) and that He 'ought not to be worshipped under the representation of any corporeal thing'. God's supreme authority, rather than His wise plan, was the ultimate guarantee of the constancy and uniformity of Nature: 'by existing always and everywhere, He constitutes duration and space, eternity and infinity'.[108] Newton's pragmatic rule of philosophising at the start of the book suggested that natural philosophers should assume that stones fell for the same reason in Europe and America; but it was the supreme government of Newton's God, underlined at the book's end, which made this assumption true.[109] The model of right government thus complemented the instrumentation of assays, which stressed the construction of reliable measurement devices to evaluate gold, then sought to instruct traders in the remote and reliable performance of tests. Newton constructed a ferocious regime of governance within the institutions of natural philosophy and the walls of the Tower. In Mint work he insisted on accurate weighings and corrected what he judged an unacceptably large tolerance of error in the average weight of coins, called 'the remedy'. The Newtonian mint became an emblematic site of administrative metrology.[110] In his 1696 Treasury report on the state of the Mint, Newton insisted on the Warden's judicial authority, 'designed to keep . . . Ministers in their Duty to the King and his people', thus neatly making the analogy between the polity of the Mint, the Kingdom and the Creation: 'nor do I see any remedy more proper and more easy then [*sic*] by restoring the ancient constitution'.[111]

Within the Mint, Newton imposed strict hierarchies, especially on the use of tools and instruments. Assaying, for example, was deemed a purely manual trade.[112] The rituals of court and law were used to sustain this new regime. Newton insisted his institution control all coining tools (see Figure 2.7). Outside the Mint's walls, he successfully lobbied to make the possession of instruments for counterfeiting into high treason deserving death.[113] These instruments, which would include blocks and rollers to make metal blanks, and crucibles, files and emery dust to turn blanks into coin, scarcely allowed forgers much mobility – they needed, rather, permanent establishments concealed in the capital's back streets.[114] Clipping – taking edges off gold coin – needed less sophisticated equipment, but in either case Newton's new Act of 1697 dictated execution. He debated whether fly-presses, common in the watch-trade and for stamping buttons, should be strictly licensed to prevent their use by forgers.[115] He noted

Figure 2.7 Eighteenth-century machinery for minting coin. In front, the coining press strikes the die against metal blanks; at rear, an edging machine to prevent fraudulent clipping. From *Universal Magazine of Knowledge and Pleasure*, vol. 7 (London, 1750), p. 69.

one judge's remarks at a trial of a clipper: 'the shears supply the place of one witness, the filings of another, and the rough clipped money of another'.[116] When persons could not be trusted to convict, things were instead forced fetishistically to speak the truth. Newton had himself made a justice to help prosecute coiners. 'These people seldome leave off and its [*sic*] difficult to detect them.'[117] By 1699 he had ten prisoners in Newgate gaol and arranged dozens of Tyburn executions.[118] A Derbyshire surgeon gaoled for debt informed him that in Newgate it was said that if the Stuarts were restored Newton would be shot.[119] 'We were placed in a Garrison,' Newton urged, 'that the Exchange and Treasury of the Nation might be guarded in our custody from all manner of invasion' – French, Stuart, Catholic, criminal or imprecise.[120]

As in Guinea, in London the key term governing Newton's polity was 'credit'. Financial credit grounded the national war effort against Louis XIV, and it was credit too which allowed the security of the coin. Thus Newton complained of widespread metropolitan laxity towards coiners and clippers: 'this vilifying of my agents and witnesses is a reflexion upon me which has gravelled me and must in time impair and wear out and ruin my credit'.[121] Credit, as the often-bankrupt Daniel Defoe explained, meant both the grounds of belief and the grounds of value. In a memoir to the Treasury defending the Mint during the credit crises of 1697 and the wartime financial difficulties, Newton explained how ''tis mere opinion that sets a value upon money; we value it because we can purchase all sorts of commodities'. This had implications for whether money appeared in the form of gold or paper. 'All the difference is that value of gold and silver is set upon their internal substance or matter and is therefore called intrinsic, and that value of paper money upon the apparent form of the writing and therefore called extrinsic, and that the value of the former is more universal than that of the latter.' Such universality and opinion meant that stable value needed, and only needed, regular administration, not necessarily an obsession with the purity of the money form. 'Let it be considered what rate of interest is best for the nation and let there be so much credit (and no more) as brings money down to that rate of interest.'[122] 'Credit' established the distinction between reliable publicly assured values and illusory, easily manipulated, ones. Newton's Mint played its part in this intricate process.

Two exemplary careers, in both of which Newton was a protagonist, illustrate how the process worked. William Challoner was a London artisan who in the early 1690s 'turned coiner and in a short time put on the habit of a gentleman'. By arranging the printing of Jacobite propaganda, then shopping the printers to the government he sought public reward and credit. Then in 1696 he tried to get Newton to hire his associates who would act as informants within the Tower, while at the same time 'he accused that worthy gentleman Isaac Newton esq as conniver at many abuses and cheats there committed'. The Parliament's Committee on Mint Miscarriages heard Challoner's allegations that the Mint was in league with forgers, and that because of the complex division of labour of its staff it was in urgent need of a general supervisor, one just such as Challoner who boasted too of the machines he'd designed to groove coin edges, thus ensuring against counterfeits. Parliament was initially convinced by the projector's schemes. Newton set out to destroy Challoner's credit and thus destroy him. The Mint's warden tried sending Challoner straight to Newgate, but he was quickly released. Then during 1699 Newton took to clandestine meetings in London pubs to gather information against his enemy. Eventually Newton had Challoner back in gaol, where the coiner 'feigned himself mad, running stark naked at midnight about the Ward . . . but now he seems more rationall'. Despite Challoner's remarkable mercy plea to his persecutor ('O God my God I shall be murdered unless you save me') Newton had him hanged at Tyburn in March 1699. The case dramatised the link between the conduct of state values and those of the London hanged, just the matter of Jonathan Swift's satire, and the means through which distinctions of true value were set up – the ceremonies and techniques of administrative law.[123]

Such, too, was the role of Newton's measures in the career of Jonathan Swift's principal target, William Wood. Just as Challoner started coining in London, Wood started in the early 1690s as a Midlands iron dealer. Like Challoner, he soon rose through ingenious manipulation of the public credit system, in this case becoming receiver for the land tax (where he profited from the interest on the money which flowed through his office) and a mechanical projector through an iron and steel marketing scheme.[124] When Newton's Mint called in tenders for a new copper coinage in 1717, Wood's iron company obtained a contract.

Like Challoner, Wood had from the early 1720s close links with the parliamentary and cabinet committees. Schemes for copper coinage prompted Newton to define an assay for pure copper. After many trials he decided that copper which did not crack when red hot would be the standard. 'There is no certain assay or rule yet known,' he told the Treasury in summer 1718, 'by wch the fineness of it may be ascertained . . . There have been blanks made of much finer copper than the money by people not imployed by me, & stamped in the Mint without my knowledge & then polished to give them a more beautifull gloss & shewn about to deceive people & bring the money into discredit.' Assaying hinged on the administration of credit. 'The goodness of copper cannot be known by the looks alone. It must be assayed . . . some judgement may be made by a skilful person: but the surest trial is by the malleability of the copper hot or cold.'[125] Newton understood that in copper coin the difference between face and intrinsic value should be maximum. With the powerful Whig patronage Wood got for his Irish project for copper coin from 1722, Newton did what he could to back the credit of this new scheme, personally drafting the terms of the Irish contract and its safeguards. The contrast with Challoner's fate was dramatic. Wood's coins were well made, of the best copper Newton knew; no parliamentary licence had ever before been required to issue copper coin anywhere in the kingdom. Despite the loud protests of Swift and his Irish allies about London colonialism, Wood's coin passed Newton's test in April 1724 in a dramatic public trial and was imposed on Ireland.[126]

Assays worked because of their theatre of proof. It was necessary to set up a 'trial plate', a standard sample of putatively pure metal. Then minted coins would be hydrostatically estimated, cupellated or applied to the touchstone to be judged against this standard. Newton knew that hydrometric tests, like any other, could be unreliable. 'I cannot undertake absolutely that there shall be no faulty barrs which may escape the assays,' Newton acknowledged, 'but I am safest in people that are afraid of me.'[127] So the drama of the assay mattered. Mint assays were checked in a terrifying, and, for Newton, often troublesome ceremonial. The test of his Mint output, including Wood's coins, involved a Trial of the Pyx, so named for the box in which randomly selected coin samples were stored before trial.[128] These boxes, the schedules of the coin made, and enough coals for the assay fires were rowed up the Thames from the Tower to Palace Yard by liveried footmen in the Mint barge. Newton and his colleagues arrived at Westminster, the Privy Counsellors and the Lord Chancellor an hour later. When a jury of metalsmiths had been sworn in, they were left alone with the mint assayers to try the metal samples, followed by a dinner at Newton's expense at a nearby inn.[129]

This assay ceremony was a ritual of state power over value and a battlefield between the many agencies involved in making value. Conflicts were endemic about the best way of running an assay. Newton's men were sometimes thrown out of the room by the smiths; the Mint's master was much concerned whenever he thought the trial jury too private. Parliament reported that Challoner may have been a victim of a conspiracy by the goldsmiths and Newton, 'nor will the pyx box or the jury prevent it as the law now stands'.[130] Newton later argued that 'the tryall of the moneys in the Pyx is to be performed . . . after the most just manner that can be made by fire, by water, by touch or by weight, or by all or by any of them'. Metrology was mixed with morality. If the gold standard were set too pure, then Newton's work would look defective. 'Fine gold' was not a self-evident type but the result of many judgements about labour and manners. Newton told the Treasury that 'by the assay I am satisfied that there are various degrees of fine gold . . . by wch means the standard of gold is rendred uncertain'.[131] In his contemporary reports to the Royal Society, Newton also emphasised the intrinsic variations in 'the weight of fine Gold', and described the trials with gold coins which definitively evinced the reality of empty space and thus the action of gravity.[132] These were not only issues of the artificiality of species and of universal gravitation, but were also vital matters of fiscal probity and credit. Assays demanded the right choreography of credit and measurement – who could be trusted, which tools should be used, who could act as judge, and who should be banned or penalised. Only thus, Newton concluded, could the assay be 'made the standing universall rule of Valuing Gold in all Nations in point of finenesse'.[133]

Attention to these rituals of instrumentation helps a symmetrical account of different societies' rules and values. In African trading posts, Akan courts, London counting houses and Whitehall chambers instrumental dramas were used to get temporary workable agreement about value. Akan traders carried individual gold-weights to measure gold dust, and balances and bags to weigh and hold it. Considerable spiritual power was vested in devices used for measurement. Akan state systems cultivated rituals of power in which this command over gold played a fundamental role. Links between individual assay instruments, spiritual ritual and sovereign authority were just as apparent in England. There coin-weights were carried by many traders and travellers because the relation between face value and intrinsic value of coin was risky and unsure. Newfangled hydrometers and other assay techniques could be used by individuals within regulated systems of calibration. The entire system was publicly warranted by ceremonies such as the Trial of the Pyx, where the state prerogative was made to look like the fount of the entire system of good values. These were matched by equally dramatic trials of coiners and clippers, where the authority of the state was publicly exercised over the bodies of condemned criminals who had subverted the system of measures.

This chapter has been about the place of assays in metropolitan and marginal sites. The Enlightenment legacy of these enterprises is what has been called 'the quantifying spirit', the urge to impose uniform weights and measures in the name of globalised economies of capital and commodity exchange. Important in this enterprise was the geography of impurity and of forgery. The instruments which discriminated between mundane stuff of real value (commodities) and others' showy but worthless materials (fetishes) were the agents of a new commodity fetishism. The Enlightenment discourse on the fetishism of aliens and primitives was the source for Marx's own account of the fetishism of commodities under capitalist production.[134] This geography also shows the development of these metrologies in museology and natural history, as well as in the testing offices and the excise laboratories of the capitalist world-economy. The exploratory and accumulative natural histories of the Enlightenment mapped and ordered the world in the name of accumulation and study in the metropolitan botanic gardens and museums of Uppsala, Edinburgh, London and Paris.[135] The geographer David Livingstone argues that the

> passion for precision characterised the whole enterprise, mathematical precision in astronomical observation, in cartographic accuracy, and in scientific illustrations. By those representational devices, the categories of European scientific ways of seeing came close to enclose – to engulf – non-Western realms. These precision tools, indeed, became the very instruments used to cut the template on which the idea of the exotic was fashioned.[136]

If anything has been added to these sage views here, it is to insist that through these assay instruments definition of value and error took place inside each metropolis as well as at their margins.

Acknowledgement

I particularly thank the British Academy for its support during the research for this chapter.

Notes

1 J. Pettus, *Fleta Minor Part 2: Spagyrick Laws containing Essays on Metallick Words*, London: Thomas Dawks, 1683.

2 N. Broc, *La Géographie des Philosophes: Géographes et Voyageurs français au 18e Siècle*, Paris: Ophrys, 1974, p. 9. M.-N. Bourguet, 'L'explorateur', in *L'Homme des Lumières* (edited by M. Vovelle), Paris: Éd. du Seuil, 1996, pp. 285–346, 289–93.

3 W. MacGaffey, 'Dialogues of the deaf: Europeans on the Atlantic coast of Africa', in *Implicit Understandings: Observing, Reporting and Reflecting on the Encounters between Europeans and Other Peoples in the Early Modern Era* (edited by S.B. Schwartz), Cambridge: Cambridge University Press, 1994, pp. 249–67. J. Law, 'Technology and heterogeneous engineering: the case of Portuguese expansion', in *The Social Construction of Technological Systems* (edited by W. Bijker, T.P. Hughes and T. Pinch), Cambridge, Mass.: MIT Press, 1987, pp. 111–34. S.J. Harris, 'Long-distance corporations, big sciences and the geography of knowledge', *Configurations* 6, 1998, 269–304. C. Withers, 'Geography, natural history and the eighteenth-century Enlightenment: putting the world in place', *History Workshop Journal* 39, 1995, 136–63.

4 J. Brewer, 'The eighteenth-century British state: contexts and issues', in *An Imperial State at War: Britain from 1689 to 1815* (edited by L. Stone), London: Routledge, pp. 52–71, 1994, p. 60.

5 P. Linebaugh, *The London Hanged: Crime and Civil Society in the Eighteenth Century*, London: Penguin, 1991, pp. 153–83. J.V. Grabiner, '"Some disputes of consequence", Maclaurin among the molasses barrels', *Social Studies of Science* 28, 1998, 139–68.

M. Ogborn, *Spaces of Modernity: London's Geographies, 1680–1780*, London: Guilford, 1998, pp. 158–85. W.J. Ashworth, 'Between the trader and the public: British alcohol standards and the proof of good governance', *Technology and Culture* 42, 2001, 27–50.

6 L. Stewart, *The Rise of Public Science: Rhetoric, Technology and Natural Philosophy in Newtonian Britain*, Cambridge: Cambridge University Press, 1992, pp. 208–11.

7 C. Bayly, *Imperial Meridian: the British Empire and the World 1780–1830*, London: Longman, 1989, S. Subrahmanyam, *The Portuguese Empire in Asia 1500–1700*, London: Longman, 1993, G. Dening, *The Death of William Gooch: A History's Anthropology*, Melbourne: Melbourne University Press, 1995, R. Blackburn, *The Making of New World Slavery: From the Baroque to the Modern 1492–1800*, London: Verso, 1997.

8 F. Braudel, *The Perspective of the World*, London: Collins, 1984, pp. 430–41.

9 D.D.C. Chambers, *The Reinvention of the World: English Writing 1650–1750*, London: Arnold, 1996, pp. 48–77. P. Brantlinger, *Fictions of State: Culture and Credit in Britain, 1694–1994*, Ithaca, N.Y.: Cornell University Press, 1996, pp. 48–87. C. Nicholson, *Writing and the Rise of Finance: Capital Satires of the Early Eighteenth Century*, Cambridge: Cambridge University Press, 1994, pp. 91–122.

10 M. Douglas, *Implicit Meanings*, London: Routledge & Kegan Paul, 1975, p. 50. Blackburn, *Making of New World Slavery*, op. cit., pp. 1, 589.

11 E. Grendi, 'Counterfeit coins and monetary exchange structures in the Republic of Genoa during the sixteenth and seventeenth centuries', in *History from Crime* (edited by E. Muir and G. Ruggiero), Baltimore, Md.: Johns Hopkins University Press, 1994, pp. 170–205. J. Styles, '"Our traitourous money makers": the Yorkshire coiners and the law, 1760–1783', in *An Ungovernable People: the English and their Law in the Seventeenth and Eighteenth Centuries* (edited by J. Brewer and J. Styles), London: Hutchinson, 1980, pp. 172–249.

12 S. Shapin, 'Robert Boyle and mathematics: reality, representation and experimental practice', *Science in Context* 2, 1988, 23–58, p. 49. Idem, *A Social History of Truth: Civility and Science in Seventeenth-century England*, Chicago: Chicago University Press, 1997, p. 348.

13 R. Boyle, *Medicina hydrostatica, or Hydrostaticks applied to the Materia Medica*, London: Samuel Smith, 1690, pp. 134–5.

14 J. Locke, *An Essay concerning Human Understanding* [first published, 1690] 28th edn, London: Tegg, 1838, p. 212.

15 Ibid., pp. 336–8. C. Caffentzis, *Clipped Coins, Abused Words and Civil Government*, New York: Autonomedia, 1989, pp. 75, 214.

16 R.S. Westfall, *Never at Rest: a Biography of Isaac Newton*, Cambridge: Cambridge University Press, 1980, pp. 491–3. M. Hunter, 'Alchemy, magic and moralism in the thought of Robert Boyle', *British Journal for the History of Science* 23, 1990 (387–410): 399–405.

17 Grendi, 'Counterfeit coins', op. cit., p. 188.

18 P.H. Smith, *The Business of Alchemy: Science and Culture in the Holy Roman Empire*, Princeton, N.J.: Princeton University Press, 1994, p. 222.

19 R. Boyle, 'An invention for estimating the weight of water in water with ordinary ballances and weights', *Philosophical Transactions* 4, 1669, 1001–3. Idem, 'A new essay-instrument', *Philosophical Transactions* 10, 1675, 329–48.

20 L. Ercker, *Treatise on Ores and Assaying* (edited by A.G. Sisco and C.S. Smith), vols. xxiii–xxv, Chicago: Chicago University Press ([1580] 1951), pp. 134–7.

21 Boyle, 'A new essay-instrument', op. cit., pp. 336–8, p. 345.

22 R. Boyle, *A Free Enquiry into the Vulgarly Received Notion of Nature* (edited by E. Davis and M. Hunter), Cambridge: Cambridge University Press, 1996, p. 5.

23 T. Hobbes, *Leviathan*, London: Andrew Crooke, 1651, p. 15.

24 Boyle, 'A new essay-instrument', op. cit., p. 338.

25 J. Beale, 'A letter relating to the contents of the tract next foregoing', *Philosophical Transactions* 10, 1675, 353–5.

26 Boyle, *Medicina hydrostatica*, op. cit., p. 217.
27 R. Ruding, *Annals of the Coinage of Great Britain*, 3 vols. (3rd edn), London: Hearne, vol. 1, 1840, pp. 64–5. I. Newton, *Correspondence*, 7 vols., edited by H.W. Turnbull, J.F. Scott, A.R. Hall and L. Tilling, Cambridge: Cambridge University Press, 1959–77, vol. 3, pp. 218–19. Boyle, *Medicina hydrostatica*, op. cit., p. 149. Hunter, 'Alchemy, magic and moralism', op. cit., pp. 404–5.
28 Boyle, *Medicina hydrostatica*, op. cit., pp. 176–7, 180–1. Notes on the statute against multipliers are in Boyle's notebook, Royal Society MSS 189 fol. 1. Notes on conversations with the governor of Cape Coast are in Boyle Papers, vol. 21, pp. 280–1.
29 J. Porteous, *Coins in History*, New York: Putnam, 1969, pp. 212–19.
30 Linebaugh, *The London Hanged*, op. cit., p. 56. Caffentzis, *Clipped Coins*, op. cit., p. 233. P. Vilar, *A History of Gold and Money 1450 to 1920*, London: Verso, 1991, p. 214.
31 K.G. Davies, *The Royal African Company*, London: Longmans, 1957, pp. 181, 360.
32 J. Barbot, *Barbot on Guinea: the Writings of Jean Barbot on West Africa 1678–1712* (edited by P.E.H. Hair, A. Jones and R. Law), London: Hakluyt Society, 1992, vols. LXXV–LXXVI. Boyle's notes on Jobson are in Royal Society Boyle Papers, vol. 39, fols. 89–91; Barbot's address is recorded at the start of Royal Society Boyle MSS 189, dated 16 September 1689; Boyle's notes about gold quality, dated to about 1686, from 'an ingenious Gentleman that was imploy'd in the French colony on the Coast of Afric', are in Royal Society Boyle Papers, vol. 21, pp. 282–3 and partly printed in R. Boyle, *The General History of the Air*, London: Churchill, 1692, p. 158.
33 Boyle, *Medicina hydrostatica*, op. cit., p. 188.
34 M. Hunter, *Establishing the New Science: the Experience of the Early Royal Society*, Woodbridge: Boydell Press, 1989, p. 93. A. Hill, 'Inquiries for Guiny', *Philosophical Transactions* 2, 1667, 472. D. Carey, 'Compiling nature's history: travellers and travel narratives in the early Royal Society', *Annals of Science* 54, 1997, 269–92.
35 R. Boyle, 'Articles of inquiries touching mines', *Philosophical Transactions* 1, 1666, 330–43.
36 M. Govier, 'The Royal Society, slavery and the island of Jamaica, 1660–1700', *Notes and Records of the Royal Society* 53, 1999, 203–17, p. 214. A. Jones, *German Sources for West African History 1599–1669*, Wiesbaden: Franz Steiner, 1983, pp. 153–4, 250–2. Oldenburg's queries are 'Inquiries for Guiny to Mr Floyd minister to the English Factory there', Royal Society Classified Papers, vol. 19, no. 56. Oldenburg's unacknowledged source is Wilhelm Mueller, 'Die afrikanische auf der guineischen Gold Cust gelegene Landschaft Fetu' (1673), translated in Jones, *German Sources*, op. cit.
37 T. Sprat, *History of the Royal Society of London*, London: Martyn, 1667.
38 S. Purchas, *Hakluytus posthumus, or Purchas his pilgrimes*, 20 vols., Glasgow: James Maclehose, ([1625] 1905) vol. 1, pp. 45, 75. L.E. Pennington (ed.), *The Purchas Handbook: Studies of the Life, Times and Writings of Samuel Purchas*, London: Hakluyt Society, 1997, p. 4. Smith, *The Business of Alchemy*, op. cit., p. 224. Anon., *The Golden Coast, or a Description of Guinney*, London: Speed, 1665, p. 67.
39 J. Ogilby, *Africa*, London: Thomas Johnston. Sprat, *History*, op. cit., 1670, pp. 408–9. Govier, 'The Royal Society', op. cit., p. 206.
40 M. Cranston, *John Locke: a Biography*, Oxford: Oxford University Press, 1985, pp. 146–56. Davies, *The Royal African Company*, op. cit., pp. 62–5.
41 J.R. Jacob, 'Restoration, reformation and the origins of the Royal Society', *History of Science* 13, 1975, 155–76. R. Boyle, *Early Essays and Ethics* (edited by J.T. Harwood), Carbondale: Southern Illinois University Press, 1991, p. 196.
42 Sir J. Craig, *Newton at the Mint*, Cambridge: Cambridge University Press, 1946, p. 10.
43 P. Laslett, 'Locke, the great recoinage and the origins of the Board of Trade 1695–1698', in *John Locke: Problems and Perspectives* (edited by J. Yolton), Cambridge: Cambridge University Press, 1969, pp. 137–64. J.O. Appleby, 'Locke, liberalism and the natural law of money', in *Past and Present* 71, 1976, 43–69.

44 Linebaugh, *The London Hanged*, op. cit., pp. 55–6.
45 R.A. Kea, *Settlements, Trade and Polities in the Seventeenth-century Gold Coast*, Baltimore, Md.: Johns Hopkins University Press, 1982, pp. 171–3. Y. Defontaine, *Guerre et Société au Royaume de Fetu 1471–1720*, Paris: Karthala, 1993.
46 T. Garrard, *Akan Weights and the Gold Trade*, London: Longman, 1979, p. 193. M.W. Plass, *African Miniatures: the Gold Weights of the Ashanti*, London: Lund Humphries, 1967, K. Arhin, 'Monetization and the Asante state', in *Money Matters: Instability, Values and Social Payments in the Modern History of West African Communities* (edited by J. Guyer), London: James Currey, 1995, pp. 97–110, p. 99.
47 A. van Dantzig, 'The Akanists: a West African Hansa', in *West African Economic and Social History* (edited by D. Henige and T.C. McCaskie), Madison: University of Wisconsin Press, 1990, pp. 205–16. Garrard, *Akan Weights*, op. cit., pp. 171–210. M. Macleod, *The Asante*, London: British Museum, 1981, pp. 122–33.
48 N. Villault, *Relation des Costes d'Afrique appellées Guinée*, Paris: Denis Thierry, 1669, p. 253.
49 Kea, *Settlements*, op. cit., p. 225.
50 Jones, *German Sources*, op. cit., p. 250.
51 Kea, *Settlements*, op. cit., p. 238.
52 P. de Marees, *Description and Historical Account of the Gold Kingdom of Guinea (1602)* (edited by A. van Dantzig and A. Jones), Oxford: Oxford University Press, 1987, pp. 60–1.
53 W. Pietz, 'The problem of the fetish – 2', *Res* 13, 1987, 23–45, p. 40.
54 Ibid., p. 36.
55 Boyle, *Medicina hydrostatica*, op. cit., pp. 186–7.
56 W. Pietz, 'The problem of the fetish – 3a', *Res* 16, 1988, 105–23, pp. 110–11.
57 Barbot, *Barbot on Guinea*, op. cit., p. 490. Boyle, *Medicina hydrostatica*, op. cit., pp. 188–9.
58 W. Bosman, *A New and Accurate Description of the Coast of Guinea*, London: James Knapton and Daniel Midwinter, 1705, pp. 73–5. See A. van Dantzig, 'Bosman's *New and Accurate Description of the Coast of Guinea*: how accurate is it?', *History in Africa* 1, 1974, 101–8.
59 Barbot, *Barbot on Guinea*, op. cit., pp. 150–1.
60 Bosman, *A New and Accurate Description*, op. cit., pp. 154–5.
61 Garrard, *Akan Weights*, op. cit., pp. 43–4. P.F. de Moraes Farias, 'Silent trade: myth and historical evidence', *History in Africa* 1, 1974, 9–24.
62 R. Jobson, *The Golden Trade* (edited by W. Rodney), London: Dawsons, 1968, p. 130.
63 W. Smith, *A New Voyage to Guinea*, London: Nourse, 1744, p. 138. H.M. Feinberg, 'An eighteenth century case of plagiarism: Smith's *New Voyage to Guinea*', *History in Africa* 6, 1979, 45–50.
64 Bosman, *A New and Accurate Description*, op. cit., p. 82.
65 Ibid., p. 85.
66 K.Y. Daaku, *Trade and Politics on the Gold Coast 1600–1720*, Oxford: Clarendon Press, 1970, pp. 28–33.
67 Garrard, *Akan Weights*, op. cit., p. 87.
68 de Marees, *Description*, op. cit., p. 60.
69 Ibid., p. 191.
70 Smith, *A New Voyage to Guinea*, op. cit., pp. 15–17.
71 de Marees, *Description*, op. cit., p. 73.
72 Pettus, *Fleta Minor*, op. cit., p. 52.
73 de Marees, *Description*, op. cit., pp. 60, 191.
74 B. Latour, *Petite Réflexion sur le Culte Moderne des Dieux Faitiches*, Paris: Synthélabo, 1996, p. 28.
75 Barbot, *Barbot on Guinea*, op. cit., p. 483.
76 L. Stewart, 'The edge of utility: slaves and smallpox in the early eighteenth century', *Medical History* 29, 1985, 54–70. Govier, 'The Royal Society', op. cit., pp. 203–17. Blackburn, *The Making of New World Slavery*, op. cit., pp. 254–6.

77 Vilar, *A History of Gold*, op. cit., pp. 220–31. I.K. Steele, *The English Atlantic 1675–1740: An Exploration of Communication and Community*, Oxford: Oxford University Press, 1986.
78 J. Brewer, *The Sinews of Power: War, Money and the English State 1688–1783*, London: Unwin-Hyman, 1989, p. 69.
79 A. Smith, *Inquiry into the Nature and Causes of the Wealth of Nations*, 2 vols., London: Grant Richards ([1776] 1904), vol. 2, p. 144.
80 Craig, *Newton at the Mint*, op. cit., pp. 1–6. C.E. Challis (ed.), *A New History of the Royal Mint*, Cambridge: Cambridge University Press, 1992, pp. 351–7.
81 Boyle, 'A new essay-instrument', op. cit., pp. 329–48, 331.
82 Challis, *A New History*, op. cit., pp. 341–8, 379–85.
83 M.-H. Li, *The Great Recoinage of 1696 to 1699*, London: Weidenfeld & Nicolson, 1963.
84 W. Cunningham, *The Growth of English Industry and Commerce in Modern Times: the Mercantile System*, Cambridge: Cambridge University Press, 1912, p. 435. A. de la Pryme, *Diary* (edited by C. Jackson), Durham: Surtees Society, 1870, p. 97.
85 Nicholson, *Writing*, op. cit., p. 47.
86 J.S. Peters, 'The bank, the press and the return to nature: on currency, credit and literary property in the 1690s', in *Early Modern Conceptions of Property* (edited by J. Brewer and S. Staves), London: Routledge, 1996, pp. 365–88.
87 G. Simmel [1907] *The Philosophy of Money* (2nd edn), London: Routledge, 1990, p. 186.
88 Ruding, *Annals*, op. cit., vol. 2, pp. 58–9.
89 R. Iliffe, 'Those "whose business it is to cavil": Newton's anti-catholicism', in *Newton and Religion* (edited by J.E. Force and R.H. Popkin), Dordrecht: Kluwer, 1999, pp. 97–199.
90 Craig, *Newton at the Mint*, op. cit., pp. 8–9.
91 I. Newton, *The Principia* (translated and edited by I.B. Cohen and A. Whitman), Berkeley: University of California Press, 1999, p. 795. A. Koyré, *Newtonian Studies*, Cambridge, Mass.: Harvard University Press, 1965, pp. 261–72.
92 I. Newton, *Mathematical Principles of Natural Philosophy* (edited by F. Cajori), Berkeley: University of California Press, 1934, p. 589.
93 I. Newton, *Philosophiae Naturalis Principia Mathematica: Third Edition with Variant Readings* (edited by A. Koyré and I.B. Cohen), Cambridge: Cambridge University Press, 1972, pp. 666–7. Idem, *The Principia*, op. cit., pp. 877–8.
94 Newton, *Correspondence*, op. cit., vol. 5, pp. 220–3, 241.
95 Newton, *The Principia*, op. cit., p. 827.
96 Newton, *Correspondence*, op. cit., vol. 5, p. 226.
97 Newton, *Correspondence*, op. cit., vol. 5, p. 233.
98 J.D. Cassini, 'Les élémens de l'astronomie vérifiez par M. Cassini par le rapport de ses tables aux observations de M. Richer (1684)', in *Recueil des Observations faites en plusieurs Voyages pour Perfectionner l'Astronomie et la Géographie* (separately paginated), Paris: Académie Royale des Sciences, 1693, p. 55.
99 Newton, *The Principia*, op. cit., p. 832.
100 Ibid., p. 830.
101 Newton, *Philosophiae*, op. cit., pp. 601–2. Idem, *The Principia*, op. cit., p. 831.
102 Newton, *Correspondence*, op. cit., vol. 5, p. 261.
103 Newton, *The Principia*, op. cit., p. 831.
104 W. Warntz, 'Newton, the Newtonians and the Geographia Generalis Varenii', *Annals of the Association of American Geographers* 79, 1989, 165–91. R. Iliffe, 'Maupertuis, precision measurement and the shape of the earth in the 1730s', *History of Science* 31, 1993, 3–14.
105 Newton, *Correspondence*, op. cit., vol. 5, pp. 245, 250–1.
106 Steele, *The English Atlantic*, op. cit., pp. 213–28.
107 Iliffe, 'Those "whose business it is to cavil"', op. cit., pp. 101–6.

108 In the third edition of the *Principia* (1726) Newton omitted the phrase 'eternity and infinity'.
109 Newton, *The Principia*, op. cit., p. 941. S. Snobelen, '"God of Gods and Lord of Lords": the theology of Isaac Newton's General Scholium to the *Principia*', *Osiris* 15, 2000.
110 Craig, *Newton at the Mint*, op. cit., p. 37.
111 Newton, *Correspondence*, op. cit., vol. 4, p. 208.
112 E.G.V. Newman, 'The gold metallurgy of Isaac Newton', *Gold Bulletin* 8, 1975, 90–5.
113 Newton, *Correspondence*, op. cit., vol. 4, p. 545.
114 Styles, '"Our traitourous money makers"', op. cit., pp. 178, 195.
115 Newton, *Correspondence*, op. cit., vol. 7, pp. 102–3, 118, 113.
116 Craig, *Newton at the Mint*, op. cit., p. 20.
117 Newton, *Correspondence*, op. cit., vol. 7, p. 289.
118 Sir J. Craig, 'Isaac Newton and the counterfeiters', *Notes and Records of the Royal Society* 18, 1963, 136–45.
119 F.E. Manuel, *A Portrait of Isaac Newton*, London: Muller, 1968, p. 233. Westfall, *Never at Rest*, op. cit., p. 570.
120 Craig, *Newton at the Mint*, op. cit., pp. 20–2, 34, 122.
121 Newton, *Correspondence*, op. cit., vol. 4, p. 209.
122 Craig, *Newton at the Mint*, op. cit., p. 42. J.G.A. Pocock, 'Early modern capitalism: the Augustan perception', in *Feudalism, Capitalism and Beyond* (edited by E. Kamenka and R.S. Neale), London: Arnold, 1975, pp. 62–83, p. 79. Nicholson, *Writing*, op. cit., pp. 45–6.
123 Craig, 'Isaac Newton', op. cit., pp. 136–45, 140–3. Manuel, *A Portrait*, op. cit., pp. 238–44. Westfall, *Never at Rest*, op. cit., pp. 571–5.
124 J.M. Treadwell, 'William Wood and the Company of the Ironmasters of Great Britain', *Business History* 15, 1974, 93–112. M.W. Flinn, 'William Wood and the coke-smelting process', *Transactions of the Newcomen Society* 34, 1961, 55–71.
125 Newton, *Correspondence*, op. cit., vol. 6, pp. 453–4.
126 J.M. Treadwell, 'Swift, William Wood and the factual basis of satire', *Journal of British Studies* 15, 1976, 82–90. Stewart, *The Rise of Public Science*, op. cit., p. 295. Newton, *Correspondence*, op. cit., vol. 7, pp. 275–8.
127 Westfall, *Never at Rest*, op. cit., p. 845.
128 S.M. Stigler, 'Eight centuries of sampling inspection: the Trial of the Pyx', *Journal of the American Statistical Association* 72, 1977, 493–500.
129 Ruding, *Annals*, op. cit., vol. 1, pp. 69–77. Newton, *Correspondence*, op. cit., vol. 4, pp. 371–3. Craig, *Newton at the Mint*, op. cit., pp. 48–9. S. Wortham, 'Sovereign counterfeits: the Trial of the Pyx', *Renaissance Quarterly* 49, 1996, 334–59.
130 Ruding, *Annals*, op. cit., vol. 2, pp. 465–70.
131 Newton, *Correspondence*, op. cit., vol. 5, p. 82.
132 Newton's earlier alchemical work raised questions about gold as the type of purity: see K. Figala, 'Newton as alchemist', *History of Science* 15, 1977, 102–37. During the week in March 1712 when Newton spoke at the Royal Society on intrinsic variations in gold density, he also completed his definitive argument in the *Principia* that the fall of gold coins in air proved a vacuum and wrote his important report to the Treasury on the dangers of over-valuing refined gold in England and Ireland. See Newton, *Correspondence*, op. cit., vol. 5, pp. 246, 248, as well as idem, *The Principia*, op. cit., pp. 810, 939; see also Royal Society, Journal Book, vol. 11, p. 279.
133 Newton, *Correspondence*, op. cit., vol. 5, pp. 84–6. Westfall, *Never at Rest*, op. cit., pp. 607–12.
134 W. Pietz, 'Fetishism and materialism: the limits of theory in Marx', in *Fetishism as Cultural Discourse* (edited by E. Apter and W. Pietz), Ithaca, N.Y.: Cornell University Press, 1993, pp. 119–51. Latour, *Petite Réflexion*, op. cit., p. 27.

135 M.L. Pratt, *Imperial Eyes: Travel Writing and Transculturation*, London: Routledge, 1992, pp. 24–37. L. Koerner, *Linnaeus: Nature and Nation*, Cambridge, Mass.: Harvard University Press, 1999, pp. 113–39. D. Turnbull, 'Cartography and science in early modern Europe: mapping the construction of knowledge spaces', *Imago Mundi* 48, 1996, 5–23.

136 D. Livingstone, *The Geographical Tradition: Episodes in the History of a Contested Enterprise*, Oxford: Blackwell, 1992, p. 126.

3 The project for a map of Languedoc in eighteenth-century France at the contested intersection between astronomy and geography

The problem of co-ordination between philosophers, instruments and observations as a keystone of modernity

Christian Licoppe

The aim of this chapter is to provide a detailed case study of an episode in eighteenth-century French astronomical cartography, namely the mapping of Languedoc, with a particular focus on the efforts of the Montpellier Société royale des sciences. Analysing the Languedoc project as a case study and putting it into perspective will allow us to identify a number of distinctive tensions that were operative in the first half of the eighteenth century, and to explore more deeply the themes of this book within a particular but quite significant historical context. The whole story takes place in the context of a rising centralised demand for information from the absolutist state, which met the usual resistance from local structures intent on keeping control over local affairs (a good example of this is the local mistrust of their operations often remarked upon and complained about by the geodetic expeditions). The status of the people involved was strained by a tension between the ideal-type exemplar of the provincial academic astronomer as an individual anchored in the Republic of Letters, and disinterestedly pursuing knowledge in the leisure time that his other activities left him (with a hint of the wish to emulate Parisian colleagues who, from the start, had a distinctive ethos), and that of the geographer as an entrepreneur who, much like a printer, operated at the crossroads between the businessman running a workshop and the culture broker, catering for the interests of a mixed audience of practitioners (the military, for instance) and avid members of the Republic of Letters. The issue of the proper cartographic methods to be followed was strained between the representation of geographical work as a painful and tedious exegesis of the work of previous geographers, as well as an assessment of the accuracy of old and new travel narratives (geography looking here like an ongoing and endless process of critical commentary), and that of a cartography founded on abstract objectivised measurements, largely independent of local context (since mediated by astronomical and geometrical instruments) and ideally providing positions of places with respect to one another that could be regarded as final. Maps could correspondingly be thought of either as combining geography and curious facts and stories attached to places, or as enhancing the separation of the geographical and the historical by focusing on the spatial layout of places of interest (for instance actualised by rendering visible a grid of parallels or meridians). Two ideal-type epistemologies can be compared: on the one hand an epistemology founded on the circulation of first-person accounts written by eye witnesses; on the other, an epistemology built upon comparable instruments (or claimed to be such) and the circulation of the tables of trigonometrically processed figures that were the outputs of the former. With respect to instruments themselves we will identify a tension between, on the one hand, a tight relationship of the natural philosopher to his tools (which was marked by eponymy, for the instrument and its user were not distinguished, and by a moral economy of prestige and ostentation, patronage and gift-giving), and, on the other hand, the instrument treated as a reified commodity exchanged for money, allegedly usable and readable by any observer. We will also show how the issue of the comparability of instruments impacted on this latter tension in the way that the role and trustworthiness of the instruments were construed.

We will see that these tensions made up the profusion of alternatives through which the promoters of the Languedoc project had to find their own path, with an additional set of practical constraints including the level of funds made available by the province, expected schedules and admissible or non-admissible delays, and various impediments to fieldwork such as seasonal constraints (work could not proceed in winter) and variable relationships with the local gentry, administration and informants, etc. But this did not simply involve shuffling out categories and concepts like cards in a new deck. The whole process was also creative, and I will dwell on three examples in particular that testify to this creativity. First at the level of organisational problems, for instance with the gradual realisation that the key to all subsequent truth claims about the map was the availability of the measured figures and the trigonometric calculations that had been made with them, with a trial-and-error approach taken to deal with this. Secondly at the political level, for contra a simplistic view of the centralisation process in France, we will see how the Etats and the Société royale des sciences of Montpellier subverted the absolutist demand by working within the frame of its political and scientific requirements simultaneously to promote provincial interests and institutions. Finally, at a cognitive level we will see novelty in the trial-and-error approach taken to the combination of geography and history; that is, in regard to the intended map and the various memoirs that would not be fitted into it but would rather accompany it.

In the light of the multifaceted inventiveness of the Languedoc project, I will approach it as an exceptional-normal case, following the controversial oxymoron popularised by the Italian *Microstoria* school.[1] Relying largely on local archives,[2] I seek to show that it was normal in so far as it revealed a fairly detailed pattern of alternatives that structured the experience of the actors in the course of the tasks in which they engaged. At times in the analysis, this perspective will guide us to perceive aspects of that very fabric. But we will see that it was also exceptional in the sense that those involved in it actively created their own recipe for the mixing of these alternatives in the course of the project. Whatever the transience of a project that in the end could not be stabilised historically by an authoritative output, such as the intended map of the province, in its own way this creative handling of cartographic, empirical, institutional and social practices along their tensile seams contributed to a shift in the whole pattern of the fabric for others to build upon, however critical of the project they may have been.

The initial stages of a province-funded, academy-inspired project for a map of Languedoc in the first half of the eighteenth century

In the 1720s, the highest local administrative body of the province of Languedoc (the Etats de la Province) agreed to fund the drawing up of a detailed and precise map of their territory. This was supposed to be the joint accomplishment of the astronomers of the Société royale des sciences of Montpellier, founded in 1706, which claimed to be a daughter of the Académie royale des sciences in Paris, and of the Paris-based royal geographer Delille, also a member of this famous Parisian institution and the first geographer to be admitted to it.

Among the several historical strands that intersected to produce this decision, we note the increasing demand for information about the provinces from the absolutist state. In 1664 Colbert asked senior state officials in the provinces, the *intendants*, for memoirs describing the current state of their *généralités*, including good maps showing administrative divisions and various politically sensitive data.[3] Similarly, in order to educate the young Duke of Burgundy by providing an accurate description of the current state of the kingdom, in 1697 the *intendants* were sent requests for memoirs describing the width and natural formations of the province, the production of its soils, and demographic and administrative data.[4] In 1716 the regent Philippe d'Orléans commissioned the Académie royale des sciences to obtain detailed memoirs from the *intendants* describing the natural and artificial products of their provinces. Réaumur, then director of the Académie, collected, analysed and even criticised the resulting documents (in some cases he would ask directly for further data).[5] A similar set of inquiries, although not involving the Académie, was to be initiated in 1730 by Orry, then Contrôleur Général des Finances.

The relevance of several state-initiated geographic and cartographic endeavours also needs to be emphasised. At the turn of the century, Cassini and Maraldi had gone into Languedoc to complete the triangulation of the meridian to the south. The president of the Etats, the archbishop of Narbonne, charged Montferrier, syndic of the province, with the administrative responsibility for the construction of the Languedoc map. When assessing the costs of similar enterprises Montferrier learned of the renowned map of the Pyrenees completed in 1719 by the military engineer Roussel.[6] The Etats of Languedoc were not the only ones to be involved in the construction of an accurate map of their province, since the Etats of Brittany had also considered raising funds for drawing up a similar map.[7] Although embedded in and organically linked to state-sponsored administrative inquiries and cartographic mapping, the map of Languedoc was to be an achievement of the province, which thus had the opportunity to show 'its zeal in the service of the king and in the good of the public', and to boast of being an example to all the other provinces.[8] That it confined itself to the institutions of the province is amply attested to by Montferrier's constant bickering over excessive costs and additional delays. We note in passing that this particular example calls for a dynamic model of the centralisation of France under absolutist rule, in which local officials interiorised and anticipated what they perceived to be the main object of the state's demands.

According to the savants of the Société royale des sciences in Montpellier, it was to be the first map of its kind. It involved the combination of local maps made by hired geographers (*cartes particulières* or particular maps) and a large-scale triangulation made by the astronomers and mathematicians of the Société taking astronomical and geometrical observations. (The astronomical observations were to be measurements of latitude by the observation of meridian heights, and geometrically the map would involve triangulation similar to that performed by the Académie royale des sciences for its determination of the length of the Paris meridian arc.) The sheer size of the province – 2,000 square leagues – justified the co-ordination of local mapping by geographers and large-scale triangular canvassing by geometer-astronomers. Moreover, by using the existing triangles and signals of the Parisian astronomers, as well as some of their astronomical determinations of longitude (that of Paris with Montpellier, for instance), the entire map was to be oriented and exactly positioned in relation to the operations of the Académie royale des sciences. It is clear that the provincial scientists saw this form of co-ordination to provide the public (and Paris) with a very concrete exhibition of the symbolic filial link that the Montpellier institution boasted with the Parisian academy. This was no mere pretence either. De Plantade (one of three men who was influential in the creation of the Société royale des sciences in Montpellier, and who would also work on the large-scale triangulations needed for the map of Languedoc with his colleagues Clapiès and Danizy) had accompanied Cassini and Maraldi in their 1700 Languedocian campaign, and performed measurements with them. Cassini's presence in Montpellier in 1701 can be considered as one of the elements that combined to provide the impetus for the creation of the Société royale des sciences in Montpellier in 1706.

The 1720s plans for a map of Languedoc were original in another respect because they involved the co-operation of geographers and academic savants. Map-making geographers were usually craftsmen, drawing and selling maps to interested practitioners or literati. Geographers could, in that sense, be compared to printers of the modern period who were both entrepreneurs and culture brokers.[9] Geography was often a family business, as the case of well-known geographers of the 1680–1750 period illustrates. Guillaume Delille, the first geographer to become a member of the Académie, entered the practice founded by his father Claude. After their deaths in 1720 and 1726 respectively, the business Guillaume had inherited from his father went to his widow (in keeping with guild customs), with whom Montferrier would conduct his dealings in the 1720s. The Sansons and the Nolins ran their geographers' business over several generations. When the Sanson geographic lineage died out in the 1730s after a century of successful geographic practice, one of the heirs to Pierre Moullart Sanson, Gilles Robert de Vaugondy, bought out the practice and launched a geographical career for himself and his son. Hardworking and upcoming Nicolas de Fer was the son of an engraving salesman, while Jaillot got into the engraving and map-making business when he inherited from his father-in-law. An apprentice could also become a master.[10] Philippe Buache started as a draughtsman in Delille's workshop before taking over part of the practice, and in the course of his geographic work succeeded Guillaume Delille at the Académie royale des sciences.

Negotiations between Montferrier and the Delille widow over the intended map of Languedoc were indeed business. After a *conférence* held in Paris with Montferrier in 1727, she sent him a draft workplan concerning the engineers she would hire to make the *cartes particulières* in the province, covering all the relevant expenses. The expected budget also provided for the cost of copper plates and their engraving, as well as paper for the maps.[11] The total amount horrified Montferrier who subsequently tried to cut all costs, abruptly halting the work of Benedictine monks who had started on a history of the province that was initially intended to be attached to the map of Languedoc.[12] In the meantime the Delille widow tried in vain to squeeze a pension out of Montferrier, as well as funds to cover expenses allegedly incurred by the people she had paid a retainer to work on the map and whose fieldwork seemed to be endlessly postponed.[13]

By comparison, the astronomers of the Société royale des sciences in Montpellier seem a pretty disinterested and gentlemanly bunch. François de Plantade was a typical *robin*, holding commissions at the law courts in Paris and Montpellier. Besides these activities he cultivated links with the République des Lettres, meeting Bayle at the end of the seventeenth century, working with Cassini and Maraldi on the meridian measurements of 1700, and obtaining the creation of the Société royale des sciences in Montpellier in 1706 with the help of another eminent local *robin* magistrate, the *président* Bon. De Plantade became its first director, and his interests remained focused on astronomical and meteorological measurements. Aged 60 in 1730 when the fieldwork for the map actually started, he became an honorary counsellor which gave him more leisure to pursue academic investigations. His academic colleague Jean de Clapiès (the third founding father of the SRS) was responsible for the public works of the province. He had an inclination for mathematics and performed his part of the cartographic fieldwork in his leisure time (that is, between the various requirements of his commission, which forced him to take many breaks in the academic operations).[14] One of de Clapiès's students with a more modest upbringing, Augustin Danizy, eventually took over responsibility for the execution and co-ordination of the cartographic operations, particularly after the death of de Plantade mid-slope in an ascent of the Pic du Midi in 1741 (at the remarkable age of 81!). Although there would of course be no contract and overt negotiation with such gentlemen-savants, they did receive monthly stipends for their fieldwork. These did not, however, interfere with an overall disinterestedness (their main allegiance being, as we have seen, to the service of the king and the good of the province) that made them stand apart from the Delille widow *qua* a typical representative of map-making craftsmen.[15]

The 1727 meeting in Paris between Montferrier and the Delille widow did lead to agreement on several important points. First, all parties agreed that: (1) it was impossible to produce the map of Languedoc without associating astronomical and geographical observations; (2) it was necessary to draw large-scale triangles connected to the Paris meridian; (3) the availability of skilful astronomers at the Société royale des sciences in Montpellier made the involvement of hired astronomical hands to perform the corresponding operations unnecessary. Astronomy and geography, astronomer-geometers and engineer-geographers were therefore to work hand in hand. It was on these grounds that the astronomers claimed originality for the project.

Although everybody seemed readily to have agreed to that co-operation, the implementation of a pattern of co-ordination for fieldwork revealed some slight but telling discrepancies. Fieldwork supposedly involved two types of operation: first, the triangular canvassing of the province by the astronomers with their instruments and methods (dividing circles, triangulation); second, the small-scale local surveying of engineers hired by the widow, also working with the tools of their trade (the *planchette*, interviews with credible experienced local people). Delille's widow requested more money in order to hire an additional workforce and create a Bureau de Correspondance in Paris, to which fieldworkers would send the results of their operations. The Bureau would co-ordinate from a distance the detailed fieldwork covering the entire province.[16] According to her, the savants of the SRS were well convinced of this need and had promised to send their astronomical and geometrical observations as soon as they had completed their triangles. The issue might, however, have been slightly more sensitive. Although the savants of the SRS did indeed confirm they would send their observations to the Bureau (though only when they judged the number of completed triangles high enough to give a clean job to the engineers), they immediately asked the engineers to send their work and the Bureau's memoirs to the SRS as soon as they were completed. They felt that the field experience acquired in the course of their triangulations would render them more competent to make decisions in the event of unexpected practical difficulties than outsiders who had never been there. The examination of data gathered by the intended bureau would be all the more accurate if performed after that first assessment by the SRS.[17]

The organisation of work pertaining to the Languedoc map thus enforced patterns of authority. Although the map was eventually to be the result of the joint operations of (allegedly interested) geographer-engineers and (allegedly disinterested) astronomer-geometers, patterns for the organisation of work and the distribution of authority were initially contested. This would be resolved pragmatically with de Plantade, Clapiès and Danizy eventually managing all fieldwork, either because their point of view prevailed or because the deal with the Delille widow never went through completely (references to her disappear from the archives when the field operations start in 1730).

The instruments for the map of Languedoc in the context of a shifting relationship between natural philosophers and the tools of their trade

The traditional instruments used by geographers in drawing up topographical maps were the compass (*la boussole*) and the plane table (*la planchette*). Things were different on the astronomical side. Crucial to Cassini's programme were the joint development of telescope-derived instruments and a strong academy, benefiting fully from the support of the absolutist state ('the commodity of an Observatoire built to that purpose, and the wealth of all things necessary to the observers that his Majesty has bestowed with a royal munificence').[18]

The ethos of co-ordinating measurements at a distance, the comparability of instruments and the enforcement of social hierarchies between practitioners

The start of the fieldwork for the map of Languedoc was dependent on the completion by the Parisian craftsmen of the list of instruments required by the SRS academicians. Foremost on their list were 'one three and a half foot quadrant circle, two other quadrants with a two-foot radius, two plane tables (*planchettes*), two second pendulums' and so on, with lesser instruments following.[19] The order of the list is of course meaningful. The quadrant circles and time-measuring devices were indeed the distinctive tools of the astronomer's kit, while the plane tables played a similar part for the engineer-geographers. Langlois, who was the pet instrument-maker of the Observatoire at the time,[20] built the quadrant circles and charged £1,500 for the larger one and £1,000 each for the smaller ones. Le Roy, who also enjoyed academic favour for time-measuring devices, built several of them (two second pendulums and two clocks) for £1,000. On the other hand, a certain Le Bas built optical material for £130, one plane table plus various smaller items for £455. Roughly, astronomical instruments cost about five times more than standard cartographic ones in the budget for the map of Languedoc. To put the instrumental budget into broader perspective, the total cost of instruments was less than £10,000 while the widow Delille's budget for the fieldwork, drawing and publishing of the map amounted to £160,000 (a figure that Montferrier still found excessive).

Instruments could indeed be ranked according to cost, but economic issues closely reflected social hierarchies in the community of practitioners involved, at least from the academic perspective. The SRS astronomers and geometers considered their observations and triangulation to be 'the first and principal work'. Once done, the tedious work of the engineers could start in an organised and constrained way that would preclude any error on their part. Again mingling the social with the economic, they added that this sequential (and hierarchical) organisation of work would be cost-effective and make it possible to reduce work, delays and expenses. Moreover, the more costly an instrument, the larger the scale on which it allowed the observer in the field to operate. The larger quadrant would be used to measure the meridian of the province and its whole north–south span, while the great triangles were thought to be 'all the more accurate when the instruments used to measure them are larger and more exactly divided'.[21] But the larger the size of the instruments, or equivalently the size of the triangles they allowed a given observer to map, the more credit that observer would be accorded. The most prestigious academician, de Plantade, used the larger quadrant, was paid more than the others, and was rarely criticised in subsequent debates. This hierarchy of observers according to the size (and price) of their instruments was also embedded very explicitly in Cassini III's 1768 project for a map of Languedoc (itself embedded in his grand plan for a complete triangulation of France), for he provided the following description for the required organisation of work (which in this later period involved only engineers):

> The engineers will be organised in two separate groups, one observing with telescopic instruments from all church towers, while the others will use the compass to map roads, rivers and woods . . . The work of both groups will be incorporated into the general map drawn by an engineer who, with a larger and more accurate instrument, will be committed to cover all the boundaries of the province in order to enable the verification everywhere of the work of the lesser engineer.[22]

The price of the instruments, their size, their accuracy, the scale on which they allowed operations in the field, and the prestige and wages of the observers were so interwoven that it is very difficult to discriminate between them as ranking criteria for cartographic practitioners.

Typical instruments for the astronomer-geometer in the field were various forms of divided circles with telescopic eyesights, time-measuring devices (clocks and pendulums), as well as meteorological instruments such as thermometers and barometers. Indeed, on the list required by the SRS for the Languedoc operations one would find (inexpensive) glass tubes for barometric instruments, which would be used by de Plantade for hypsometric purposes. Although they featured on the list required for the Languedoc fieldwork, these instruments also lay at the core of the work done in observatories. At the Observatoire Royal in Paris, for instance, astronomers were to perform daily observations of the stars and of the state of the atmosphere; that is, of its degree of heat or cold and of its weight (actually more than daily measurements). A more detailed picture of the ethos of an observatory astronomer in France in the eighteenth century can be obtained by considering the heated dispute between the Jesuit, Father Esprit Pezenas, and his successor Saint Jacques de Silvabelle in the early 1760s. Pezenas was a Jesuit astronomer, in charge of a Jesuit astronomic observatory in Marseilles for over thirty years. When the Jesuits were ousted from France by royal decree in 1761 a more than reluctant Pezenas had to relinquish the keys to his cherished observatory, as well as to most of the instruments it contained, to his chosen successor Saint Jacques de Silvabelle. Pezenas fought back, hoping to retain a few of the instruments (we shall return to this point), and voiced some virulent criticism, emphasising the many inadequacies he perceived in the sloppy management of the *observatoire* by his successor.

He noted that Silvabelle did not perform measurements every day, did not educate a circle of students (*élèves*) as he (Pezenas) used to do, and that when asked about the greatest degree of cold that winter in Marseilles, 'he answered in my presence that he never observed it before nine in the morning, because he had other business to attend to at that time'.[23] Moreover, he preferred to use his astronomic instruments to amaze a lay and most impolite audience ('he invites all passers-by, even women of the lowest rank, to come and watch the church tower of N-D de la Garde through the telescope').[24] This caused the instruments to be unfit for proper astronomical use, so that Pezenas arranged for the mathematician Bezout to come and testify to the tarnishing of the mirror of that telescope, which consequently led to a great loss for the Royal budget: £6,000 according to Pezenas's figures.[25] Keeping the instruments and performing daily measurements were crucial to the observatory astronomer's trade. Should he fail in some respect to do so, it would be normal for his wages to be withheld. But it is extremely important to note that daily observations were deemed essential because they were required to render meaningful those made elsewhere, particularly by travelling observers. Pezenas stressed that the observations of de Chabert, who had left to perform observations in the eastern Mediterranean, would be useless without corresponding ones taken in fixed observatories in France. Moreover, because of favourable weather conditions, the observatory in Marseilles was more fit for that purpose, provided of course a skilled and responsible astronomer manned it. Saint Jacques de Silvabelle considered his status as an observer so threatened by this barely concealed stricture that he felt compelled to counter attack. He accused Pezenas, because of his irresponsible possessiveness in regard to the instruments, of having deprived him of the very clock that would have enabled him to make the simultaneous astronomical observations in Marseilles required for the relevance of those of de Chabert, thus causing him to fail the public good.[26]

A strong justification for observatories and for performing daily measurements was therefore the co-ordination of measurements performed at a distance, either within faraway observatories throughout Europe or by travelling observers. In the seventeenth century simultaneous observations of the paths taken by celestial bodies had been a building block for the formation of the astronomical community, while such collective observations (e.g. the controversial sightings of the comet(s) of 1664)[27] had also provided part of the rationale used to obtain the royal patronage necessary for the construction of the Observatoire Royal.[28] This co-ordination was obviously a cornerstone of geodetic practice and of the incursion of astronomical observations into geographical fieldwork in the first half of the eighteenth century. As shown above, establishing the definitive relative location of specific places depended on the time-synchronised observation of the same celestial events by observers in the field and the astronomers of the Observatoire Royal (for instance the emersion of the Jupiter satellites favoured by Cassini, for those furnished with the proper astronomical instruments). The same idea underlay the way meteorological observations were performed. Temperature and atmospheric weight measurements in places of interest had to be compared with similar measurements made in the Observatoire, and field observers within the sphere of influence of astronomical practices were always keen to indicate how the comparison of field measurements with those at the Observatoire allowed one – ideally within the frame of a single well-done measurement – to determine the relative height of places compared to Paris, the level of the sea in the case of barometric measurements (a programme to which Cassini, Maraldi and the hydrographer-astronomer Laval in the south of France devoted a great deal of work), or the differences between maximum degrees of heat or cold in various places in the case of thermometric ones. This objective alliance between 'a public good justification' for astronomers and observatories and the usefulness of co-operative simultaneous measurements at a distance was so pervasive as to infuse the instrumental programme at the Académie royale des sciences in the first half of the eighteenth century with its rationale. All the work on thermometers and barometers from La Hire and Amontons to Réaumur was obsessed with the construction of measuring devices that would allow for the comparison of figures taken simultaneously in separate locations. Amontons pointed out that with the help of his thermometers calibrated on the heat of boiling water, which he judged to be a fixed point everywhere in the world, one could consider it possible to know 'temperature in all the climates of the earth' by sending proper thermometers (and often observers) there.[29] Speaking of the status of incomparable thermometers before his own work, Réaumur argued thirty years later that since they used to speak in different tongues, 'one can only understand the tongue of a thermometer one has been observing for years, and can understand nothing of any other'.[30] The credibility of the instruments of the modern period was here indissolubly linked to that of its user, through the familiarity that was bred with repeated use over a long period. This was particularly so when truth was asserted through circumstantial narratives of local and contextualised events by authorised and authoritative voices. Instruments were almost always described in the République des Lettres in an eponymous manner, as the machine of Mr So-and-so, whose credit was thus inextricably merged with that of his machines. The comparability of instruments advocated by Réaumur and needed for the co-ordination of measurements at a distance clearly allowed some looseness in this link between the virtuosi and their scientific instruments.

Users, instruments and truth claims in the first half of the eighteenth century: contrasted regimes of empirical authority

That kind of experiential link that comprised the authority of scientific tools extended to their making. De Plantade asked Maraldi, for instance, to choose the best craftsmen and to supervise the production of the astronomical instruments that would be required for the map of Languedoc. Maraldi accepted, but it seems he never got any material gratification for this call on his expertise. This was not an unusual practice; Pezenas did the same for a sector in the 1730s in Marseilles. The exact role of Maraldi was clearly to select the craftsmen and pay them directly from funds provided for that purpose by the Etats of Languedoc. But the part played by Cassini or Maraldi was never described in a way that would directly relate it to the material details of the work or the quantitative performance of the instruments, and even less so to their engagement in such practical aspects (though the province academicians mentioned once that they had sent drawings and models to the craftsmen in order to obtain multipurpose, easy-to-use instruments). In a very distinctive and aristocratic manner it was said that the instruments were made 'under their eyes', which bore testimony to the worth of instruments within the rhetoric of patronage rather than by criteria based on performance.

It was, moreover, noted later that Maraldi and de Plantade were in-laws. Whatever the part played by kinship we may deem it contingent, for after the untimely death of Maraldi in 1729 Cassini smoothly took over the task where his unfortunate colleague had left off. Maraldi said he engaged in the Languedoc project because he found it a 'grand' one. But since what contributed substantially to its being grand was that it was carried by the province academicians as a pledge to Cassini's geodetic programme, and thus by a provincial institution that claimed to be the daughter of the Académie royale des sciences, we see that the reciprocities involved in patronage ran deep. Maraldi and Cassini were faithful to their word for they apparently persuaded some craftsmen to interrupt their current commissions in order to work on the Languedoc assignment. The choice of the craftsmen also exhibited the emerging tensions between a scientific patronage modelled on its aristocratic counterpart, and bureaucratic requirements. Trying to cut on costs, and irked by a remark by Maraldi that only one craftsman in Paris was skilled enough for such instruments – a dependency which in Montferrier's opinion slowed down work that might have gone much faster with several working in parallel[31] – Montferrier suggested to Cassini that he organise a competition between several craftsmen, arguing that rivalry would ensure better and faster work at a lower price.[32] Cassini replied haughtily that craftsmen worked well because they longed for his protection, since only he had the power to 'make them fashionable (*leur donner la vogue*)', and their being aware of that fact provided good cause for them 'to hurry for completion, and give us perfect instruments'.[33] This time, scientific patronage was to prevail over bureaucratic efficiency this time.

The strength of the link between the instrument and its user also shaped the way in which the objects related to natural philosophy circulated. One could of course buy an older instrument, but capturing some of its scientific aura and asserting ownership on that basis involved, at least to some extent, the inscription of the exchange and circulation of instruments in an economy of gift-giving and reciprocity in the sense of Mauss.[34] When trying to retain the ownership of the instruments he had used in the laboratory from which he had just been expelled, the Jesuit Father Pezenas claimed that a large and thus expensive sector was his because it had been used by his predecessor, the hydrographer Laval, who had given it to him. In his mind – and without a shadow of doubt – gift-giving superseded the bureaucratic allocation of the instruments to the observatories. His claim over the ownership of the Laval sector would eventually be vindicated in court. When not offered as gifts, instruments cost money. Pezenas and his successor Saint Jacques perceived the link between money and ownership very differently. To Pezenas some instruments were his because, as a sure sign of his dedication to the public good, he had paid for them from his own funds and had sometimes not even been reimbursed afterwards by the minister. These gifts to the public good cost hundreds or even thousands of pounds, and provided him with grounds to argue for the ownership of the aforementioned instruments. Saint Jacques de Silvabelle, on the other hand, claimed he had spent a few pounds from the budget given him by the state to repair some instruments, and that these should therefore be considered the property of the observatory. Thus, we have personal engagement and sacrifice, with instruments seen as an extension of oneself, on the one hand, and bureaucratic accounting and objectivised commodities directly linked to an institution on the other. There was evidently a tension between scientific objects as gifts brokered in patronage networks and their status as material commodities.

This tension operated in a larger frame, which made the eighteenth century something of a watershed following the era of curiosity cabinets and theatres of nature, whose ostentatious decorative economy incorporated the ambition of an exhaustive display of nature interwoven with the patrician ethos of its owner and the literati circles in which he moved.[35] The instruments and machines of the Scientific Revolution had found their way into these cabinets, provided their aesthetics fitted the decorative economy of the place as a whole. Although it is impossible to sustain it any further at this stage, I suggest as a working hypothesis that in regard to aesthetic issues the cartographic instruments of the Languedoc map may have reflected the complex and polymorphic dividing line that manifested in the 1730–50 period. Two of the quadrant instruments used by the Société royale des sciences of Montpellier for its operations in the 1730s have survived and can be identified.[36] One, made in 1730 by Cadot, involved an ostentatious brass ornamental pattern, reminiscent of seventeenth-century instruments of Hevelius as depicted in his books. The other, made by Langlois in 1736, had a much more sober appearance and was stamped with the arms of the province as planned by de Plantade, and was therefore directly related to that institution.

To summarise this analysis, we have been able to put the construction of the Languedoc map in the perspective of a complex tension that operated at several levels. First, trust in a given instrument and its output was largely confounded with trust in the person who had used it for a long time – and opposed the distinctive regime of trust that Réaumur advocated for his thermometers. In the latter case, trust was to a large extent carried by the instruments themselves, which could ideally be used anywhere, be read by anybody and still provide meaningful comparable measurements. Such trust in things without men was allegedly incorporated in the instrument through the rules of construction and calibration proposed by Réaumur. Second, we found a similar 'objectivised' trust in regimes of exchange and ownership. There we saw an often simultaneous and occasionally contested involvement, on the one hand, in a gift-giving economy of reciprocity that was infused with the patronage values pervading the elite République des Lettres, and on the other hand an engagement in the cold and much more anonymous marketplace for commodities. Third, when considering the making of instruments under the supervision of the Parisian academicians, we saw a tension between the aristocratic-like patronage of craftsmen fiercely advocated by Cassini, in which scientific fashion, a direct result of Cassini's protection, ultimately guaranteed the skills of the craftsmen and economic competition, the active staging of which was suggested by Montferrier to meet bureaucratic needs to cut costs, and in which the price in a competitive marketplace (still built by a limited community of craftsmen, and hence not entirely competitive in a 'liberal' sense) became the guarantee of skills.

I argue that these tensions were a characteristic manifestation of a transition period. A very strong current underlying these shifting practices stemmed from the requirements of research programmes such as early eighteenth-century cartography or meteorology, which emphasised the necessity of comparable measurements made simultaneously at a distance. What I would like to do now is follow up on this key moment in the separation of space and time, of history and geography, of experiential narratives and descriptive tableaux. The early stages in the Languedoc project have guided us to recognise these separations both at the level of disciplinary boundaries and of instrumental practices. I will now try to evoke the outcome of the whole enterprise, which takes us to mid-century.

From the demise of the provincial project for a map of Languedoc to a paradox of modernity

The later debates about the Société royale des sciences project for a map of Languedoc

The fieldwork for the map of Languedoc, which started in 1730 and was expected to last four to six years, was still not completed in 1748. Clapiès had slowly faded out of the picture and de Plantade had died on the Pic du Midi in 1741 after doggedly triangulating his due share of the province, leaving Danizy responsible for the operations performed by the Société royale des sciences of Montpellier. Lacking proper documentation for the 1730–48 period due to a lapse in the available archives, in 1748 we discover Danizy under a great deal of pressure from the provincial Etats to provide adequate material for the geographer Philippe Buache. Buache had taken over Delille's private practice and academic status as a geographer, and needed the material for the construction and printing of the whole map. Trying to retain some degree of control over the project, Danizy and the engineer Barthès walked a tightrope, having to justify the fact that only the maps of a few dioceses had been completed, and to provide Buache with suggestions about the proper way to process the collection of maps (proposing a single scale, advising against drawing the triangles on the final maps, etc.).[37] After this interaction with observers in the field and the evaluation of the state of their work, Buache drew up a draft contract between himself and the Etats that covered his reworking of all the material obtained in the field and the drawing and printing of the maps in the form of an atlas of the province.[38]

The Etats readily accepted this draft, albeit with various amendments suggested either by Danizy or Montferrier.[39] However, some doubt about the accuracy of the work performed by the Montpellier academicians was apparent. While praising the efforts of Danizy and his colleagues to link their triangles to those of the Parisian astronomers, and expressing his will to believe Danizy ('for he is a skilful and hard worker'), Montferrier nevertheless explicitly refused to personally guarantee the quality of Danizy's work, and even less so the maps to which Danizy did not contribute. But whatever his reservations, he adamantly rejected the idea that one could consider 'starting again, with new methods, work that has lasted so long and already cost too much, and which the Etats absolutely want to be brought to a conclusion'.[40] Feeling threatened, Danizy also objected to Buache's intended title. In its first version, it favoured Buache's trigonometric and geometrical processing of the empirical data, particularly with respect to the resulting overlap of their triangles with those made by 'ces Messieurs de l'A.R.S.'. Danizy argued that since he and his colleagues had drawn the triangles astronomically and geometrically, there was no reason for the title to state that they had been drawn and subjected to trigonometric and geometrical observations 'by Philippe Buache'.[41] This was a clear mid-century instance of the authorship debates allowed by the ethos of astronomical and geometrical cartography mentioned in the first section, in which observers in the field could argue against the geographers for the right to authorship pertaining to maps.

Cassini de Thury's project for a bureau in charge of drawing the map of France was also taking shape at the time. In 1749, the president of the Etats, the archbishop of Narbonne, refused to submit the Languedoc project in its current state to the control of Cassini de Thury. He argued that the fieldwork had been done competently and with proper methods, and that if there still were some significant errors left this could only be attended to by new fieldwork for which the province did not want to provide funding. He proposed informing the relevant minister, Trudaine, that once duly corrected in its present state and rendered compatible with the work performed by Cassini de Thury on his own, this work would be as good as possible, and certainly should not require new expenses from the province.[42] Cassini was, however, mandated to give expert advice on Buache's drafts, and he intervened directly by proposing the same scale as his intended maps of France ('drawn by order of the king'), thus asking 'that everything that had been done until now should be redesigned thus'. His most visible concession was vindicating Buache's intended budget.[43]

The following years were to see Cassini de Thury's cartographic projects gaining momentum. The guidelines to his planned map of France were printed in 1756 and the ripples were to reach the case of Languedoc in the 1760s. His first draft for a map of Languedoc contained a severe attack of the previous work. The link with the Parisian meridian triangles were dubious because: (1) the original signals (natural or artificial milestones visible from far enough to take alignments on large triangle summits) had disappeared; (2) the practice of the Montpellier academicians that led them to favour the top of available hills for their alignments left many parts of the countryside, particularly the bottom of valleys, unseen; (3) he had noticed that some alignments had been performed with the compass, a far less appropriate geographic tool than the *planchette*.[44] Clouds also entered the relationship between Buache and Cassini de Thury, the former refusing to collaborate with the latter, and Cassini trying to dismiss the impression that he was inducing the Etats to suppress Buache's commission and turn it into his own hands.[45] But whatever justice we may give Cassini's protest (and I would not be inclined to accord it much) this was precisely what the Etats were to do a few weeks later. This was certainly not a simple negotiation, for it required several meetings of the directors of Cassini's maps with the representatives of the Etats, with Trudaine himself attending. The only concession they obtained from Cassini de Thury was that he should verify the previous operations performed by the Montpellier Académie and correct them whenever possible.[46]

Cassini de Thury grudgingly acceded to this, voicing his opinion that when compared to his own work, some of the previous triangles of the Société royale des sciences were wrong by one to ten degrees, and noting that he would give the necessary elements for his engineers to ascertain whether the discrepancies were their fault or that of their predecessors.[47] A few weeks later he felt emboldened enough to state to Montferrier that the previous map was useless to him, because of the scant data available about its computing.[48] Danizy made a last and somewhat pathetic effort, claiming that work that had already been done, the correction of which he had just completed (thirty-nine years after it was started!) and which was satisfactorily linked to the Paris meridian triangles, had no reason to be done all over again: 'It is sad, at the moment I expected to triumph and satisfy the wishes of the Archbishop, those of the Academy and your own, to have endured so much pain in vain and see my work rejected.'[49] And Danizy's last-ditch attempt was indeed rejected. After a final meeting with the archbishop, Buache and Cassini in Paris, he was informed that although his measurements were considered 'very exact and accurate'(!) his map could not be made to fit the other maps of France (that is, the Paris meridian triangles and Cassini de Thury's set of co-ordinated and triangulated maps all over France). The king wanted all maps of his kingdom to be done in the same way (that is, Cassini de Thury's), which required new fieldwork in Languedoc to provide a map within this format. This was meant to be final: 'I think it would be as improper as useless to criticise or complain about this.'[50] And final it was, for the 1770s and 1780s were devoted only to the dialogue of the Provincial Etats with Cassini de Thury's engineers. The map that the Montpellier academicians and the Etats themselves had dreamt of in the 1720s, and on which they had worked for nearly thirty years, was to be stillborn.

Establishing truth in numbers, organisational problems for the cartographic projects and a new dialogic relationship with local representatives of the higher orders

Danizy and Cassini de Thury agreed on one point at least: that a map should not be trusted on its own. It needed to be accompanied by the results of the instruments that had been used in the field; that is, by tables of measured angles, distances and relevant computations. The Etats of Languedoc had requested that all the fieldworkers 'report yearly on their operations and that they be submitted for examination'.[51] The integration of all triangles into a whole was at stake. Danizy complained that he could not link de Plantade's triangles with his own in spite of the exactness that was a well-known trademark of de Plantade's practice, because although de Plantade's map of his triangles had survived, all the records of his computations had been lost. Cassini de Thury stressed this weakness, remarking that a 'map that was unaccompanied with observations and triangles was useless to him'.[52] While earlier it had been common to supplement travel narratives ultimately founded on the authority of the author having been there and seen things with his own eyes (or occasionally quoting credible local witnesses) with detailed particular maps of interesting places, maps founded on geodetic and co-ordinated practices could not stand by themselves without a separate memoir describing instrumental practices, measurements and calculations. In the background of the map itself, trust in numbers only seemed to have replaced trust in people. This was certainly not so. Danizy stated the paradox of the matter well. On the one hand, co-ordinated astronomical and geographical observations and measurements allowed a form of mapping that astronomers represented could be performed from a distance, without requiring the engineers to visit every nook and cranny:

> One is even surprised that, according to this method [geometrical observations subordinated to astronomical observations], it would be useless to transport oneself generally to each place, that this would encourage poor work, and that a sequence of triangles drawn from mountain to mountain, or from places elevated enough for the countryside to be visible, provides the location of each place with all the required precision.[53]

On the other hand, experience showed that the greater the claim to be able to proceed by measurements that involved instruments and practices foreign to local life and lore, the greater the need for this local life and lore, for numerous problems were encountered on high places. Danizy actually noted that since the engineer did not know the detail of the remarkable places and spots he could discover from hilltops, he needed an informant to make sense of the view from above in terms of standard ground-level local experience.[54] This is absolutely and, somewhat paradoxically, the crux of the matter: 'The parts where we have found such informants (intelligent and trustworthy) have been perfect, those where we have found untrustworthy informants necessarily contained some omission or inversion.' The importance of local support led Danizy to a complex psycho-epistemological typification of his informants. He tried to distinguish between untrustworthy informants who led the fieldwork astray because they jeopardised the mapping project and feared for their own interests, and honest informants that were too confused by the view from above to identify familiar objects. When it was possible to perform several alignments with instruments, such an impediment could easily be circumvented; otherwise the astronomers and engineers had to rely fully on informants, irrespective of the potential drawbacks. This was a generic problem at the time. In a textbook on map-making, the engineer Dupain de Montesson advised cartographers not to ask direct questions, but rather to start with questions about different places or roads (so that the informant would be kept unaware of the real goal of the investigator); to ask many interrelated questions several times in different orders, carrying out a kind of cross-examination of the information offered; and to cajole and flatter the informant every time he gave a visible token of goodwill, such as correcting a previous mistake.[55]

Our actors thus needed the contribution of local inhabitants. In the 1730s the astronomers of the Société royale des sciences recruited peasants to guide them and act as local informants.[56] After the fieldwork had been performed and triangles drawn covering the dioceses, the maps were shown to local officials, asking them to certify to their accuracy.[57] Accuracy of spelling also had to be checked *ex post* in this way.[58] The officials concerned clearly dragged their feet, for Barthès complained that 'it is difficult to gather several persons for a business that does not interest them'. When trying to collect the data into a single map, Buache also planned to send a list of geographical questions with local relevance for the parish priests to answer. Danizy recalled that he had such a list sent to the local priests with the mediation of the diocesan bishops. But here again the feedback was disappointing, with answers 'too vague to be useful', so that they had to send them back with the maps themselves once these had been completed. They then received a great deal of criticism, most of which Danizy felt could easily be attended to. Interaction with locals in the fieldwork of the Languedoc map could be broken down into: (1) negotiating access to visited places through mobilisation of local officials and recruitment of peasants as help and informants; (2) once the maps had been drawn, checking with noblemen or priests (that is, literate local notables) to ascertain that they contained no glaring mistake immediately apparent to an eye accustomed to the lie of the countryside. In such a process, the business of observations and measurements was made to look as though it had been largely independent from the opinion or engagement of the literate, trustworthy and knowledgeable inhabitants (though of course a lot of polite socialising could – and usually did – take place during fieldwork).

Part of the problem was in fact organisational. Enthusiastic, in a typically 1730s fashion, about the power of astronomy and astronomy-based triangulating geodesy to obtain once and for all the location of all places within triangles without going there (that is, from a distance), with an output made of measurement figures given by comparable instruments and disengaged both from the context of their production and local knowledge (therefore arguably able to travel without further negotiation from the field to the geographer's workshop), the Montpellier academicians had strayed as far as possible from traditional travel practices. In those one would gain authority by going to the place, seeing things with one's own eyes, collecting testimonies that were as trustworthy as possible from local people, and building up from there a compound of narratives and detailed maps to elicit the attention of the République des Lettres back home. Here, going in person to the most interesting spots was deemed useless, for a lot could be worked out from a distance, while informants were used mostly for access, and the local gentry and bureaucracy called upon to confirm *ex post* the validity of the maps. Apart from all the criticism he levelled at the previous project for a map of Languedoc, Cassini de Thury introduced a new organisation and relationship with the diocesan and parochial establishment that would remedy many of the shortcomings he felt his predecessors had been unable to avoid.

First, he remarked that locations for which the intersecting alignments could not be taken from the high places that served as triangle summits had to be mapped with standard geographical tools such as the plane table. He was adamant that after verifying the details of the operations of the previous team, he could detect that many angles for those difficult places were measured with the compass, which he considered far less suitable for the purpose than the plane table. To avoid this particular pitfall he asked the engineers working out the smallest-scale *cartes particulières* (therefore the lowest rank of engineers in his social division of work) to visit the priests and local gentry *during their fieldwork*. This was not only to check that nothing of interest was being omitted, but also that the local notables could testify to Cassini in writing that 'the engineers have observed all visible points not with the compass but with the *planchette*'. They were thus to be deeply involved in the cartographic project as discipline enforcers, for Cassini de Thury eventually asked that 'everything be done under the eyes of the gentry and the priesthood' while mapping was under way. This intervention was required during the process of mapping and not afterwards, as was the case before, because then errors could be compounded and thus made irretrievable: 'The engineers only build their maps after their return to Paris. They may have represented the field on drafts where points were not set correctly, and the engravers may not have copied their drawings faithfully.' But a solution was at hand: 'Here are many causes for error, which can be remedied only by showing each lord or priest the map of his land or parish.' One must read 'draft' rather than 'map' here. This was to be done while the fieldwork was under way and while that which had been said could be checked by measurements: 'They [gentry and priesthood] will tell it [whatever is faulty] to the engineer; the latter, *carrying his instruments*, will correct it and add what is missing' (my emphasis). The more abstract and objectivised the instrumental work, the more it needed to be performed in a dialogic relationship with the local elite. Here lay the path to perfection for 'by this means each map being approved by all persons that *alone* are entitled to criticise it, will be as perfect as one might expect'.[59] Danizy himself appeared to concur, suggesting later to Montferrier that to work in this way it would be better if the mandates of his engineers appeared to be given by the Etats of the province, for this might ease the dialogue in the field.[60] Of course the dialogic construction of the map in the field would not preclude *ex-post* verifications of the final proofs by the parish literate.[61] However, at that later stage there would be no detailed dialogue between local experts and instrument-laden engineers; the local experts would merely have to indicate blatant inversions of places with respect to one another, other errors falling into the span of intervention of trigonometric computations.

To summarise, we see that producing a map founded on triangulation measurements only, and integrating three scales (the smallest one of the detailed *cartes particulières*, the intermediate one of the triangular canvassing of the province, and the larger one of the French territory or even the globe, which is the scale of the Parisian meridian measurements), involved more than simply taking distant measurements with comparable instruments, plus the circulation and trigonometric processing of the resulting figures. It directly engaged the dialogic intervention of the local gentry and priesthood during the measurement campaigns themselves. There is indeed a paradox embedded here: the more the Languedoc cartographic process seemed to rely on instruments, measurements and their trigonometric processing, rather than on the testimonies of travelling individuals, the more it involved engaging credible people in the dioceses and parishes in a dialogue that had to occur during the fieldwork, when the engineers were using their instruments (and not afterwards). The condition for trusting the figures was a mobilisation of local people while the measurements were under way. It was therefore a social and moral problem, as historians have argued.[62] I think, however, that the crucial issue here is the co-ordination of measurements taken by comparable instruments at a distance. It apparently disengaged the fieldworkers from a form of travel, the accounts of which were founded on eye witnessing. But where the latter meant the construction of an authoritative discourse about the spaces and sites it described that did not require explicit confirmation by the natives, the co-ordination of distant measurements required the representation of encounters in the field – in other words, an ongoing dialogue with the 'natives' combined with instrumental measurements. The more 'objective' and disengaged from local social context the measurements (that is, the more separated in appearance the abstract space provided by the instruments and the time and history-loaded space of local experience), the more some forms of encounters would become useful resources to vindicate the trust in the map that was to be the outcome of the cartographic project. The enhanced sensitivity to local context necessary for the construction of knowledge would not solve all the moral and political implications of such encounters; they would have to be worked out in the goal-oriented setting of the scientific activity and within each specific context. It would, however, provide distinct new frames for assessing distant social spaces and establishing channels of trust with the mediation of 'objectivised' instruments and measurements.

Conclusion

I argued in the introduction that the project for the map of Languedoc could be construed as an 'exceptional-normal case', in the spirit of *Microstoria*. I have tried to show how it rendered visible some of the alternatives that constituted the web of meaning in which the practices of our various actors made sense. It remains to be seen now how their uneasy path helped to carve the shape of things to come. A good example of this influence in shaping what was to follow in a path-dependent way was Cassini de Thury's criticism of the way the Montpellier academicians used their signals. One of the core criticisms Cassini de Thury addressed to the Languedoc project concerned the signals they built:

> they judged they should make a choice of the higher places in the mountains that could be seen everywhere, and since such places did not offer any determined objects, they had signals erected that could only be used during their observations there, and that one could not find when working out the details of the province.[63]

The transient nature of the signals built and used by the Montpellier academicians was thus exploited in two distinct ways by Cassini de Thury, both relating to the integration of the different scales of the project. Because the signals for the triangles were temporary, the link between the triangles and the *cartes particulières* was threatened; that is, the link between the fieldwork of astronomers/geometers and engineer geographers. For the very same reason, any certainty about the fact that Parisian signals and provincial ones erected for connecting triangles between the meridian and the province were one and the same thing was precluded, and the overlap of the provincial canvassing and the Paris meridian was deemed at best very dubious by Cassini de Thury. As we have seen, the latter point was relayed by the Etats and harshly conveyed to Danizy definitively to dismiss the provincial cartographic project.[64]

For a whole century of astro-geodetic cartographic triangulation, however, the geometers of the Académie royale des sciences had been climbing the mountains encountered in the course of their triangulations, and building signals on them. Mountaintops were previously considered inhospitable and improper locations for the literati, and in the first half of the century astronomers engaged in geodesy and cartography had been the only ones to climb them in a systematic manner.[65] This was a component of a public ethos for the astronomy-inclined and observatory-wise geometer, cartographer or hydrographer. One need only recall the place given by La Condamine to various ascents in his accounts of the expedition to Peru. Even accounts of more ordinary trips feature characters such as the hydrographer Laval climbing mountains in southern France in the 1710s and advising the young noblemen who wanted to accompany him that such a place was unfit for them. The Montpellier academicians had been following a path that French astronomers (particularly those of the Observatoire Royal) had trodden repeatedly, and they plied a trade learnt essentially from them (one must remember that de Plantade worked with Cassini and Maraldi in the 1700 extension of the meridian measurements).

But very often the signals built on the mountains did not last long. La Condamine complained bitterly of the dumb Peruvian Indians who destroyed their signals on the Andes. Cassini de Thury himself acknowledged this point, recognising that his own signals did not exist any more, for he used to work in the very same way he was now criticising. The issue was now that the context did not simply involve triangulating for measuring large bases, whether meridians or parallels; rather, it meant drawing a map integrating the local and the global on three scales, of which triangles were intermediate. This, in turn, required repeated use of known signals for the practitioners working on each scale so as to share a fixed reference point for their angles. Without that there was simply no way to posit smaller maps within larger frames. Mapping was not a definitive set of measurements performed by a tight-knit unit of practitioners, it was a process involving the co-ordination of several groups working at different scales and different times – groups of engineers, each with instruments of a given size and scope, advocated by Cassini de Thury. But this idea of a process went further. Independent groups had to be able to co-ordinate their work (which the Montpellier and Paris astronomers might have done, had signals survived long enough to be shared by them). A stable structuring of the landscape with milestones that were useful for instrumental triangulation was an essential part of the co-ordination within this larger framework. While astronomers in the first half of the eighteenth century, including those in Montpellier, had proceeded with the idea that a single measurement or expedition would settle the geographic issue of fixing the relative loci of given places (which made the proposal to use co-ordinated observations of astronomical events at a distance so appealing), in this particular context of an existing provincial mapping project Cassini de Thury was pushed towards the necessity to allow for the repetition, replication and co-ordination of measurements and field itineraries, supported by permanent signals. (The problem was exacerbated even when compared to his own mapping of France because, being unable for several years to dismiss entirely the calculations of his predecessors, he had to find a way to assess them in relation to those of his engineers.) This, in turn, involved his advocating less hill-climbing in favour of using well-known and permanent visible signs such as church or castle towers.

Notes

1 G. Levi, *Le pouvoir au village. Histoire d'un exorciste dans le Piémont du XVIIème siècle*, Paris: Gallimard, 1989 (English translation: Giovanni Levi, *Inheriting power: the story of an exorcist,*

Chicago: University of Chicago, 1988); J. Revel (ed.) *Jeux d'échelles. La micro-analyse à l'expérience*, Paris: Gallimard, 1996.

2 Most of the remaining letters and drafts concerning the Montpellier project can be found in Archives de l'Hérault, folders C7909/7910, folders C 4671 and 7909/7911. In the following notes and to avoid further repetition, I use the abbreviation Archives de l'Hérault.

3 While not every memoir survived, and those mentioned here were not published in their entirety, useful references can be found in L. André, *Les sources de l 'histoire de France au XVIIème siècle*, t.VII, Paris, 1938, pp. 180–207.

4 E. Esmonin, 'Les mémoires des intendants pour l'instruction du duc de Bourgogne (étude critique)', *Bulletin de la Société d'Histoire Moderne*, LV, 1956, 12–21.

5 A few were published, but scattered manuscripts and corrective memoirs can be found in the Fonds Réaumur, at the Archives of the Académie des Sciences in Paris.

6 H.M.A. Berthaut, *Les ingénieurs géographes militaires, 1624–1831; étude historique*, Paris: Service géographique de l'armée, 1898.

7 Montferrier to the widow Delille, letter dated 6 July 1728, Archives de l'Hérault, folders C7909/7910.

8 Mémoire de la Société royale des sciences au sujet de la carte du Languedoc, undated, probably 1729, Archives de l'Hérault, folders C7909/7910.

9 W. Eamon, *Science and the Secrets of Nature. Books of Secrets in Medieval and Early Modern Culture*, Princeton, N.J.: Princeton University Press, 1994.

10 M. Pastoureau, *Les Atlas français XVIème–XVIIème siècle*, Paris: Bibliothèque Nationale, 1984.

11 Veuve Delille, Devis envoyé à Mr de Montferrier, syndic des Etats du Languedoc, Archives de l'Hérault, folders C7909/7910.

12 Montferrier, Lettre au père Vaisselle, Archives de l'Hérault, folders C7909/7910.

13 Delille widow to Montferrier, dated 22 June 1728; Montferrier to the Delille widow, letter dated 6 July 1728, Archives de l'Hérault, folders C7909/7910.

14 Nouveau plan pour la levée de la carte géographique de la province, 1732, Archives de l'Hérault, folders C7909/7910.

15 The case of her husband Guillaume Delille is less clear-cut, as shown below, Archives de l'Hérault, folders C7909/7910.

16 Devis de la veuve Delille envoyé à Montferrier, syndic des Etats du Languedoc, *c.* 1728, Archives de l'Hérault, folders C7909/7910.

17 Mémoire de la Société royale des sciences au sujet de la carte du Languedoc, probably from 1729, Archives de l'Hérault, folders C7909/7910.

18 J.D. Cassini, 'De l'origine et du progrès de l'astronomie et de son usage dans la géographie et dans la navigation' [1693] *Mémoires de l'Académie royale des sciences depuis 1666 jusqu'à 1699*, t. 8, 1730, p. 37.

19 Note from de Plantade and Clapiès, 1 May 1730, Archives de l'Hérault, folders C7909/7910.

20 J. Bennett, *The Divided Circle: A History of Instruments for Astronomy, Navigation and Surveying*, Oxford: Phaidon Press, 1987, pp. 84–7.

21 Mémoire de la Société royale des sciences au sujet de la carte du Languedoc, undated, probably 1729, Archives de l'Hérault, folders C7909/7910.

22 Cassini, Projet pour l'exécution de la carte du Languedoc, 1768, Archives de l'Hérault, folders C7909/7910.

23 Letter from Pezenas dated 24 May 1766, Archives de la Marine, folder G90/91.

24 Letter from Pezenas to the Duke of Choiseul dated 29 May 1766, Archives de la Marine (Paris), folder G90/91.

25 Letter from Pezenas dated 28 May 1766, Archives de la Marine, folder G90/91.

26 Letter from Saint Jacques de Silvabelle dated 26 May 1766, Archives de la Marine, folder G90/91.

27 S. Shapin, *A Social History of Truth*, Chicago: Chicago University Press, 1994.
28 C. Licoppe, *La formation de la pratique scientifique. Le discours de l'expérience en France et en Angleterre (1630–1820)*, Paris: Éditions La Découverte, 1996.
29 Amontons, 'Discours sur quelques propriétés de l'air & le moyen d'en connaître la température dans tous les climats de la terre', in *Mémoires de l'Académie royale des Sciences pour 1702*, 1704, pp. 161–80.
30 Réaumur, 'Règles pour construire des thermomètres dont les degrés soient comparables', in *Mémoires de l'Académie royale des Sciences pour 1730*, 1732, pp. 452–506, p. 453.
31 Maraldi à Montferrier, 19 January 1729, Archives de l'Hérault, folders C7909/7910.
32 Montferrier à Cassini, 3 January 1730, Archives de l'Hérault, folders C7909/7910.
33 De Plantade à Montferrier, 8 March 1730, Archives de l'Hérault, folders C7909/7910.
34 Marcel Mauss, 'Essai sur le don. Formes et raison de l'échange dans les sociétés archaïques', in *Sociologie et Anthropologie*, Paris: Presses Universitaires de France, 1950. (English translation: Marcel Mauss, *The gift: the form and reason for exchange in archaic societies*, Norton, 2000).
35 P. Findlen, *Possessing Nature: Museums, Collecting, and Scientific Culture in Early Modern Italy*, Berkeley: University of California Press, 1994.
36 F. de Dainville, *Les cartes anciennes de Languedoc, XVIème–XVIIIème siècle*, Montpellier: Société Languedocienne de Géographie, 1961.
37 Letter from Danizy to Buache dated 21 July 1748; letter from Barthès to Buache dated 20 August 1748, Archives de l'Hérault, folders C7909/7910.
38 P. Buache, Mémoire sur la carte de la province de Languedoc from 12 October 1748, Archives de l'Hérault, folders C7909/7910.
39 Letter from Montferrier to Buache, dated 21 February 1749, Archives de l'Hérault, folders C7909/7910.
40 Ibid.
41 Danizy, Observations et réponses aux mémoires de Mr Buache pour servir à perfectionner la carte générale du Languedoc et de chaque diocèse, undated, probably 1749, Archives de l'Hérault, folders C7909/7910.
42 Letter from the archbishop of Narbonne to Joubert dated 23 August 1749, Archives de l'Hérault, folders C7909/7910.
43 Cassini de Thury, Mémoire sur les changements désirés par les Etats au projet de M. Buache, 1753, Archives de l'Hérault, folders C7909/7910.
44 Cassini, Projet pour l'exécution de la carte du Languedoc, 1768, Archives de l'Hérault, folders C7909/7910.
45 Letter from Cassini to Montferrier dated 30 July 1768, Archives de l'Hérault, folders C7909/7910.
46 Map of Languedoc, about the treaty of the Etats with Cassini dated 29 September 1768 amended on 28 May 1771, Archives de l'Hérault, folders C7909/7910.
47 Letter from Cassini dated 25 October 1768, Archives de l'Hérault, folders C7909/7910.
48 Letter from Cassini to Montferrier dated 5 November 1768, Archives de l'Hérault, folders C7909/7910.
49 Letter from Danizy to Montferrier, dated 3 January 1769, Archives de l'Hérault, folders C7909/7910.
50 Letter from Montferrier to Danizy dated 9 January 1769, Archives de l'Hérault, folders C7909/7910.
51 Danizy, Mémoire sur les opérations géographiques pour la levée de la carte générale du Languedoc, Archives de l'Hérault, folders C7909/7910.
52 Letter from Cassini to Montferrier dated 5 November 1768, Archives de l'Hérault, folders C7909/7910.

53 Danizy, Mémoire sur les opérations géographiques pour la levée de la carte générale du Languedoc, Archives de l'Hérault, folders C7909/7910.
54 Ibid.
55 D. de Montesson, *L'art de lever les plans de tout ce qui se rapporte à la guerre, et à l'architecture civile et champêtre*, Paris, 1763, p. 158.
56 Mémoire de la Société royale des sciences, 1 May 1730.
57 Letter from Barthès on 20 August 1748, Archives de l'Hérault, folders C7909/7910.
58 Buache, Tables et mémoires, December 1748, Archives de l'Hérault, folders C7909/7910.
59 Cassini de Thury, Projet pour la perfection de la carte de la France, undated MS, Archives de l'Hérault, folders C7909/7910.
60 Letter from Danizy dated 3 January 1769, Archives de l'Hérault, folders C7909/7910.
61 Letter to MM. Les syndics des diocèses dated 6 March 1777, Archives de l'Hérault, folders C7909/7910.
62 T.M. Porter, *Trust in Numbers, the Pursuit of Objectivity in Science and Public Life*, Princeton, N.J.: Princeton University Press, 1995.
63 Cassini de Thury, Projet pour l'exécution de la carte du Languedoc, 1768, Archives de l'Hérault, folder C4671.
64 Letter from Montferrier to Danizy dated 10 January 1769, Archives de l'Hérault, folders C7909/7910.
65 N. Broc, *Les montagnes vues par les géographes et les naturalistes de langue française au 18ème siècle*, Paris: Bibliothèque Nationale, 1969; M.-N. Bourguet and C. Licoppe, 'Voyages, mesures et instruments: une nouvelle expérience du monde au siècle des lumières', *Annales. Histoire, Sciences Sociales* 52 (5) 1997, 1115–51.

4 The travels and trials of Mr Harrison's timekeeper

Jim Bennett

Introduction

The story of John Harrison and his struggles to win the Longitude Prize in eighteenth-century England has come to be treated as a single, linear tale with a consistent moral: the just and heroic struggle of a virtuous individual against the calumnies of powerful vested interests.[1] While this account seems naive in its one-sided appropriation of virtue, it is also much too simple and one-dimensional to cope with the complexities of the historical record. The different interests at work exemplified both reason and prejudice, and even within the Harrison faction he and his supporters waged their campaign to win the prize for some forty years and over that time adopted different stratagems in different circumstances.

It is helpful to identify three episodes in the complex story.[2] The first covers Harrison's reception in London society, his early support from the Royal Society and the Board of Longitude, and the making and publicising of his four timekeepers. This covers the period from 1730 to 1761. It was in the latter year that Harrison requested a trial for his watch, known as H4, under the terms of the Longitude Act of 1714. The second episode, the one to be considered here, concerns Harrison's dealings with the Board, and to a lesser extent with Parliament and with the Royal Society, during the course of the trials and their aftermath. The third episode saw Harrison treating with the King and Parliament for the payment of the second half of the reward he believed was his due but had been denied by the Board, who were scarcely involved with this final phase of his story. The end of the second episode is not so clearly marked as its beginning, but is perhaps best dated to the period between Harrison completing his second watch, H5, in 1770 and its trial at George III's observatory at Richmond in 1772. The final grant was made by Parliament in 1773 and Harrison published the last of what was by then a long series of tracts and pamphlets in 1775,[3] which might reasonably be taken as the end of the whole turbulent business.

From 1761 onwards, the Board of Longitude faced some of the difficulties and questions that arise at the point where instruments meet both travel and science. They were called upon to arrange a very public test of the ability of an instrument to travel, and by travelling successfully this instrument would itself enable successful and advantageous travel more generally. John Harrison presented them with the problem of testing a watch on a voyage to the West Indies, so as to measure the longitude at sea by comparing standard time carried by the watch with local time discovered from the sun or stars. According to the Longitude Act of 1714 that they were charged with administering, if such a feat could be performed to within an accuracy of half a degree, the deviser of the method would be entitled to a reward of £20,000, lesser sums being stipulated for lesser accuracy.

Along with various admirals and other naval officials, the Board included the Speaker of the House of Commons, the President of the Royal Society, the professors of astronomy at Oxford and Cambridge and the Astronomer Royal. Their situation was complicated by the fact that Harrison's was not the only invention claiming a reward, and the competing methods were fundamentally different in character. Along with devising and administering adequate testing procedures, they found themselves wrestling with such fundamental issues as the distinction between discovery by method and by accident, and then the very nature and meaning of 'discovery' itself.

A great deal hung on the result of the Board's determinations and the competence of their procedures – both large sums of government money and profound developments in navigational practice and, consequently, in economic and political development. There were two possibilities of failure in the Board's competence and either would represent dereliction and disgrace. They might give the prize away for a less than satisfactory technique, and thereby waste a great deal of public money in a very public manner. Or they might fail to recognise the importance of the technique, fail to capture what control over it might be possible and thus lose a profound technological and economic advantage that should fall to British seamen.

Voyage to Jamaica

Establishing a proper procedure for testing the reliability of an instrument thousands of miles from their supervision was the challenge facing the Board in 1761 when John Harrison applied to them to arrange a trial of two timepieces.[4] By this stage Harrison had received many years of funding from the Board and, whatever doubts they may have had about the likely performance of his machines under the demanding and adverse conditions of a long voyage, they were effectively committed to such a trial. But how was it to be carried out when there were no precedents available to them? The Commissioners for the Longitude sought the advice of the Royal Society.

The Society responded with 'A Plan for an adequate Trial of Mr Harrison's Time Keepers, as recommended by the President and Council'.[5] The essential components of this plan were as follows. Local time at Portsmouth would be determined by John Robertson, then first master of the Royal (Naval) Academy there, and the timekeepers set accordingly under his supervision. Robertson was a credible witness, known to the London mathematical circle, having been mathematical master at Christ's Hospital. He would later be appointed clerk and librarian to the Royal Society. Some of the ship's officers were to be appointed 'Witnesses who are always to accompany Mr Harrison' when he had access to the timekeepers during the voyage, and the times shown by the timekeepers would be recorded at least once a day. A competent person would be sent on board the same ship to determine the local time on arrival, to note down 'Before Witnesses' both this result and the time given by the timekeepers, and to send these results under seal to the Admiralty. The same person 'or some other sufficient Person' would remain in Jamaica to determine the longitude by eclipses of Jupiter's satellites. A similar determination would be made at Portsmouth, and Nathaniel Bliss, Savilian Professor of Geometry at Oxford, was willing to undertake this task.

One telling stipulation concerns the instruments appropriate to the two measurements of longitude by Jupiter's satellites. The telescopes to be used at Jamaica and at Portsmouth were to be 'of the same focal Length and magnifying power'. Indeed, the Society stipulated sending to Jamaica 'an equal altitude Instrument, a reflecting Telescope of two feet focal Length and a good Astronomical Clock', and 'a similar set of Duplicates' to Portsmouth.

This provision reflects a problem with the accuracy of this method for finding longitude on land. It was not clear how the moments of immersion or emersion of the satellites with respect to the planet were to be defined; the phenomena were not clear-cut and the determination depended on the judgement of the observer, the seeing conditions and the quality of the telescope. The magnification in particular was critical. It was therefore important to have the telescopes used at the different stations as similar as possible, so as to minimise differences in the observing conditions. Duplicate instruments left the other variables unchecked, but even this level of control was not straightforward, as reflecting telescopes, such as were commonly used for such observations to improve definition with their wider apertures, had not been generally associated with precision instrumentation. Their manufacture was a hand-craft skill and the figuring of the mirrors in particular a trade secret, while the use of reflectors was linked more with popular star gazing than with professional measurement. It is not clear that 'duplicate' instruments could be made, any more than duplicate observers or conditions of seeing, yet the whole method depended on the identification of simultaneity by observing an instantaneous astronomical event from separate continents.

Harrison heard that the Society's proposal for the person appointed to travel to Jamaica was Captain John Campbell, one of the Royal Navy's most experienced officers, who had accompanied George Anson on his circumnavigation of 1744. Anson was now first lord of the Admiralty and as such the recipient of Harrison's various memoranda. Between 1757 and 1759 Campbell had tested, on the Board's behalf, Tobias Meyer's design of the reflecting circle, made by John Bird. This was, of course, connected with the rival, lunar method of finding longitude at sea, and Campbell is generally credited with having the first sextant made by Bird to the same end. It is unlikely, given his association with the method of lunar distances, that he seemed a 'sufficient Person' to John Harrison.

Harrison seized an opportunity in October 1761 to propose a different protocol and on the apparently innocent grounds of inconvenience with respect both to Jupiter (it was now too late in the year for the observations) and Campbell (he had been given another public commission) both were absent from his suggestions.[6] The planned departure of the newly appointed governor of Jamaica, William Henry Lyttleton, coincided with his watch being 'ready for making the Experiments of the Longitude'. He no longer wanted his third timekeeper included in the trial. Local time at Portsmouth would be determined by Robertson, as suggested before. But Harrison now proposed that his son William, who it had already been decided should accompany the watch on account of his father's age, might be supplied with an equal altitude instrument for finding local time at Jamaica 'before Witnesses', and comparing this with the time given by the watch.

This much, Harrison said, was sufficient under the Longitude Act: 'This Experiment, as your Memorialist apprehends, is what the Act of Parliament, concerning the Longitude, demands and requires to be made.' But he was willing to go further and suggested that the procedure could be repeated on the return voyage to Portsmouth. Harrison now referred to 'these Two Trials'.

The Board were not to be so easily moved. However, they were under pressure to discharge their public responsibility without undue delay, and the departure of a new governor was a very visible opportunity to do so. Failure to take it would certainly attract criticism from Harrison's supporters. But William Harrison could not be responsible for the time determination – even under the scrutiny of witnesses, that would scarcely be impartial. The Board's conditions were that Robertson should find the time and see the watch set as before, witnessed by Captain Richard Hughes, Commissioner of the Navy, Captain Dudley Digges, who was in command of the *Deptford*, the ship bound for Jamaica, and William Harrison.[7] The watch was to be secured by four locks with different keys held by Governor Lyttleton himself, the Captain and first Lieutenant of the ship and William Harrison. Thus all four would need to be present for the daily routine of winding the watch to ensure that no illicit tampering took place.[8] John Robison was to be taken to Jamaica on the *Deptford* and would find the local time on arrival, witnessed by Lyttleton, the Captain, the first Lieutenant and William Harrison, and the same group would also attest to the time given by the watch.

Campbell's replacement, Robison, was a graduate of Glasgow and would later become Professor of Natural Philosophy at Edinburgh. No replacement was offered for the Jupiter's satellites method of finding the longitude of Jamaica, an omission that is perhaps the greatest anomaly of these modified proposals. Under them the longitude would indeed be found by the watch, by comparing time carried from Portsmouth with local time determined in Jamaica; but against what standard would this be tested, so as to discover the discrepancy – the difference that would determine the level of any award? What faith could be placed in existing measurements or estimates of the longitude of Jamaica? There is an even more fundamental point here. If there were doubts over the methods of finding longitude on land – arising from the determination of simultaneity – what was the point of finding longitude at sea, so as to determine position with respect to landfall? The legislation was drawn up in terms of a discrepancy with a standard which, it suddenly seemed, may not exist. This first trial simply ignored the problem.

'Astrophilus', writing in the *Gentleman's Magazine* in October 1761, in advance of the voyage, was a well-informed and exacting commentator.[9] His understanding, in the absence of a determination of a standard longitude, was that the method depended on keeping the longitude account by the watch in both directions and comparing the values found on the outward and return voyages, but he regarded this as inadequate. In fact he points, reasonably enough, to an inadequacy even in the Royal Society's proposed method, the one that had included independent determination by Jupiter's satellites. The equal-altitude measurement of local time was to have been made on land, whereas of course in any real navigational exercise this would be performed at sea. His point was that what was being tested was the longitude method, not simply the going of the watch, and finding local time at sea was essential to the method. It was not at all clear how well local time could be determined at sea with a Hadley quadrant, but it would certainly be much less reliable than on land with one of Bird's fixed instruments.

Astrophilus thought there would be a future role, perhaps a complementary one, for lunars, pointing out that that was, as distinct from the watch, a 'method which is founded in nature'. His account also mentions a third competitor – the 'marine chair' designed by Christopher Irwin. This was a form of suspension intended to give an observer a sufficiently steady platform for observing Jupiter's satellites at sea. The possibility of this land-based longitude method developing into a technique for use at sea may be part of the explanation of Harrison's antipathy to the use of Jupiter's satellites.[10]

After hearing the conditions stipulated by the Board, the Harrisons visited the workshop of John Bird to see the instruments that were to be used for the trial. Why should they have done this, when William Harrison was no longer to be the observer? The inspection is indicative of a climate of caution and even mistrust, confirmed by Harrison's account of meeting at the workshop James Bradley, Savilian Professor of Astronomy and, as such, a Commissioner for the Longitude: 'The D^{r}. seem'd very much out of temper and in the greatest passion told M^{r}. Harrison that if it had not been for him and his plaguy Watch Mr Mayer and he should have shared Ten thousand Pounds before now.'[11] Harrison noted that he had formerly considered Bradley one of his best friends on the Board, 'but now self Interest seemed to be the Principle'.

The voyage went ahead as planned, but in Jamaica the equal altitude observations were slow to get under way, and then, so as to catch a ship returning unexpectedly to England, were hurried and inadequate. One pair of measurements were taken on only one day. In addition, Harrison's account reads as though they were the work of Robison and he working together, with no other witnesses. A further problem emerged at the subsequent consideration of the results by the Board. They insisted that the pair of sights – before and after noon – had to be taken from the same station, and apparently this had not always been done, at least not at Portsmouth. Harrison disputed the necessity for this as well as the obligation to take a series of measurements. The Board called on testimony from John Bird: 'he said the more Observations to be sure it must be the better, But if the Instrument had been moved from one place to another it could not be depended upon'. According to the manuscript record prepared by the Harrisons and known as the 'Harrison Journal', 'this very much surprized Mr. Harrison as he always took Mr Bird to be better acquainted with the Instrument'. Bird was probably the leading instrument maker in Europe at the time, and the Board would have had no difficulty deciding where the greater authority lay on this question.[12]

The Board arranged for three mathematicians to make independent calculations from the observations and considered the results at a meeting on 17 August 1762. They concluded that 'the Experiments made of the Watch, have not been sufficient to determine the Longitude at Sea'.[13] Harrison's own calculations concluded that the watch had 'lost' only 5.1 seconds on the voyage to Jamaica, but this figure was arrived at only after applying a losing rate of 2.66 seconds per day. The legitimacy of applying the rate was denied by the Board. It is not clear whether they had previously reached any understanding about this, but certainly Harrison had not declared a rate before the voyage so could not, it was determined, declare and apply one retrospectively. The point was an important one, for the difference in distance determination with and without the rate was that between 52.5 miles and 1.25 miles.[14]

The Board were beginning to appreciate some of the problems attached to any real, practical trial – a component in the reward scheme that sounded straightforward enough when stated in the legislation but which in practice was becoming beset with difficulties. They realised that they were encountering two levels of difficulty, which they now differentiated. First there were the technical difficulties of an adequate trial, of testing whether a device met the conditions laid down in the original Longitude Act. But there was the further question of whether satisfying these conditions was anything like the same thing as finding the longitude. The Act stipulated that the reward would be paid

> when a Ship by the Appointment of the said Commissioners . . . shall thereby actually Sail over the Ocean, from Great Britain to any such Port in the West-Indies, as those Commissioners . . . shall Choose or Nominate for the Experiment, without Losing their Longitude beyond the Limits before mentioned.[15]

But it also said that the reward was to be paid when the Commissioners had found a longitude method to be 'Practicable and Useful at Sea'. The Commissioners who were now actually faced with the question were finding that these conditions might not be equivalent, even though they were enshrined in the same Act. Little wonder that they would seek to resolve their dilemma through further, clarifying legislation. The Cambridge mathematician William Ludlam, who was sufficiently close to the Board to be appointed by them as a witness when Harrison eventually explained his watch, wrote that the Act was 'very defective', since 'The enacting part, taken exclusively of the preamble, seems to consider the whole reward on the accidental success of one trial.'[16]

If we return to the record of the Board's meeting on 17 August 1762 we see these problems at work. Their dissatisfaction with the rigour of the trials was presented to the Harrisons, who first thought they were still dealing with objections to the equal altitude observations, but they were immediately reassured by Bliss, who was now Astronomer Royal, that this was not the stumbling block. In fact the problems went much deeper. Harrison's record of the response from Bliss is telling:

> Supposing the watch had answer'd in this Trial is that any reason it would do so again, but we do not Know that it has answered, for we do not know the Longitude of Port Royal in Jamaica.[17]

It is not clear from the record that Harrison had appreciated the significance of these points, but Bliss had succinctly stated two fundamental and quite different problems, which, since this was a verbatim record of the meeting of the Board, must have been exercising them at the meeting. One (the second stated by Bliss) concerned the implications of the uncertainty over the longitude of the target, and perhaps the uncertainty generally over longitudes of target landfalls. After Harrison's objections to using Jupiter's satellites, he was finding that the absence of a standard was now a problem for this particular trial, but Bliss's other point addressed the efficacy of trials in general and the need for greater reassurance than could be delivered by a trial with respect to any solution to the longitude.

How, the Board wondered, were they to know whether a single determination, such as that stipulated in the Act, was not a fluke? How could they be sure that the watch could be reproduced in numbers by the generality of competent workmen? For much of his early years in London Harrison had been presented to the public, who came to see him and his clocks, as a very special case, as one whose natural ingenuity had not been dulled and distorted by education. What guarantee, now, could there be that his methods were generally applicable? The Board did not know what these methods might be, but if Harrison's skill really was intuitive and unlearned, could he formulate and articulate it even if he wanted to? These questions concerned the 'Practicable and Useful at Sea' side of their responsibilities under the Act, and they were to come to greater prominence later, once a properly regulated trial had been completed. For the present, the Board had learnt much about the kind of rigour that must accompany such trials. They had entered unfamiliar territory, but were learning fast, and the conditions for the new trials they realised would have to be made, reflect this.

The meaning of 'discovery'

Between the two trials, the turn of events was such as to precipitate some further thinking about how a watch might represent a solution to the longitude problem.

Ironically this consideration was prompted by Harrison or, as he was to claim, by the strong advice of his friends. This may be significant, for these friends were undoubtedly closer to the objections of the mathematical members of the Board than Harrison was himself. He was always inclined to see the matter simply in terms of compliance with the letter of the 1714 Act; others could appreciate the argument that a longitude solution could not rest on a watch having once, or even twice, performed the feat stipulated in the Act. It was necessary for there to be an account of how it worked, an understanding of this and the possibility of replication by other makers. It was here that the lunar method had the advantage, and to match it there would have to be, literally, a process of 'disclosure' or 'discovery'. Harrison would have to open his watch in appropriate and carefully regulated company and explain how it worked.

The increasing influence of Harrison's 'friends' is indicated also by the beginning of a propaganda campaign, signalled first by the appearance of three broadsheets in December 1762 and then by two editions, the second greatly expanded, of the pamphlet *An Account of the Proceedings in order to the Discovery of the Longitude at Sea.*[18] None of these was written by Harrison himself, and the pamphlet at least has generally been ascribed to the telescope maker James Short. Short made only reflecting telescopes, so had no professional interest in lunar distances, which would depend on the kind of instruments for precision measurement that were the prerogative of John Bird.

Harrison's friends were sure that a second trial should not be contemplated 'before he had made the discovery of his invention', and in a memorandum to the Board Harrison offered to 'disclose' its principles, given certain safeguards. This led to Harrison petitioning Parliament to the effect that he was willing to 'disclose the manner and principles of Framing it [the watch]' for an appropriate reward, the remainder to be paid on further trial. Harrison surely did not write this petition, which responded to the Board's more general worries by explaining that the disclosure would be 'in such a manner as that other Workmen may be able to Execute the same so that it may in a short time become Serviceable to this Kingdom and to all who use the Sea'.[19]

In the short term this petition met with success. After a committee of the House of Commons had taken evidence from various witnesses and made its report, an Act was passed appointing 'commissioners' to arrange and supervise disclosure. Their declared satisfaction would bring Harrison a reward of £5,000.[20] This sum would be deducted from any subsequent longitude prize awarded after further trial arranged in accordance with the original Act. There were now, for a relatively short period, two sets of commissioners, the original 'Commissioners for the Longitude' or 'Board of Longitude', and the 'Commissioners for the Discovery of Mr Harrison's Watch'. The latter numbered eleven and included mathematicians and fellows of the Royal Society alongside five working watch and clockmakers. One member, in the former category, was James Short.

The problem faced by these Commissioners was rather different from the one that had confronted the Board. It was no longer a matter of arranging and supervising adequate trials, but of arranging and certifying adequate 'disclosure'.

At a meeting on 13 April 1763 they determined what they considered would amount to 'a full and Clear discovery of the Principles of his Watch for the discovery of Longitude'.[21] Eight members were present: Lord Charles Cavendish, FRS, the Earl of Morton (who was to become President of the Royal Society the following year), the mathematician George Lewis Scott, James Short and four clockmakers – Alexander Cumming, Thomas Mudge, William Frodsham and James Green. The influence of these four is evident in the set of resolutions that emerged. Not only were the expected requirements of explaining the different parts of the watch and providing drawings of them included, but the tools used by Harrison were to be disclosed and explained as well as the watch; in addition, the methods of tempering the metals were to be explained and performed before the committee, workmen were to be employed to make watches under Harrison's supervision and subject to the committee's inspection and questioning, and these new watches were to be tested to ensure 'that the discovery made by M^{r}. Harrison is sufficient to enable other Workmen to make Watches of equal goodness'.

This was not the outcome to his initiative Harrison had been expecting. He thought he would be engaged simply in explaining the workings of his watch to an appointed committee. But the Commissioners were anxious that this disclosure should be adequate, which could only mean adequate to permit the making of similar watches by other workmen, which in turn they could only be sure of by experiment. They took the inclusion of practical mechanics among their number as an indication of this, and one of them, Alexander Cumming, kept a useful record of the meetings and of his own concerns.[22] It had not, for example, escaped Cumming's notice that if they declared themselves satisfied with Harrison's 'discovery', he would be awarded £5,000 despite the fact that the watch had not yet passed an adequately regulated sea trial.

It is clear that the Commissioners were very exercised by their public responsibility. The lengths to which they were prepared to go to regulate disclosure is seen in the proposal that when the Commissioners found they had questions to ask of Harrison, as making the new watches progressed, 'questions and answers shall, if required, be taken down in Writing to be signed by M^{r}. Harrison and the person or persons who shall ask such question or questions'. Cumming could imagine a time when, should Harrison die or his watch be lost in a future trial at sea, he might actually be called upon, in the public interest, to live up to an earlier affirmation and make such a watch.

Harrison objected to the making and testing of new watches. When he pleaded that he could not afford to engage workmen to make duplicates, Lord Morton replied, 'if you cannot do it of yourself, you must get your friend M^{r}. Short or some other Friends to assist you'.[23] Harrison was prepared to show and explain the necessary tools, explain and show drawings of the parts of the watch, and explain how to temper the metals, but he would not engage in workshop demonstrations, replications and subsequent trials of replicated watches. The Commissioners responded that the trials need not be the same as those obligated on the Board of Longitude by the original Act, but that they had to be satisfied that his disclosure was sufficient 'to enable other Workmen to make other such Watches or Timekeepers'. This could only be determined by practical trial and only then could they sign a certificate 'consistently with the Trust reposed in them'.[24] Cumming recorded separately that the Treasury could hardly be expected to part with £5,000 on the basis of a certificate that the committee 'believed' that Harrison's disclosure would enable workmen to duplicate his watch.[25]

Short had been present at both meetings of the committee. He had objected that the second meeting had not been properly summoned, and Harrison had been told that the resolutions did not reflect the sentiment of those present. On writing to the members, again offering disclosure, the replies he records are generally bland, but Short makes the specific point that he would not consider it necessary for Harrison to take the watch to pieces in the course of explaining its principles, though he suggests that a majority of the meeting may insist on its dismantling. So it proved: at a final meeting of these Commissioners, though there was disagreement among those present, the majority insisted on the terms of disclosure already formulated.[26]

Harrison withdrew from this argument and returned to a request for a second trial under the terms of the original Act, addressed to the original Commissioners – that is, the Board of Longitude. Although there were objections to this now being a legitimate request, since Harrison had not complied with the new Act and had refused to meet the conditions set by its Commissioners, a second trial was allowed by the Board. It went ahead without disclosure, but the discussion surrounding that issue, though it had no immediate consequence, did have a lasting influence. The new set of Commissioners had been concerned with discharging their responsibilities under the new Act and with the consequences to their reputations of a public blunder. This led them into a focused consideration of what was meant by 'full and complete' disclosure and how assurance could be given that this had been achieved. These concerns were directly related to the fundamental question of what could constitute a longitude solution, and in this respect their deliberations and conclusions would have an enduring effect. The questions that had been raised could not easily be set aside.

Voyage to Barbados

Thus the Board of Longitude returned to the question of a trial for the watch under the original Act and reverted to the scheme proposed by the Royal Society. Harrison objected to determining the standard longitude difference by Jupiter's satellites, but the Board decided it was the best method available. John Bradley, nephew of the late Astronomer Royal James Bradley and purser in the Royal Navy ship *Dorsetshire*, was to make equal altitude observations on at least three days at Portsmouth before departure. Nevile Maskelyne was now named as one of the astronomers to be employed in the trial: he was a Cambridge-educated mathematician, who had recently been sent by the Royal Society to the island of St Helena to observe the transit of Venus, and in the course of his voyage had had some success with finding the longitude using lunar distances. He and Robison were to make measurements of equal altitudes and of

Jupiter's satellites in Jamaica. The former observations were to occupy at least three days, and more if possible. The reflecting telescopes and equal altitude instruments that had been prepared were to be sent to the Royal Observatory to be examined by Harrison and whomever he chose to bring along, in the presence of Bliss, Anthony Shephard (Plumian Professor at Cambridge) and Thomas Hornsby (Savilian Professor at Oxford).

This examination of instruments took place on 6 August 1763, when Harrison took with him to Greenwich Colonel William Roy, Dr John Bevis and James Short, together with a two-foot reflector provided by Short; he proposed to match the Board's three nominees with Roy, Bevis and Charles Green, but since Green was assistant to Bliss at the Royal Observatory the situation was difficult for him. Harrison's nominees compared their telescope with three two-foot reflectors by Bird, declaring Short's to be superior. They also examined a portable transit instrument by Bird for taking equal altitudes, and declared themselves satisfied. Another equal altitude instrument was available for inspection, but there was not time to examine it also. Although Bliss, Shephard and Hornsby, for their part, were prepared to agree that the Short telescope was better, it was essential for the longitude determination that the different stations had equivalent instruments. Harrison objected that 'he thought the most correct instruments that could be had should be sent', but Bliss countered that Bird had already made these ones for the purpose and at the public expense.[27]

The voyage was to be to Barbados. On this occasion Harrison was to be allowed to announce a rate and have it applied to the times given by the watch, and this meant additional formality at Portsmouth. Three more locks were to be applied to the box containing the watch, with the keys in the possession of the Commissioner for the Port, Richard Hughes, the Captain of the *Tartar*, bound for Barbados, and John Bradley. Bradley was also to be supplied with instruments for taking equal altitudes at least three days before departure, in the presence of the other two keyholders and William Harrison. On the basis of these observations, which would be communicated to the Admiralty, William Harrison was to determine and declare a rate. Hughes's key would be handed to the First Lieutenant of the *Tartar*, Bradley's to the Second Lieutenant, and these officers, with the Captain, would be present whenever William Harrison had access to the watch during the voyage.

It was decided that Maskelyne and Robison should go on an earlier ship to Barbados, and would be responsible for the longitude determination by Jupiter's satellites and the comparison between local time from equal altitudes and time from the watch on Harrison's arrival. Again officers from the ship were to witness this. Corresponding observations of Jupiter's satellites were to be made at Portsmouth and at Greenwich 'by Instruments exactly similar'.[28] In the event Robison declined to go and was replaced by Charles Green.

One reason for the separate voyage of Maskelyne and Green, which is often overlooked, is that this whole enterprise should be seen as the trial not of one method but of three. This was the general public perception of the venture, and may be one reason why the Board agreed readily to a second trial for Harrison, despite the inconclusive outcome of the consequences of the new Act of Parliament. The other two methods were that of lunar distances, using the lunar tables of Tobias Meyer, and 'Irwin's marine chair'.

Again readers of the *Gentleman's Magazine* were treated to a well-informed and critical commentary, sceptical but not entirely hostile to Harrison, this time above the name of 'Nauticus'.[29] The central technical issues are examined again – Jupiter's satellites as the method to be used on land, the application of a rate and, more forcefully than before, the need to find local time at sea with a Hadley quadrant. The conclusion offered once more is that lunars will be required as a complementary method, but now the successful experience of Maskelyne can be adduced, together with his recently published handbook on lunars, *The British Mariner's Guide*.

One of the first things William Harrison did, on his arrival in Barbados on 13 May, was to challenge the integrity of Maskelyne as a witness. With Green's assistance Maskelyne had, as he was instructed, taken the longitude by lunar distances on his own voyage in the *Princess Louisa*. For the purpose he had a brass sextant by Bird, loaned to him by the Board, and his own mahogany quadrant by Bird. Evidently he had been speaking of his success and of his enthusiasm for the method, and when Harrison arrived, by his account, he was told that Maskelyne considered himself a candidate for the longitude prize. A very public and acrimonious quarrel ensued in front of Sir John Lindsey, Captain of the *Tartar*, who, again according to Harrison, agreed with his assessment that Maskelyne's interest made him 'a very improper Person to take the Observations of equal altitudes'.[30] Indeed, Harrison's account makes clear his belief that Maskelyne accepted the commission to act in Barbados in order to ruin the prospects for the watch. Naturally Maskelyne took this as a serious reflection on his integrity: 'he alleged that if he was not to observe it would be a great dishonour to him and therefore he insisted upon it'. In the end he observed first, with he and Green then taking alternate days.

No trials were intended for the return journey, and at the next meeting of the Board all the measurements and Harrison's declaration of the rate of the watch were referred to three mathematicians – John Campbell, Bevis and George Witchell. As a result of this meeting, William Harrison was at least able to report that 'The Chair is dead', and from then on his father was able to focus his venom on what he was to call 'the improper, troublesome, erroneous, tedious Method' of lunar distances and the 'infamous Nautical Almanac'.[31]

At a meeting of the Board on 9 February 1765 the processed results were considered and it was clear that the watch had passed the trial stipulated in the Act. This being so, if the Board wanted to reserve the question of a practicable longitude method, the point had to be made immediately. They moved swiftly to separate carefully and precisely the two issues of succeeding in a trial and solving the longitude. It was agreed unanimously that Harrison's watch had kept time during the voyage beyond what was required under the original Act to qualify for the highest award – 'even considerably within the same' – but that Harrison had not yet made a Discovery of the Principles upon which his said Timekeeper is constructed; nor the method of Carrying those Principles into Execution, by means whereof other such Timekeepers might be framed, of sufficient correctness to find the Longitude at Sea within the Limits by the said Act required, Whereby the said Invention might be adjudged Practicable and useful in terms of the said Act and agreeable to the true Intent and meaning thereof.[32]

Formerly the Board had been perplexed by the ambiguities in the Act; now they decided to separate the different senses of 'discovering the longitude', to define them, and to consolidate the separation by allocating a financial identity to the components, which taken as a whole would complete the value of the full reward. They ruled that the moneys paid to Harrison would be made up to half of the maximum reward, that is to £10,000, when he had discovered the principles and manner of making his watch to certain persons to be named by the Board, but that payment of the remainder would depend on 'Proof being made to the Satisfaction of this Board, that his Method will be of common and general utility in finding the Longitude at Sea'.[33] It was clear that this proof would involve having watches made by others and their trial on appropriate voyages.

The Board determined to consolidate through Parliament their resolution of the former ambiguities by applying for a new Act. Ludlam characterised their action as follows: 'An application was therefore made to parliament for an act to explain that of Queen Anne, and to settle the meaning of this wonderful word *discovery*.'[34] Harrison countered by petitioning Parliament for payment in full under the original Act and continued his programme of propaganda. But after a parliamentary committee was convened and witnesses examined, the Board's conditions were eventually enshrined in 5 Geo. II c. 20, passed in May 1765. It was 'An Act for rendering more effectual an Act, made in the 12th year of Queene Anne . . . ', and it sought, at least in this instance, to legislate for 'discovery'.[35]

Ten thousand pounds would be due to Harrison when the principles of his watch had been explained; the remainder would depend on the making of similar watches and their successful trial to the same limits. The Board chose six people to witness Harrison's disclosure. They were John Mitchell and William Ludlam, both mathematicians with strong practical interests in instruments, three watchmakers (Thomas Mudge, William Matthews and Larcum Kendall) and John Bird. After much argument and prevarication, or as Ludlam put it, 'after boggling some time about discovering the principles of his watch', Harrison finally complied.[36]

Part of Harrison's 'boggling' concerned one of the conditions stipulated by the Board for this 'discovery'. The principles of the watch were to be explained by drawings and written explanations, but also by word of mouth and by 'experimental exhibitions'. This last expression would have had a ready meaning for the members of the Board: they wanted Harrison to take his watch apart and explain its working and the methods of manufacture before the eyes of their nominees by practical manipulation and demonstration. Harrison said he found the expression dangerously meaningless: he 'endeavoured to shew the absurdity of the latter part of the second resolution, as there was no knowing what might be demanded from it'. He wanted to be sure of the first half of the reward 'when he had done, but . . . he might never know when he had done from the words "Experimental Exhibitions"', and he pleaded with the Board 'to tell him what they meant by these words'.[37]

The question of the meaning of 'experimental exhibitions' continued in a fruitless manner for some time and became the final sticking point between Harrison and the Board, he refusing to sign an oath he did not understand, the Board refusing either to alter it or to restrict its meaning by a more precise formulation. The Board were under some pressure from the public discussion of the whole matter and at Hornsby's suggestion resolved to publish the minutes of these meetings, 'that the world might not be imposed on'. Once the oath required of Harrison had appeared in print, there was no possibility of the Board withdrawing 'experimental exhibitions'.

There was substance to Harrison's concern here. The older sense of 'experiment' concerned trial of the practical viability of a device or a project. It was particularly used in this sense in relation to voyages of discovery that tested speculative geography and the possibilities for navigational projects. To Harrison it sounded suspiciously like an opening for new trials. The Board's sense was much more open and ill-defined: it could not be formulated in a particular protocol, even though they had a sense of the general direction such experiment would take. As the argument continued, Harrison asserted that

> in his opinion the words might mean the making of a hundred more Timekepers or nobody Knows what; Lord Egmont then told Mr Harrison that they only meant by it the manner of Tempering Metals to which Mr Harrison replied then please to put it so, No, says My Lord we will not be confined.[38]

The argument blundered on without progress, to the increasing exasperation of the Board, until Harrison, protesting to the end, felt obliged to sign, the 'discovery' could proceed, and the nominees gave him the certificate he needed to claim half the reward. Four of the signatories – Ludlam, Mitchell, Kendall and Bird – later, in different ways, expressed doubts over the real practicality of Harrison's watch as a model for others. Ludlam was particularly outspoken in print. In a letter to the *Gazetteer*, reproduced in the *Annual Register*, he averred that Harrison's methods were not readily transferable, that disclosure had not been complete, that his adjustments were made simply by trial and error rather than on principle, and that the temperature adjustment was inadequate.[39] It must have been frustrating for the Board to find that even after all their care over precisely defined and regulated conditions, backed by an Act of Parliament, published by the Board and adhered to through a prolonged argument, the 'discovery' procedure could not be made to stick.

'Voyages' at Greenwich

Another clause in the oath Harrison had signed, where 'experimental exhibitions' had caused him such problems, was the surrender of his watch to the Admiralty. The Board were now able to conduct their own trials, and they decided to test its timekeeping at Greenwich. Thus there was a third attempt at a trial, but in this case the watch did not go to sea. It was transferred to the Admiralty on 5 May 1766, and John Ibbetson, Secretary to the Board, Larcum Kendall and Maskelyne conveyed it to the Royal Observatory. It was wound daily and compared with the transit clock, generally by Maskelyne himself, sometimes by an assistant, but always in the presence of one of the officers of Greenwich Hospital, where a key was kept for unlocking the box. An officer had to climb Greenwich Hill every day: even the Astronomer Royal's access to this instrument had to be regulated and witnessed. The trial continued for ten months, generally with the watch horizontal and face up, but occasionally in other orientations.[40]

In absolute terms the watch's performance was far from satisfactory. In the early days it was gaining nearly twenty seconds per day. Maskelyne acknowledged that this would not be a serious problem, only a slight inconvenience, if the gaining or losing rate were fairly constant, but in fact it was erratic, and not clearly related to temperature. If the performance of the watch was fundamentally unreliable, had it after all succeeded on its trial to Barbados merely by chance? Maskelyne hit on a way to assess the chance of success from the figures he now possessed.

What were the odds in favour of keeping the time on a voyage to the West Indies, given the discovered level of irregularity and granting the application of a rate? Maskelyne identified six successive six-week periods when the watch was horizontal, and he treated these as six virtual voyages to the West Indies. The rates calculated for the six 'voyages' differed widely, but if each rate in turn was applied as a correction to the timekeeping on the subsequent 'voyage', over the six voyages the probability of keeping the longitude to within half a degree to the probability of not doing so was 1 to 4. The probability of keeping it to within two-thirds of a degree to that of not doing so was 2 to 3. Further, in one week the timekeeping of the watch had been particularly badly affected by the temperature sometimes falling below freezing. If this week was disregarded, Maskelyne found that the respective probabilities were even and 3 to 1 respectively.

So, even though the timekeeping was inherently unstable, the method of applying a rate deduced from observations taken in the period before the voyage greatly reduced the odds against a successful performance.

> These considerations are sufficient to explain the motives which might have actuated Mr. Harrison, as a man of prudence, in desiring to send his watch upon two voyages to the West Indies, upon the idea that he should be intitled to the large rewards prescribed in the act of the 12th of Queen Anne, in case his watch kept time within the limits there mentioned, whether the method itself was or could be rendered generally useful and practicable, or not.[41]

Of course, in general use such chances would not be worth taking, neither would it be practicable to revise the rate frequently, nor always to avoid extremes of cold.

Maskelyne's verdict on the trial results was that, although Harrison's watch was a valuable invention and could be usefully applied in conjunction with the lunar method, even if it were not exposed to extremes of cold it 'cannot be depended upon' to keep the longitude within a degree on a voyage to the West Indies, nor to keep it within half a degree for more than a fortnight. In extreme cold these estimates would be much worse. Of course the watch had already kept the longitude to within half a degree on a real voyage to the West Indies where all the appropriate protocols had been maintained. Maskelyne's trials were being used to test the second component in the reward. He was claiming that the virtual voyages showed that the success achieved in the single, real voyage had been by chance – a chance whose odds Maskelyne claimed he had measured.

Naturally Harrison was stung to reply, though the pamphlet he published was written by someone else.[42] He challenged the protocols: the officers were old and infirm, and were particularly enfeebled after climbing the hill to the Observatory, the locks were insubstantial, the transit clock that provided the standard for comparison was not protected from tampering, and so on. He also challenged Maskelyne's integrity in every respect, including his truthfulness in the former trials of lunar distances:

> I am not satisfied with the Truth of his reporting other Observations relative to the Longitude, as I do maintain that in both his Voyages the Observations which he said he made the Land by, were not calculated till after he had seen the Land: and I am certain those he has given, in the Publication now before us, are not genuine.[43]

He accused Maskelyne of being 'deeply interested' in a scheme for gaining a reward by the provision of lunar tables. In spite of, and perhaps in contradiction to, these protests Harrison claimed that his calculations on Maskelyne's published figures showed that he could extract a rate that, applied to any six-week period of going in a horizontal position and outside extremes of cold, would yield the longitude to 'within the Limits of the Act of Queen Anne'.[44] He does not say which limit he referred to and, more importantly, he does not give his calculations.

Harrison quotes the passage reproduced above about his being 'a man of prudence', retorting:

> this Insinuation (published under the Authority of the Commissioners of Longitude) that I had contrived a Trial which I knew the Watch would fulfil, whilst I was conscious that it would not answer the general Purposes of the Act of Queen Anne, and consequently that I had formed a villainous Scheme to rob the Publick of the Reward without really and effectually performing the Conditions, strikes me as a Charge of so atrocious a Nature,

> that I think myself not only justified in publishing to the World what has been done with respect to Trials of the Merit of my Invention, but even indispensably obliged so to do.[45]

If Harrison had not previously appreciated the nature of the gap between himself and the Board, it had at last become clear to him. The Board were still not convinced that the performance of his watch on the voyage to Barbados had not been achieved by chance, and so, although they had a certificate to say that his watch had been 'discovered', the *longitude*, in their understanding of the matter, had not been discovered. That would require a discovery of a different order, a discovery in nature or of some mechanical principle applicable to universal horology: it was not a thing that could be represented by a single working machine. It was here that the move from the local to the universal was invoked. The method that would deliver the longitude, whatever it was to be, to rise above accidental success would have to be secured by virtue of a natural law or a universal principle.

In April 1767 (the Greenwich trials had ended in March), the *Gentleman's Magazine* carried a contribution from the Harrison side of the argument that shows that this move had been understood. It is headed 'The Principles of Mr Harrison's Time-Keeper', and, after a short and general description of the working of the watch, concludes, 'So that it is plain from this, that such a time keeper goes entirely from principle, and not from chance.'[46]

Harrison's own tract continued with generalised complaints about his treatment by the Board and about the demands of the new Act, saying that by this frankly worded account he was probably forgoing the second half of the reward. The Board, for its part, backed by the new Act, began to arrange the making of duplicates and of new watches by Harrison. Their paths had begun to diverge and they had little more to say to each other. Harrison moved into the third phase of his campaign, appealing past the Board to Parliament and the King.

Conclusion: three ambiguities

The trials of Harrison's timekeeper illustrate a range of difficulties associated with regulating and testing the movement and distant performance of a measuring instrument, and of other instruments and procedures related to it. It was a very public example and a well-documented one. There were problems with identifying and securing standards, and with the reliability and integrity of operators, or at least with the general acceptance of their credit. No one was in any doubt that there were 'sides' and 'interests' at work here; the question was how they were to be managed. The reliability of other instruments, on which the trials depended, was open to question and challenge, as were the protocols for their use. All the critical operations in the process required precise regulation and credible witnessing. The social management of these procedures was enshrined in the resolutions of official committees and even in Acts of Parliament, and yet it proved all but impossible to secure agreement and assurance.

One plausible reading of the outcome was not that Harrison had solved the longitude problem and received his just reward, but that, as they had feared all along, the Board had, despite all their efforts, failed in their public responsibility. After Harrison had been awarded by Parliament a sum equivalent to the second half of the reward, William Emerson felt able to write in 1770 that the longitude was

> still a secret, and likely to continue so, for tho many thousands of pounds have been paid for the pretended discovery thereof; we remain just as wise as we were before the discovery, except the ill success of it happens to teach us so much wit as to take better care of our money for the future.[47]

The intensity of the contest and the importance of the prize served to magnify attention to the social interactions involved. With so much at stake, it had to be accepted and admitted publicly that instruments were variable and required inspection by both sides, that access to instruments on trial had to be controlled, and that operators could be fallible or biased and their observation and recording had to be witnessed. Practically nothing seemed free of 'interest'. Harrison was even offered the opportunity to witness calculations commissioned from mathematicians to extract results from observations – to witness the *process* of calculation, not simply to examine its results. A social apparatus of locks, keys, witnesses, officers, sealed envelopes, committees, commissioners, societies, Acts of Parliament, etc., was deployed to control these trials. In the end, universal principles seemed to offer the only possibility for transcending the contingencies and fallibilities of local testing regimes.

In conclusion, we can note three difficult, contested and closely related terms, whose ambiguities point to some of the critical issues involved. These are 'experiment', 'discovery' and 'accident'.

Harrison claimed not to be able to assign any sense to 'experimental exhibitions'. To him 'experiment' seemed suspiciously close to 'trial', and the conjunction with 'exhibition' was absurd. He demanded a closer definition before he could sign the conditions for the discovery of his watch, but that was just what the Board were unwilling to give. For them, this was a move to open up the discussion, to escape from the bounded and 'accidental' conditions of a specified trial and to question the matter in an unconstrained way, in search of a universal understanding, a resolution of their doubts that could not be ascribed to chance. As Egmont told Harrison, 'we will not be confined'.

Similar considerations surround what we have seen one commentator call 'the meaning of this wonderful word *discovery*'. At the height of the argument in 1765, following the second voyage, when the Board sought a precise definition of what it could mean to discover the longitude, went to Parliament to enshrine it in legislation, and placed a value on its different components, an anonymous letter – 'the private sentiment of a bye stander' – appeared in the *Gentleman's Magazine*.[48] It was a well-informed and closely argued comment. We have already seen at least three senses to 'discovery' at play in this story. The 'discovery of the longitude' had two meanings: finding one's longitude at sea or devising a method for finding it. The 'discovery of the watch' meant, literally, uncovering it and explaining how it worked. Now a fourth sense was made explicit, one that sought to encompass and reduce the others, in the manner of the natural sciences.

The original longitude legislators in 1714, the author of the letter asserted, must have meant to reward 'the discovery of some general principle by which the longitude could be ascertained and applied to practice'. The case of 'an ingenious artist' spending a lifetime making and adjusting a single watch so that it managed to keep time on a single voyage did not come close to this standard: 'The bare construction of one single machine can never be said to discover the longitude.' To say that something more than watchmaking was required did not mean that the solution would not be mediated through watchwork:

> It remains, therefore, to be known, whether, besides the construction, there be any new and general principle discovered by which longitude is to be ascertained by watches, made upon that principle, with the common care of good and approved workmen, in a reasonable time, and at a moderate expence.

Without a universal discovery in this sense, a longitude solution could never be secured. Even if a practical method were to be found in mechanical horology, it was not the trial of successful watches that would substantiate the solution, but the discovery of a universal principle of nature:

> The artificial construction of a timepiece can never be accounted a discovery of longitude, even though it should go to Barbadoes and back again 1,000 times without varying; any more than a chemical process that will produce silver from lead, can be said to be a discovery of the philosopher's stone, or the transmutation of metals, even tho' the same process should produce the same effect 1,000 times repeated.

On this understanding, the performance of Harrison's watch was 'accidental' as distinct from essential: it depended on the local and contingent circumstances inside its box, brought to a nicety of adjustment by the obsessive attentions of its maker. It depended on no universal principle that could be formulated, and by being accidental in this sense it had performed its artificial feat 'by accident', or by chance. By this account, the watch had not really travelled at all: rather it had carried its artificial world with it, isolated in its box by however many locks were currently prescribed by the Board of Longitude. It could be said that the ultimate difference between Harrison and the Board was that he held that by travelling successfully his timekeeper had discovered the longitude, while they, invoking a universalist standard of knowing by means of a different concept of discovery, held that it had to discover the longitude before it could be said to have travelled.

Notes

1 D. Sobel, *Longitude: the True Story of a Lone Genius who Solved the Greatest Scientific Problem of his Time*, London, 1995; W.J.H. Andrewes (ed.) *The Quest for Longitude*, Cambridge, Mass., 1996, contributions by Andrew L. King, William J.H. Andrewes, Anthony G. Randall, Martin Burgess.

2 The best account remains H. Quill, *John Harrison: the Man who Found Longitude*, London, 1966. Note also R.T. Gould, *The Marine Chronometer: the History and Development*, London, 1923.

3 J. Harrison, *A Description Concerning such Mechanism as will Afford a Nice, or True Mensuration of Time*, London, 1775.

4 The basic narrative of the trials was established in Quill, op. cit., chapters 10–18, drawing principally on the minutes of the Board of Longitude (now Cambridge University Library MSS RGO 14, 5) and the Harrison 'Journal', which exists in several copies, one being in a public collection, State Library of Victoria, Melbourne, Australia, MSS Box 110/9, H 17809. For the 'Journal' (hereafter cited as 'Harrison Journal'), see Quill, op. cit., pp. 83–5, 234, 244. Both these sources are used generally in supplying the narrative of the present account, with references given only to quotations and other specific documentary matters.

5 *An Account of the Proceedings in order to the Discovery of the Longitude at Sea*, 2nd edition, London, 1763, appendix vii; Harrison Journal, ff. 3–4.

6 Ibid., ff. 5–7.

7 Quill, op. cit., pp. 91–2; Harrison Journal, ff. 7–8v.

8 During the voyage the necessity of having all four keyholders in attendance seems to have slipped, as on the certificate Harrison obtained from them they would attest only that the watch was not opened during the voyage 'except in the presence of One or more of us' (Harrison Journal, f. 12v).

9 *The Gentleman's Magazine*, 31, 1761, pp. 437–9.

10 For longitude by Jupiter's satellites, see A. Van Helden, 'Longitude and the Satellites of Jupiter', in Andrewes, op. cit., pp. 86–100; for the lunar distance method, see D. Howse, 'The Lunar-Distance Method of Measuring Longitude', in Andrewes, op. cit., pp. 150–62.

11 Harrison Journal, f. 9.

12 Ibid., ff. 11v–14v. Harrison later published the opinion that from this objection raised by two astronomers on the Board, 'it is plain that these great Astronomers did not understand either the Principles or Use of one of the most simple Instruments in Astronomy', J. Harrison, *Remarks of a Pamphlet lately published by the Rev. Mr. Maskelyne*, London, 1767, p. 18.

13 Harrison Journal, f. 20v.

14 Quill, op. cit., p. 113.

15 Ibid., p. 227.

16 *The Monthly Review; or Literary Journal*, 53, 1775–6, p. 321.

17 Harrison Journal, f. 22v.

18 Quill, op. cit., pp. 111–12, 117–18; *An Account of the Proceedings in order to the Discovery of the Longitude at Sea*, 2 editions, London, 1763.

19 Quill, op. cit., pp. 118–20; Harrison Journal, ff. 23–8v.

20 *Journal of the House of Commons*, 29, 1761–4, pp. 546–53, 601.

21 Guildhall Library MS 3973/2; Harrison Journal, ff. 30–1v.

22 Guildhall Library MS 3973.

23 Harrison Journal, f. 31v.

24 Ibid., ff. 32–5; Guildhall Library MS 3973/3.

25 Guildhall Library MS 3973/10.

26 *A Narrative of the Proceedings Relative to the Discovery of the Longitude at Sea*, London, 1765, pp. 1–2; Guildhall Library MS 3973/12; Harrison Journal, ff. 36v–42v.

27 Harrison Journal, f. 50v.
28 *A Narrative of the Proceedings*, op. cit., p. 6.
29 *The Gentleman's Magazine*, 33, 1763, pp. 230–4.
30 Harrison Journal, f. 57.
31 Harrison, *A Description Concerning such Mechanism*, op. cit., pp. 66, 53.
32 Harrison Journal, f. 64.
33 Ibid., f. 64v.
34 *The Monthly Review; or Literary Journal*, 53, 1775–6, p. 322.
35 Guildhall Library MS 3973; *Journal of the House of Commons*, 30, 1765, entries for 19 February and 6, 8 and 20 March.
36 *The Monthly Review; or Literary Journal*, 53, 1775–6, p. 322; note also *The Annual Register*, 5, 1765, pp. 113–33.
37 Harrison Journal, ff. 69–71.
38 Harrison Journal, ff. 76v–77.
39 For Ludlam, see *The Annual Register*, 8, 1765, p. 130; for Ludlam and Mitchell, see Quill, op. cit., pp. 151–2, 154–5; for Kendall, see Quill, op. cit., p. 181; for Bird, see J.A. Bennett, 'Science Lost and Longitude Found: the Tercentenary of John Harrison', *Journal for the History of Astronomy*, 24, 1993, 281–7, see p. 286.
40 N. Maskelyne, *An Account of the Going of Mr. John Harrison's Watch, at the Royal Observatory*, London, 1767.
41 Ibid., p. 23.
42 *Remarks on a Pamphlet Lately Published by the Rev. Mr. Maskelyne*, London, 1767.
43 Ibid., p. 12.
44 Ibid., p. 13.
45 Ibid., pp. 15–16.
46 *The Gentleman's Magazine*, 37, 1767, p. 156.
47 W. Emerson, *The Mathematical Principles of Geography, Navigation and Dialling*, London, 1770, p. 172.
48 *The Gentleman's Magazine*, 35, 1765, pp. 34–5.

5 Landscape with numbers

Natural history, travel and instruments in the late eighteenth and early nineteenth centuries

Marie-Noëlle Bourguet

On the passport the Spanish government issued him before he set off for America in June 1799, Alexander von Humboldt had asked that the name of the botanist who accompanied him – 'mon ami (secrétaire) A. Goujau-Bonpland' – be written down, as well as a special mention of the equipment he was taking on board, referring to the extraordinary array of portable 'instruments de physique et d'astronomie' he had assembled – barometers, thermometers, hygrometers, electrometers, compasses, dipping needles, magnetic needles, microscopes, timekeepers, etc.[1] In themselves, those few words sounded like the announcement of a programme: completing the catalogue of nature through the discovery of new botanical species, on the one hand, and understanding the face of nature through the observation of plant geography and the making of multiple measurements, on the other. Both were to be part and parcel of the scientific travellers' endeavour (Figure 5.1).

In the history of science and travel of the late eighteenth century and the early nineteenth, Humboldt stands as an extraordinary figure, epitomising the idea of the naturalist as a heroic traveller, going to the most remote and desolate places in order to encounter nature *in situ* and scrutinise it through all kinds of instruments. His trip to America instantiated what Susan Faye Cannon eponymously dubbed 'Humboldtian science', which developed during the Victorian period and opened a new phase in the practice of natural sciences, based on series of observations and on accurate and repeated measurements, gathered from all over the world by a network of scientists and travellers.[2] However unique its scope and accomplishment might be, Humboldt's attempt to merge botany and physics in a single programme has yet to be accounted for within the historical and cultural context which framed it: how did it happen that, in the course of the eighteenth century, it became a requisite for naturalist-travellers to carry instruments and make measurements along the way, as well as to collect and classify samples? To what kind of botanical programme would numbers and measurements add some epistemological value? In turn, to what extent did these quantifying practices contribute to reshaping the travellers' encounter with nature and their perception of it?

With this set of questions in mind, I will try to chart the landscape in which such changes occurred through a set of case studies, located in the mountainous

Figure 5.1 Humboldt and Bonpland are depicted with some of the tools and instruments they used during their Andean expedition (a sextant, a barometer, a flora, etc.).

Painting: Friedrich Georg Weitsch, *Humboldt und Bonpland in der Ebene von Tapia am Fuße des Chimborazo, 1810*, Stiftung Preussische Schlösser und Gärten Berlin-Brandenburg, Potsdam.

provinces of France and focused on a few travellers and naturalists whose work Humboldt happened to come across. In the shape of a journey, the story will follow in the footsteps of Ramond de Carbonnières in the Pyrenees, Giraud-Soulavie in the Vivarais, Latourrette and Villars in the Dauphiné and, lastly, Humboldt and Bonpland in the Cordillera. Its variety of places and characters aims at delineating the ways naturalists responded to the mountaineering and quantifying tradition that physicists had developed in France as early as the mid-eighteenth century, and at highlighting both the diversity of situations and programmes and the common values or assumptions that were in play in the spreading of quantitative practices in the field of natural history from the 1770s onwards.

Why should naturalists travel with instruments?

On 10 August 1787, while climbing the Pic du Midi de Bigorre in the Pyrenees, the young Louis-François Ramond de Carbonnières encountered near the lake of Oncet two travellers, the physicist and mineralogist Henri Reboul, a member of the Academy of Toulouse, and his colleague, the astronomer and geometer Jean Vidal. Whereas Ramond had no equipment other than a walking stick, a compass, a mineralogist's hammer, an abstract of Lamarck's *Flore française*, and a notebook, the two travellers were slowly and painfully carrying down the mountain a heavy set of measuring rods, levels, graphometers, barometers and thermometers they were using for their geodesic operations.[3] This episode stages, emblematically, two types of practices that coexisted in scientific travel at the end of the eighteenth century. For Vidal and Reboul, their arduous and minute work belonged to a mountaineering tradition, rooted in the development of geodesy and cartography in France, and closely linked with astronomy and physics.[4] Completing previous expeditions by Parisian and provincial savants like Jacques Cassini (1700) and his son Cassini de Thury (1739), de Plantade (1741), d'Arcet and Monge (1774), Flamichon and Palassou (1780), they had undertaken to ascertain the exact height of the mountain by a direct levelling measurement, made with instruments they had specially designed.[5] Moreover, their operation was to make the mountain itself into a standard, a kind of gigantic and embodied rule ('une montagne toute graduée') which, finely graduated into vertical degrees, would provide an accurate scale for testing the variations of the barometer and experimenting with its use in height measurements. For both issues, instruments had a crucial role to play.

Ramond, for his part, had a different perspective while walking towards the summit. Born into a Protestant family settled in Alsace, he had studied law and medicine at the University of Strasbourg in the mid-1770s, and soon asserted his taste for natural history. In the meantime, through his friendship with the German dramatist Jacob-Michael Lenz, he was introduced to the *Sturm-und-Drang* movement and to the writings of Johann Wolfgang von Goethe, whose *Werther* deeply impressed him.[6] In fact, literature appealed to the young man as much as science: a tour of the Swiss Alps with Lenz in 1777 was to earn him some literary notoriety via the footnotes and extensive additions that enriched his French translation of William Coxe's *Sketches of the natural, civil, and political state of Swisserland*.[7] To be sure, when he arrived in Barèges in midsummer 1787, as a secretary of the Cardinal de Rohan whom he accompanied in exile, ambition as well as curiosity impelled him to scale the nearby peaks: if the Alps already had the names of writers like Haller and Rousseau, and of scientists like Deluc and Saussure, attached to them,[8] the Pyrenees, as yet only partially explored, still offered the traveller a chance for scientific discovery and literary renown. 'However lacking in the means that can make for the success of a campaign of observations, I could not envision myself living in the core of that famous range without exploring at least some of it.'[9]

Yet, Ramond had to find his own path to fame. That awareness made him interestingly self-reflexive and anxious to assess his accomplishments, in a way that sometimes resulted in detailed, even verbose, arguments. In the account of his first ascents, *Observations faites dans les Pyrénées*, published in 1789, as in later notebooks and unpublished manuscripts,[10] Ramond repeatedly portrayed himself as a heroic and solitary mountaineer, travelling with hardly any scientific accoutrements. Speaking somewhat disingenuously of his meagre travelling outfit, he proudly claimed the legitimacy of an unmediated encounter with nature for a devoted observer:

> Travelling alone, lacking for my observations all the resources that only instruments can afford, I had to overcome all the difficulties that can be encountered in a survey expedition, and to define for myself a programme that could be fulfilled with the only means which are always available, in any place and time, if only one is close enough to nature and takes some pleasure in its contemplation.[11]

Beyond the literary topos, Ramond's vindication helps to delineate the issues that were at stake for late eighteenth-century naturalist travellers around the question of how to approach nature.

Nature through the looking-glass of the senses

During his first sojourn in the Pyrenees, as well as during the Revolutionary period when the Jacobin regime confined him in Barèges as a semi-prisoner restricted to daily trips to the nearby mountains, Ramond indeed had to make his senses, primarily his eyesight, a privileged medium for approaching nature. 'I wandered throughout the Pyrenees, with no instruments other than my own senses.'[12] In part, this statement had a practical meaning, alluding to the cognitive skills all travellers – sailors and explorers, as well as mountaineers – develop in order to make their way in an unknown and dangerous environment. But it also had an epistemological meaning, since Ramond's ambition was to make his journeys the basis of a new knowledge about the range structure and natural history. In his writings, cast in the genre of a travel narrative, Ramond never directly refers to the long-lasting debate that the question of sensory evidence had provoked among natural philosophers from Francis Bacon and Locke to Kant and Condillac. Nor does he cite any works like the essay by the Swiss Jean Senebier on *L'art d'observer* (1774), which discussed the uses of the senses in medicine, natural history and all the sciences based on empirical observation. Yet his remarks on the matter indicate that his views were neither naive nor simplistic, and are to be read within this broad cultural context.

Like any mountaineer, Ramond knew all too well how much the environment can delude a traveller's senses:

> This country is a chaos, where the eye cannot find its way; at every step, the various objects separate from one another or merge together; their appearance and respective location change; a new sight is only a new uncertainty, and doubts grow along with the many perspectives.[13]

Such experiences did not however lead him to dismiss sensory evidence: rather, his emphasis on the frailty of perception sounds a call for the training and disciplining of the senses. Yet that process itself is less described than suggested, through expressions like 'une longue habitude', 'un œil exercé', 'un coup d'œil' (bird's-eye view), which Ramond did not bother to explain in any detail, as if the observational skills and cognitive techniques they pointed out belonged to a tacit knowledge, if not to an innate talent. One must browse through his narrative and notes to bring to light some of the bodily and sensory aspects of his practices of 'cognition in the wild',[14] as well as the social and cultural context that constituted them.

Although Ramond complacently presented himself as a solitary hero, he did not usually travel alone. His mountaineering excursions were accomplished with one or two guides, hired in Barèges or the nearby villages – like Jacou, a descendant of the guide who introduced Tournefort to the Pyrenees, or Laurens, from Barèges, who was for many years his favourite. He would also often take along with him as 'compagnons de promenade' some local notable or naturalist whom he had met in Tarbes, Barèges or Bagnères – such as the botanist Saint-Amans, the mining engineer Duhamel, or the physician Borgella, who became his brother-in-law. Moreover, on some occasions, Ramond introduced to the Pyrenees travellers from Paris or foreign botanists who visited the region – like the German Friedrich Stromeyer, or the Danes Jens Wilken Hornemann and Niels Hofman Bang.[15] The local guides or fellow travellers, however, playing the role of the 'invisible technicians' in many experimental accounts of early modern science,[16] are rarely named or even mentioned in his reports: only incidentally does one glimpse their role in helping Ramond to confront the mountainous environment and in teaching him how to interpret meteorological phenomena or where to find certain plants.[17] More often, their presence is used as a *faire-valoir*, allowing Ramond to emphasise his own superiority: when climbing the Maladetta, for instance, he was anxious to get to the summit before his guide and the hunter who had joined them, because he alone, as a naturalist going for the sake of science, was 'driven by a personal will'. A few years later, reporting a discussion he had had with Laurens about the better way to reach the Mont Perdu, he boasted: 'I was right, since the knowledge I could gather from my observations was so superior to mere routine.'[18]

The only travelling and cognitive practices Ramond described at some length are those in which he used his body or his sensory experience to get an estimation of space and distance, with some kind of measurement and computation at stake. In the history of travel, the human body has commonly been the first standard to be used: land surveyors or explorers made up for the lack of instruments by making the regular pace of a man – or of an animal, a camel or a horse – into a measuring device.[19] Similarly, a mountaineer could calibrate his walk and calculate the distance travelled: a note in Ramond's diary, recording a set of trials made in 1788, reported that he could walk a third of a toise per footstep, and cover from 50 to 100 toises per minute, depending on the conditions.[20] Other techniques were closer to the art of drawing: here, the standards were to be found in nature itself, through the perception of the colours, shape or size of objects seen in the distance. In a footnote of his translation of Coxe, Ramond explained how he was able to give an accurate estimation of the depth of a narrow gorge by taking as a measuring unit the size of a familiar object – the fir trees that grew on its slopes. A few years later, on the top of Mont Perdu, a similar procedure helped him to confront a landscape which, at first glance, defied any attempt at measurement: the presence of some izards bounding across a frozen lake, at the bottom of the scene, enabled him to get a sense of size and proportion, and to estimate the height of the peak.[21]

By the words he used on such occasion, like 'module' or 'étalon', Ramond underlined the continuity between natural or corporeal standards on the one hand, and instrument-based ones on the other. Far from conforming to a simple schema that would see the use of instruments as a radical break from sensory evidence, Ramond's cognitive practices indicate that they should rather be regarded as part of a continuum. As a matter of fact, when some instrumental measures were at hand, Ramond relied on them to check and calibrate his own perceptions: for the Gavarnie Falls for instance, which, by eyesight, he estimated to be 900 feet high at most, 'the measures recently made by MM. Reboul and Vidal [gave] 1266 feet as their exact height'.[22] The interplay between sensory evidence and instrumental measurements could also work the other way around: from a few visual clues – 'a certain light blue and sharpened line, . . . seen from a distance' – Ramond was able to forecast that some glaciers would be found in the highest part of the Pyrenees as in the Alps; this discovery, in turn, allowed him to support the computations that some other travellers had made from afar about the height of the range.[23] In the end, this back and forth movement confirmed Ramond's confidence in his own perceptual skills. Years of travel, of observations, of comparisons had made his eyesight a reliable tool for reading the book of nature and deciphering its language:

> When visiting the upper part of the range, I could trust all the more my ability to calculate the height from looking at the aspect of the everlasting snow, taking into account the orientation and the declivity, that in some places I could . . . confront my own observations with the results of the geodesic operations.[24]

Out of bravado, he went so far as to recommend, in a footnote, his method as an accurate substitute for the use of instruments, for any traveller able to confront nature with both sensitivity and knowledge:

> Whoever likes the mountains and knows them, if he happens to find himself in a new range without any instrument, will know how to apply my method, and perhaps will even draw resources from it, that no instrument could afford.[25]

Nature's standards: the botanical barometer

Ramond's self-fashioning as a solitary and true observer of nature struck the German traveller Wilhelm von Humboldt, when he met him in Barèges in 1797. Observing that Ramond '[had] no instrument, and [did] not work in relation with other scholars' – a practice so vividly opposed to that of his own brother, Alexander – Humboldt wondered whether his attitude was grounded in 'a true love of nature'. Casting a penetrating glance over Ramond's persona, he pointedly commented: 'He does not stand simply in front of nature but, at the same time, in front of a mirror . . . He throws himself between his own senses and nature.'[26] As critical as this remark is, it draws attention to a crucial point: the representation of nature which grounded Ramond's claims, even in his most extreme or arrogant pose. For his confidence in his own senses was based upon the conviction that nature speaks for itself and makes its order visible through signs that an observer can perceive, if attentive and careful enough. Ramond's attitude may be compared with the position Goethe adopted in the same years, when asserting that objective empirical knowledge can be grounded on the senses and observations of individuals: 'As soon as an observer gifted with acute senses happens to pay attention to objects, he becomes both inclined to make observations, and excellent at them.'[27] Because of his views on the direct relation between nature and the observer, Ramond could express an offhand disdain for the use of instruments and stand as a simple observer in charge of registering the language of nature. Thus, when on his first ascent he reached the upper region of the Pic du Midi, he observed that there was 'no tree or bush in that part of the valley, . . . except for the rhododendron, which appears at 200 toises above Barèges'. Drawing on his previous observations in the Alps, he added: 'In the Pyrenees as in the Alps, the rhododendron signifies to the naturalist that he has reached an absolute height of 8[00] or 900 toises above the sea level.'[28] With no need for Vidal and Reboul's graduations, the naturalist relied on another scale, divided according to the ordered succession of the plants that grow on the slopes of a mountain.

The representation of the mountain as a natural scale, divided into degrees according to the various plants living on its slopes, was by no means a novelty at the end of the eighteenth century. Going back to Tournefort's travels to the Near East and to Linnaeus's *Oratio de telluris incremento*, the powerful image of the mountain as an ideal microcosm, displaying all types of climates and plants in successive strata, was a common theme in the literature of natural history, and a scheme that framed travellers' perceptions. Ramond had met this image already when he translated Coxe who, observing the changing of vegetation as he was walking up an Alpine valley, echoed the proposal made by 'a French writer' to take this natural display as a measuring instrument in its own kind: 'One could build a floristic scale, or at least imagine it.'[29]

The author quoted by Coxe–Ramond was probably the secretary of the Academy of Lyon, Marc-Antoine Claret de Latourrette, a lawyer who devoted his life to botany, herborised with Jean-Jacques Rousseau, and corresponded with Carl Linnaeus, Albrecht von Haller, Bernard de Jussieu, and other botanists of his time. In his *Voyage au Mont-Pilat*,[30] a mountain near Lyon where he often went for herborising, and the height of which had not yet been ascertained, Latourrette had evoked the debates of his time about the measurement of mountains:

> If one wants to use the angular method, the atmospheric refraction will necessarily make the result inaccurate . . . The barometer . . . could offer a better and shorter way, if only one had found the precise relation between the mercury level in the tube, and the absolute height of the place.[31]

Because the results of savants like Mariotte, Cassini, Needham or Deluc were so uncertain, the naturalist felt entitled to search for some reliable resource that nature itself could supply. Building on Linnaeus's remarks upon the distribution of plants according to altitude, Latourrette suggested that one could read the succession of strata as the graduations of a hidden instrument, with a different plant for every degree of height. As the 'Flora's calendar' counts the time according to the cycle of the plants, the Flora's barometer ('le baromètre de Flore') would measure 'the height of a mountain by giving the respective locations and names of the various plants that grow on its slopes'.[32]

The notebooks where Ramond recorded his courses and listed his samples illustrate the use he made, for himself, of the idea of a botanical scale ('l'échelle de la végétation'). First, the plants were arranged taxonomically, each one being given a number according to Jussieu's natural classification that he had learnt at the Jardin du Roi in Paris in 1788. Then they were given a second number, corresponding to their location: 'The figures inscribed on the right of the name of the plant, going from 1 to 7, follow the order in which they were found: 1 stands for the height of Barèges, 7 for the Pic summit.' The description was completed with a few remarks upon the aspect of the plant and its vegetative cycle. As an example of this system of double indexation, which arranged the botanical samples according to both a classificatory and a topographical order, one can read, for instance, at the entry of 8 August 1792:

On the road from Barèges to the Pic, . . . 46: *Gentiana verna*, 4, one single plant with flowers, albeit faded . . . – 50: *Rhododendron ferrugineum*, 2, a single plant still in bloom . . . – 55: *Artemisia glacialis*, 6, near the door of Reboul's cabin and downwards . . . – 62: *Papaver alpinum*, 7, the only perfect flower I found on the summit.[33]

Except for the rhododendron which he gave as the equivalent of 'an absolute height of 800 toises or so, in the Alps; 100 toises higher in the Pyrenees',[34] Ramond did not make any systematic attempt to translate the divisions of his topo-vegetal scale into altitudes above sea level, even for places where the Vidal and Reboul measurements were at hand. Why not? Here, the absence of measures probably had less to do with technical motives than with a theoretical argument: for Ramond insisted that height was not 'an object of crucial importance' in itself, but only so far as other circumstances were tightly linked to it.[35] In fact, as far as the growth of plants is concerned, temperature was a more significant element to take into account: 'The distribution of plants on the slopes of a mountain follows, mostly, the local temperature of their respective climates.' Since the heat of a given place can be considered as a result that integrates the influences of a complex set of factors (height, latitude, exposure, winds, etc.), a botanical graduation should rather be indexed on temperature, the everlasting snow-line being taken as a kind of 0 degree isotherm line. 'The upper level of the floristic scale is not to be found at a constant height above the sea level, but at a given height under the everlasting snow limit.'[36] From that line downwards, each plant could be given its specific location (the rhododendron was to be found at '300 or 350 toises below the everlasting snow-line', and so on[37]), and, following that reverse order, a complete scale of the distribution of plants could be conceived.

In the report they presented in 1789 at the Académie royale des sciences on Ramond's *Observations sur les Pyrénées*, the two savants d'Arcet and de Dietrich picked up this aspect of his work – the study of the geography of plants in relation to heat – as a new and important one: 'He draws from his observations the general law that the position of plants on the slopes of mountains mostly conforms to the temperature of their respective climates.' What seemed to them most original in Ramond's method and style was the way, 'from observations that seem at first merely local and geographical', the author engaged himself in theorising, and 'offers conclusions entirely his own'.[38] The academicians actually did not make any comment or discussion about those very conclusions, which were thus given as the personal judgement of an individual, waiting for further demonstration and proof. Yet, by awarding its *privilège* to the book, the Académie acknowledged that Ramond's account of the Pic du Midi vegetation offered, beyond its local interest, a suggestive model for approaching the question of the geographical distribution of plants.

Natural history with 'weights and measures'

A few years earlier, the members of the Académie had already broached the topic of plant distribution, in the context of the work of a young scholar from the province of Vivarais, Abbé Jean-Louis Giraud-Soulavie, whose *Histoire naturelle de la France méridionale* came out, in the first two volumes, in 1780.[39] There is no evidence that Ramond ever read or heard about Giraud-Soulavie's work, the publication of which was never completed. Had he read it, he would have probably felt both sympathy and antipathy for the stance taken by Giraud-Soulavie, and for a programme which made measuring instruments the very tools of his investigation and the ground for his argument.

Born in 1752 into a well-to-do family from Largentière, Giraud-Soulavie had studied in Avignon before becoming a priest at Antraigues, a village in the upper part of the Vivarais, in the mid-1770s. It was in fact natural history that interested him most: a self-taught and widely read naturalist, a persevering traveller despite an ailing constitution, he sought to give a complete description of his native province, whose ancient volcanoes had recently brought it to the attention of the learned world. At first glance, Giraud-Soulavie's approach to nature resembles Ramond's desire to stand as a true and empathic observer. A passage of his book, shaped as an epistemological story, tells how in his youth he arrived at the intuition of a geo-botanical science and decided to 'look at the distribution of vegetation in the southern part of France'. As he was going to the upper Vivarais in order to restore his health, his mother had him observe the slow changing of the vegetation, each zone being separated from the others by 'natural divides':

> While we were climbing, we could see passing before our eyes, successively, the domain of the fruit-trees, the upper part of the vineyard zone and its disappearance, the chestnut-trees continued, and finally the highest climate . . . where only alpine plants can grow.[40]

Endowed with both emotional and demonstrative power, the mountain scenery appealed to the traveller's sensitivity, while displaying all the elements that could help him decipher the order of nature. With this model in mind channelling his energy, Giraud-Soulavie methodically combed his native province, until he settled in Paris in 1780. To keep 'close to the course of nature', he had chosen to walk up the rivers, a travelling practice meant as a heuristic strategy: by following the valleys – 'the true instrument that nature offers' – he was able to observe, step by step, the different types of minerals, the minute changes of the vegetation, the boundaries of their successive climates.[41]

At this point, however, the parallel with Ramond's travels stops short. Far from claiming he could do without instruments, Giraud-Soulavie proved to be extremely concerned with measurements, which he considered a crucial part of his programme. He tells how, before going on his hikes, he strove to procure the best barometer he could with the help of a scholar from Lyon: 'I tried to get one that did not have the faults these instruments usually have. Mr Le Camus, a member of the Academy of Lyon, well skilled in this field, was kind enough to have mine checked by the best artisan in Lyon.' From then on, he proudly portrayed himself travelling everywhere with a large-sized barometer on his back, and made all his observations with 'a barometer and a thermometer in hand'. Only in most difficult situations would he leave his equipment aside:

'Climbing on top the mountain near Crussol, I had to drop my barometer and hold on to a shrub.' Above all, to assess the accuracy of his measures and make them part of a cumulative series of data, he carefully compared them with the ones made by local observers (the monks of the Chartreuse de Bonne-Foy, at the bottom of Mont Mézenc) and also with those made by renowned naturalists like Faujas de Saint-Fond and Adanson, who visited the area and climbed the Mont Mézenc in 1779.[42]

The 'physics of plants'

This equipment, and the social and bodily practices that go with it, suffices to indicate how different his endeavour was from that of someone like Ramond or previous naturalist-travellers. Indeed, Giraud-Soulavie regretted that Tournefort had collected plants without determining with precision the altitude of their climates or mapping their geographical distribution. Furthermore, he considered the way most botanists, Linnaeus included, had approached the question of plant distribution to be too mechanistic and narrowly limited to the link between vegetation and height: 'The height of the climates of plants vary like the climates of the globe. The summer temperature is higher in a mountain in the Vivarais, at 800 toises above the sea level, than in a mountain in Sweden, albeit at a similar height.' Referring derisively to Latourrette as the 'respectable author of a memoir on the Mont Pilat', he mocked his proposal of a botanical barometer, based on the idea that 'one can use the stations of plants as barometric graduations'. Nature does not allow for such a direct reading: 'Since the upper limit [of a plant] varies according to latitude, therefore the graduation of the so-called Flora barometer should vary with every degree of latitude, that is with every twenty leagues or so.'[43] Instead, like Buffon and Adanson[44] (whom he referred to in his work), Giraud-Soulavie considered nature as an integrated and complex unity: 'No element in nature can be isolated from the rest.'[45] There were so many entangled factors that one could not hope for a significant mapping directly drawn from nature. In particular, if the distribution of plants along the slopes of a mountain was to be considered as a measuring scale, a natural instrument incarnate in the landscape, it had to be understood as a very complex and sophisticated one, whose every single graduation had a global and synthetic meaning, integrating a set of elements – altitude, latitude, temperature, exposure, hygrometry, soil, and so on. It was the scientist's endeavour to analyse them and find out the laws of nature.

Hence the need for instruments, and the justification for the traveller's obstinate carrying of his barometer and thermometer up and down the Vivarais: the thermometric measurements were meant to calculate the quantity of heat necessary for each plant, whether wild or cultivated, to mature and fructify; the barometric ones aimed at determining the lower and upper boundaries of each species, taking the height where a plant can no longer reproduce itself as its natural limit.[46] In the end, a whole series of 'climates' of plants were identified, characterised by their respective temperature and barometric height. Botany, as

Giraud-Soulavie understood it, was not a science concerned only with description and classification ('la botanique systématique'), but with the geographical distribution of plants and their conditions of life as well ('la botanique physique' or 'la physique des plantes'). This physical and geographical dimension made it a quantitative science, based on instrumental measurements: 'The botanical system by climates is to be understood with weights and measures, because the plants are subject to variations induced by so many different factors.'[47]

Giraud-Soulavie's determination to link natural history, particularly botany, to the physical sciences and to their quantifying and measuring practices is worth stressing here, for it strongly asserted a new approach of nature. Some fifty years earlier, during the French expedition to Peru in the 1730s, the natural philosopher Charles-Marie de La Condamine had expressed his doubts about the possibility of accounting for plant distribution by some mathematical law: 'One clearly feels that the diverse nature of the soil, the variety of exposures, the winds, the seasons, and other physical circumstances necessarily modify the limits of plants, which cannot be geometrically calculated.'[48] At the end of the century, a traveller like Ramond could still propose an empathetic and unmediated reading of nature, based on visible phenomena like the everlasting snow limit, as a legitimate substitute for precise measurements. It is thus remarkable that as early as 1780 this attitude seemed no longer up to date or appropriate for the provincial priest.

It is not that he was ignorant of the complexity of the phenomena to be accounted for, as his criticism of Latourrette's barometer makes clear. Nor was he unaware of the 'uncertain history' of barometric measurements, which he evoked at length from Cassini and Bouguer to Saussure and Deluc. But the lack of a definite law, enabling one to read the barometric tube graduations as height figures, did not persuade him to abandon his instrument. Rather, he narrowed the scope of his endeavour, and contented himself for the moment with giving the results of his observations as 'barometric heights', expressed in *pouces* and *lignes*, allowing for no generalisation or comparison in terms of absolute height.[49] Despite the general principles which grounded it, his science still had to be linked to a given site, and it would take time for its locality to be made commensurable. It was left to later scientists to translate his local barometric figures into exact measures of height, once that became possible, and, by so doing, to make his work on the Vivarais commensurable with others and inserted into a global picture of the geography of nature. Thus, with a mixture of modesty and assurance, Giraud-Soulavie firmly stood for a physical approach to nature: his barometer and his thermometer were in charge of laying the milestones for a new science. Whatever the short-term limitations of its results, his measuring endeavour sought to analyse natural phenomena, which were too complex to be seen by direct observation and had to be disentangled into discrete elements, through series of distinct observations and measures. A new idea emerges – namely, that only through minute and multiple measurements can nature be apprehended in sufficient detail and its laws understood. For the naturalist, physics was to replace theology: 'Physical geography is the only true basis for a history of nature. It shows the natural distribution of all living beings on the surface of the earth . . . This distribution has its own laws, as well as its own boundaries.'[50]

Yet that was not all: in a daring move, the young abbot enlarged the scope of his programme so as to comprehend the past history of nature together with its present state. The instruments for that undertaking were easy to find, if one was imaginative enough: to reconstruct the meteorology of ancient times it was enough to look at the fossil plants held within the mineral strata of the volcanic mountain, and consider them as 'thermometers of the past', true natural instruments that made it possible to compare the ancient climates of plants to the present ones. 'We will examine the difference between the temperature of the ancient and modern times. The fossil plants will be our thermometers.'[51] Since the laws of nature were universal in time and space, both the history of the earth and the geography of living beings were to be subjected to the scrutiny of measurement.

Mapping the landscape: a geo-botanical diagram

To describe the spatial organisation of plants, Giraud-Soulavie usually borrowed expressions like 'frontiers', 'lines of demarcation', 'natural boundaries' from the political domain. He did so consciously since, as he explained, those terms revealed the origins of his idea of a visual and cartographic language, and of his project of mapping nature: 'Just as political geography is represented through maps, similarly the geography of plants allows for the drawing of botanical maps, which show the various climates of plants.'[52] Martin Rudwick has convincingly argued that the usual paucity of illustration in most eighteenth-century natural history books had not so much to do with the cost of engraving as with the fact that illustration was not then perceived as necessary, or even useful, by most traveller-naturalists.[53] That very paucity makes all the more interesting Giraud-Soulavie's early intuition of the potential of a visual language, and his attempts to conceive, with the collaboration of the engineer-geographer Jean-Louis Dupain-Triel,[54] the maps, drawings and graphic techniques that could convey his observations and ideas on the geography of plants.

Two documents, he explained, were sufficient for him to 'reduce' his work into a comprehensive representation. The first one, a 'Carte géographique des plantes, où l'on trouve la situation et les limites naturelles de leurs climats depuis le bas Vivarais jusqu'au Mt Mezin',[55] offers a bird's-eye view of the province. Like the sheets of Guettard and Lavoisier's *Atlas et description minéralogique de la France* (1780), it adopted conventions derived from a well-established cartographic tradition, such as the use of spot symbols to represent the distribution of minerals and plants, of colour washes to denote the major rock-types, and of dotted lines to signify the boundaries between different bedrocks and plant climates. The second, most innovative, document – 'the first map of its kind, since the appearance of the science of plants', as its author described it – is a side view of the mountain: 'Coupe verticale des montagnes vivaroises avec les limites respectives et les mesures barométriques de leur hauteur' (Figure 5.2). The two-dimensional

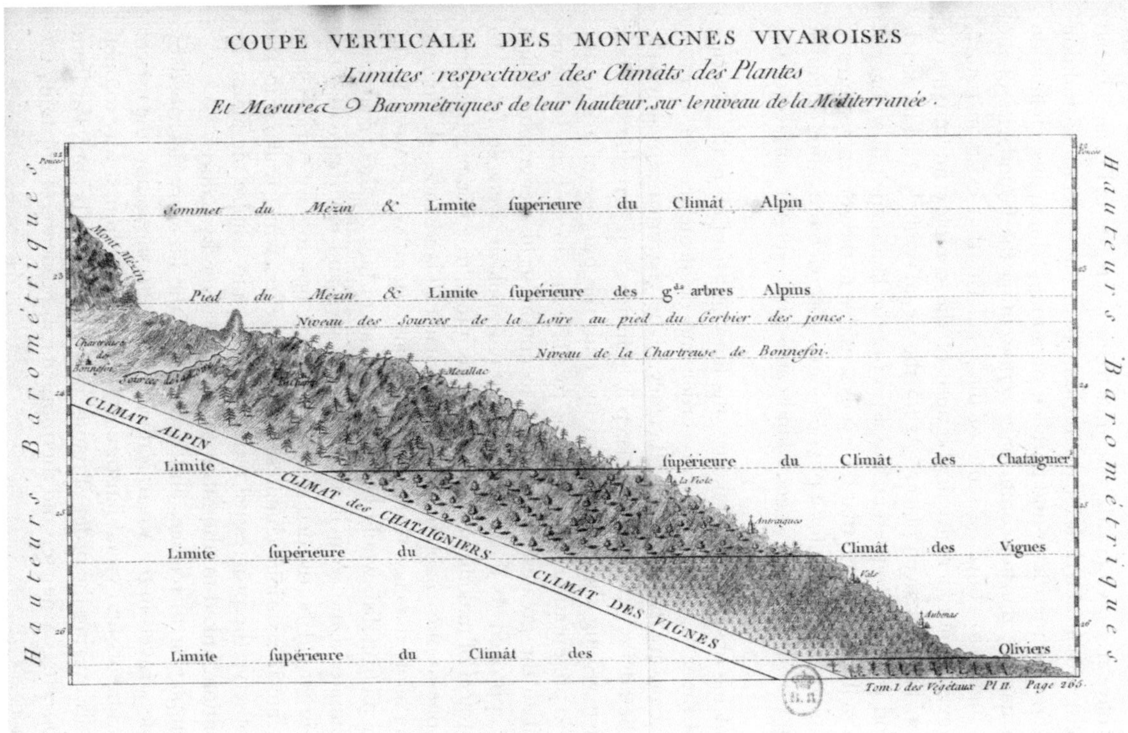

Figure 5.2 The climates from the Mediterranean sea to the upper region of Vivarais, indexed to their respective 'barometric heights' and represented by their corresponding trees or cultivated plants (orange trees, olive trees, vineyards, chestnut trees, fir trees).

Source: Abbé Jean-Louis Giraud-Soulavie, 'Coupe verticale des montagnes vivaroises avec les limites respectives et les mesures barométriques de leur hauteur', in *Histoire naturelle de la France méridionale*, Part II, 1, *Les végétaux*, Paris, 1783, facing p. 265 (Bibliothèque nationale de France, Paris).

diagram displays the superposition of climates from the Mediterranean sea to the upper region of Vivarais, indexed to their respective 'barometric heights': 'This idealised profile of the Vivarais, perpendicularly cut, allows one to comprehend at a glance the upwards succession of the climates of plants and the aspect of the landscape.' Five strata, corresponding to different trees and cultivated plants (orange trees, olive trees, vineyards, chestnut trees, fir trees), had been selected by the naturalist as characteristic of the region. As in the geographic tradition which materialises the meridians and parallels on a map, 'although such lines do not appear in nature', the horizontal lines representing the boundaries of the climates were meant to convey a sense of the regularity that exists in nature, although the eye cannot perceive it, because of the complex intermixing of plants that grow in every given place.[56] As for the visual symbols used to sketch those plants and trees, and represent 'the aspect of the landscape', they suggest, as crude and simple as they are, that the eye is sensitive to the general features of a vegetation, and open the way for a physiognomic approach of nature, attentive to the character of the various vegetal landscapes laid over the surface of the earth.

Whilst distancing observers from nature, the lines and symbols of the drawing in turn helped them to visualise its very order and delineate its face: Giraud-Soulavie was so aware of the visual efficiency of his 'barometric map of the climates' that he advised his readers to keep it in hand when perusing the book or, better, when visiting the area. For in the end, the purpose of a scientific description or representation of nature was to allow one to go back to nature and perceive it to the greatest extent. As if to epitomise his geographical and physiognomic approach to the landscape, Giraud-Soulavie went so far as to conceive the project of making his native Vivarais into a 'true botanical garden'. In this ideal one-to-one mapping of a geo-botanical landscape, the plants, instead of being allotted a class within the Linnaean system, as they were 'in the gardens of our metropolis', would be distributed according to 'their respective heights and climates', and seen in their natural sites and environment. 'The whole mountain would be divided from the bottom upwards by parallel lines that will delineate the various climates.' Every traveller, while walking to the summit, will 'observe the progressive order of plants' and admire the hierarchy of their distribution, linked to 'the more or less heat given by nature'.[57]

From contemplation to science, and back again: the itinerary described by Giraud-Soulavie's travels in the Vivarais during the 1770s shows how the uses of measuring instruments and visual illustration had been, for him, a way to open new paths for discovering the order of nature and admiring its beauty. Far from limiting his scope, his barometer and thermometer were means of enlarging the naturalist's vision:

> Once the love for physical sciences had driven the scientific travellers to the top of the highest mountains, they could contemplate the magnificent scenery of nature, displaying before their eyes the main features of the past and present times of the earth.

Combining the tradition of natural history with the practice of quantification, Giraud-Soulavie dreamed of a science that would bring together all fields of knowledge and encompass the deep unity of nature: he waited for a time to come, 'when physics, electricity, mechanics, geometry, mineralogy, chemistry, physical geography, the science of fire, air and water, all together' would jointly explore the inner parts of the earth as well as its surface. Then, at last, 'we will have a true system of the world'.[58]

Measuring variations: nature to its greatest extent

In a virulent pamphlet, the Jesuit Barruel – a compatriot from the city of Villeneuve – attacked Giraud-Soulavie's computations about the history of the earth, based on the study of mineral strata and a thermometric reading of fossil remains, as contradicting biblical chronology.[59] The polemics brought to an end the scientific career of the naturalist, who left his encyclopaedic *Histoire naturelle* uncompleted and, from the time of the Revolution, started on a new life as a diplomat and historian. Controversially received by the French scientific community (Buffon dismissed his views on the age of the earth; the Académie royale des sciences withdrew its privilege after the publication of the first two volumes), Giraud-Soulavie went into semi-oblivion during his own time, except for the reception some botanists and German natural philosophers gave his book. In his earliest scientific publication, in 1790, the young Alexander von Humboldt had already praised him as a founding-father of plant geography, a notion that he was himself to coin as a scientific field in his *Géographie des plantes* (1805).[60] Not that he made him the inventor of a new concept – the notion of plant distribution was already present in Tournefort's and Linnaeus's work. Rather, Humboldt gave Giraud-Soulavie credit for the novelty of his approach, which joined fieldwork observations and meteorological measurements. As an undertaking to understand what 'determines the plants' preference for one rather than another place',[61] the travels in Vivarais thus encapsulate the moment when, through a concern for the local conditions of the life of plants in terms of height, soil, temperature and exposure, natural history encountered with physics, and began to develop an instrumental and quantifying attitude towards nature.

The case of a contemporary of Giraud-Soulavie, a naturalist who lived in the nearby province of the Dauphiné, a section of the Alps, confirms the claim that an interest in meteorology played a crucial role in the cultural shift that was then taking place. As a physician at the army hospital of Grenoble, Dominique Villars was a skilled botanist; years of experience ('One needs to spend a lot of time before becoming familiar with the physiognomy of plants') had enabled him to evaluate the constitution of a place from a glance at its vegetation: 'Some well-made observations, a coup d'œil on the appearance of the plants, on their size and colours, . . . give the skilful physician a clear idea of the salubrity of the region.' So, like Ramond, Villars was in the habit of making his senses, trained by long practice, his usual instruments. Observing the succession of wheat and vineyards, rye and oats, woods, meadows, then glaciers and everlasting snows that occupied the slopes of the Alps, he could claim: 'These five different climates are so well inscribed in the landscape as to be perceived by any naturalist accustomed to observe the higher mountains.'[62]

Concurrently, his occupational medical concern for the local climatic conditions made him curious about meteorology:

> While visiting the mountains of this province to survey its botanical riches, I often regretted not having a barometer, a thermometer and a hygrometer with me, in order to collect the meteorological observations which interest physics and which are often so necessary in medicine.

For a long time, these two domains, as he presents them, were kept distinct. An excursion he made in July 1781, when he had a barometer and a thermometer, illustrates how Villars's interest for botany and for meteorology once happened to converge, when some unexpected and puzzling barometric observations disturbed his usual notions on the distribution of plants:

> How surprised I was when I found that, in places distant by only ten or twelve *lieues*, the barometric height of a stratum of plants differed so much as to make the woods where I expected meadows, and the rye where I expected woods, and so on.

To account for these discrepancies, the naturalist had to survey all kinds of factors at play, such as the exposure, the declivity, the winds, the hygrometry: the curious phenomenon thus opened to him a new domain of investigation. 'Our *époque* cannot content itself with rough data; it asks for details about the various causes and their consecutive effects.' But, to allow for systematic comparisons between series of commensurable data, instruments were required. Villars was aware his own work still stood at the beginnings of such an undertaking, since he had neither the equipment nor the skills that would give his observations a sufficient precision: 'I could not use M. Deluc's thermometer which I did not know at the time, nor M. de Rets' hygrometer which I did not possess.' Yet as approximate as they were, Villars was confident his measures were good enough to prove the utility of his programme and to engage some future skilled and well-equipped physicist to replicate them. Like Giraud-Soulavie's journeys in the Vivarais, Villars's excursions in the Dauphiné called for the development of a systematic measuring and quantifying approach of nature.[63]

A barometric craze: the second career of Ramond de Carbonnières

It seems a trick of history that Ramond de Carbonnières, whom we saw exploring the Pyrenees with no equipment in the early 1790s, was the one to take up the problem of height measurement at the very point where travellers like Giraud-Soulavie or Villars had to leave it uncompleted in the 1780s. In fact, Ramond was first introduced by chance to the use of instruments during the spring of 1796: recently recruited as a natural history teacher at the École centrale of Tarbes, he was asked by one of his colleagues, the astronomer and mathematician Jean-Auguste Dangos, a corresponding member of the Institut national, to take part in an expedition to the Pic du Midi. Here again, the measure of altitude was the issue: while a telescope was set up in Tarbes to make angular measurements from afar, the mountaineers were to take barometers to the summit and compare the variations of the mercury level in the tube with the geodesic measures made by Vidal and Reboul in 1787.[64] However, the barometric exercise would probably have remained an episode of no consequence for Ramond, if it had not found an echo in his own work by offering him a new and unexpected standpoint from which to observe nature, as well as giving him an opportunity to make his way into the scientific circles of post-revolutionary France.

Be that as it may, the 1796 expedition opened a new period in Ramond's travels. After his apprenticeship with Dangos, the enthusiastic neophyte – 'un passionné positif', as Sainte-Beuve described him – suddenly became as devoted to his instruments as he had previously been indifferent.[65] From then on he would no longer travel without carrying a set of meteorological devices – at least one barometer and two thermometers (one attached to the mercury tube, the other for measuring the outside temperature). Observing and measuring now went hand in hand:

> I wanted to reach its summit [of the Mont Perdu], in order to check with my barometer the elevation of this mountain, which seemed the highest of the range, and to observe the nature and arrangement of the mineral strata it is made of.

As always, a rapid glance at the panorama would come first, like a bird's-eye view, bringing him both aesthetic pleasure and intellectual understanding: 'At a quarter past eleven in the morning, I reached the summit and enjoyed seeing the entire range of the Pyrenees, displayed before my eyes.' The measuring operations were to follow: 'Soon, I put my instruments to work. A strong wind, coming from the west-south-west direction, made that operation difficult. I observed the barometer and the thermometer at noon.' Ramond no longer pictured his work separated from others: his endeavour now required matching measures. 'Meanwhile, the citoyen Dangos was doing the same observations in Tarbes, with instruments . . . that had been carefully compared to mine.'[66]

His objective during these measuring trips was to observe the variations of the barometer and find a reliable technique for reading its graduations into measures of height. Like previous physicists before him – Cassini, Bouguer, Deluc, Saussure, etc. – he then launched into long campaigns of measurements, making hundreds of barometric and thermometric observations, first in the Pyrenees ('I climbed the Pic du Midi sixteen times with meteorological instruments'[67]), later in the mountains of Auvergne, where Napoleon appointed him prefect of Puy-de-Dôme in 1806. Remarkably, Ramond was successful enough in his experiments and calculations to finish up in 1804 with a sophisticated mathematical formula, which made it possible to calculate the altitude precisely from a set of barometric and thermometric data.[68] Laplace acknowledged the coefficient as a useful correction to the formula he had proposed in his *Mécanique céleste*. Alexander von Humboldt, to Ramond's great pride, adopted it to rectify the barometric observations he had collected in America and make new calculations.[69] As for Ramond, he had become so obsessed with his barometer that his person almost became identified with the instrument – a 'préfet-baromètre', as a biographer would later describe him![70] – and Humboldt himself, as passionate as he was about instruments and quantification, was to mock this monomania, writing to his friend the Genevan physicist Marc-Auguste Pictet: 'Ramond read a new memoir on his inevitable barometers . . . He will end by measuring conscripts with his instrument, which will make his method quite commendable indeed.'[71]

As impressive as it is, Ramond's achievement in quantitative science is only a part of the story. While he was busy recording the minute variations of his instruments on top of the Pic du Midi or making tedious calculations in his cabinet, in order to turn the barometer into a black-boxed device, simple to use in any place or condition once the right mathematical formula had been found, he was also losing track of his immediate objective. In his attempt to distinguish between the many influences that caused the day and night ups-and-downs of the barometer, his attention was directed more and more towards these variations themselves. From measuring devices, the instruments were progressively made into objects of scrutiny, and their variable and unpredictable patterns, no longer taken as parasitic epiphenomena, became the focus of his investigation:

> On 30 August 1805, at the summit of the Pic du Midi, . . . the mercury was constantly moving in the tube of the thermometer, going up and down according to the temperature of the air that was brought upwards to the summit.

Such fluctuations offered the traveller a new, unexpected standpoint from which to contemplate nature at work. Through the recording of meteorological data he had the intuition of the atmosphere – 'the invisible ocean of air', as he later designed it – as a new world to explore:

> Thus, what was at first the focus of my research became in the end a subsidiary point; the story of the precise measure of altitude is, now, only the frame which brings together a number of other considerations.[72]

Remarkably, here Ramond was again confronting, with his barometer, a question which had been of great concern to him from the beginning of his travels: that of the link between height and other natural phenomena. At the time of his first trips in the Pyrenees, as we saw, he considered the height an object worthy of no attention in itself (hence, his lack of interest in measurement) but only because many other circumstances – temperature, hygrometry, winds, light – were closely dependent on it. His experiments with the barometer were now confirming that insight, by making visible through the variations of the mercury the multiple and invisible elements that were constantly at play and were influencing all living beings, humans as well as animals and plants. A new field was thus opening before his eyes: 'One rarely studies nature with some care without finding more in it than what he was looking for.'[73]

But Ramond, now a baron d'Empire and a member of the Institut, would not embark on exploring the unknown domain his instrument had revealed to him. In the concluding section of his memoirs on the barometric formula, he summed up his personal scientific experience as the slow discovery – probably for him a reluctant and somewhat discouraging one – of the complexity of nature. After 1810 he would never go back to the Pyrenees, and he did not resume his botanical researches so as to document the influence of atmospheric and climatic conditions on the distribution of plants and the physiognomy of landscape. Only at the end of his life, then retired in Paris, did he make an attempt to merge together the two sides of his travelling experience. In a 'Mémoire sur l'état de la végétation au sommet du Pic du midi de Bagnères', the last paper he read at the Académie royale des sciences in 1826, he went back to the herbal he had collected on the mountaintop, and tried to relate the botanical samples to the meteorological observations he had compiled there for many years. Here again, as throughout his life, Ramond felt torn between his commitment to empirical observations and his quest for a unique principle of explanation: viewing the peak, standing high in the air far from any other mountainous environment, as an ideal laboratory of nature, he pictured it as a site where the climate was 'a mere result of the combined influence of height and latitude', and the vegetation 'a clear and simple expression of the joint influence of these two factors'.[74] Ramond, however, was no longer confident he could assert any generalisation: whereas he had hoped nature to be transparent to his senses or to their instrumental supplements, the barometer had proved to be no more simple to read than the rhododendron or the everlasting snow limit. 'Confusion begins for us, whenever the effects of various causes happened to merge in a single point, and when, however simple by themselves, they are made complex because of their concourse.' The language of nature was neither a simple nor a direct one. 'Let us be patient in observing, comparing and compiling data, and try to stop ourselves in front of the darkness of nature, which our suppositions can hardly cast any light upon.'[75] Beyond the routine task of collecting series of observations and measures, there was still room, in Ramond's credo, for contemplation and silence.

Figuring nature: science and aesthetics

In 1821 already, writing to Alexander von Humboldt, Ramond had expressed his distaste for the abstract 'general formulas' which seemed to appeal to the 'system builders' of his time. One cannot help thinking that this defensive remark had, in fact, his German 'illustre confrère' as its immediate target. For

Humboldt, far from renouncing as Ramond had the hope for a global understanding of nature through the processing of quantified data, was actually pushing ahead the introduction of precision and accuracy into the field of natural history. Not that he had a mechanistic vision of nature, quite the contrary: coming from the German natural philosophy tradition, Humboldt could share with Ramond a view of nature as a unified whole. But his training as a mining engineer and his interest in the quantitative methods developed, by French scientists in particular, in chemistry and physics had soon distanced him from the idealistic vision of a Schiller and made him, instead, champion an empirical and instrumental approach of nature.[76] Humboldt was convinced that the most sensitive instruments were the best way to assemble material for a global and unified science; he had no doubt about the possibility of encapsulating the most minute changes and unpredictable variations into a grid of precise and accurate measurements, which would encompass all kinds of phenomena and unravel the laws of nature that the human eye and senses could not perceive. His departure to America, equipped with a complete cabinet of portable instruments, mentioned in the introduction, epitomised his view that nature has to be studied *in situ* and systematically investigated and registered by series of measurements. 'My objective is . . . to study the links that relate all the living beings to inanimate nature, and that makes it necessary for me to embrace many objects at once', Humboldt wrote to Lalande before arriving in America.[77]

The botanical observations and collections Humboldt made with Bonpland during their tour of America retained, indeed, a strong descriptive and taxonomic component: back in Europe, the two travellers would proudly endow the Muséum national d'histoire naturelle with a herbarium of more than 6,000 samples. Yet his researches were primarily directed towards the elucidation of other matters, namely the question of plant geography:

> Preferring the connection of facts that have been long observed to the knowledge of insulated facts, . . . the discovery of an unknown genus seemed to me far less interesting than an observation of the geographical relations of the vegetable world, or the migration of social plants, and the limits of the height which their different tribes attain on the slopes of the Cordillera.[78]

In Humboldt's view, any specimen or observation that could not be referred to a precise determination of the corresponding height and other local conditions was useless, and to be discarded: the description of plants, the observation of their distribution, the precision of height–barometric measurements and the like were now merged into a single enterprise.

This objective of a systematic collecting and mapping brought important changes for the naturalist in its wake. First, in the practice of travelling itself, since sparse observations or discrete measurements could no longer be considered sufficient: 'I was sorry to see', he recalled later, 'that whilst the number of precise instruments was growing, yet we still were ignorant about the exact height of so many mountains, the periodic tides of the atmosphere, the everlasting snow limit, the variation of magnetism, and so many other phenomena.' Humboldt regretted, for instance, that Bouguer and La Condamine, the two academicians who had preceded him to the top of Chimborazo, had restricted their work to 'measuring heights, air pressure, degrees of heat or cold', with no attempt to link those measurements with their botanical observations. The quest for the unity of nature seemed necessarily to go together with a global and co-ordinated approach, able to encompass and measure all types of phenomena. For the naturalist-traveller, this aim justified spending a lot of time in the upper mountains, and motivated the making of extensive collections of numerical data and material samples, botanical as well as mineralogical, which were all to be indexed to a precise location through geodesic and height measurements. 'The merit of this collection', he told the Muséum naturalists, 'is that any of its sample can be given its native place and absolute height above the sea level.'[79]

No local fieldwork, however, as systematic as it might be, would suffice for the building of such a comprehensive system of knowledge. To fulfil his programme, Humboldt had to rely on a distributed network of local observers or travellers in the field, in charge of making series of observations and measures that were to be gathered from all over the world and processed into a unique and global science. Once back in Europe, whether in Paris or Berlin, Humboldt became the centre of a vast network of informants from whom he expected accurate and precise series of observations and measures; he also sent his own data to some scientists. Not surprisingly, Ramond de Carbonnières happened to be one of these many correspondents. Besides special instructions about the barometric formula, Humboldt asked Ramond to collect data on temperature, which he would integrate into his work on isothermal lines: thus, on 25 October 1816, Ramond sent him a note about the thermometric observations he had collected 'every day at noon, for seven years', while he was prefect in Clermont-Ferrand. 'I believe these computations to be accurate, within two-tenths of a degree', he asserted, before adding a post-scriptum in which he deferentially asked Humboldt to send him in turn some barometric information about the Cordillera. On another page, undated, which Humboldt entitled with the heading 'Pyrenees, Ramond', the French naturalist had listed the alpine plants to be found on the slopes of the Pyrenees, with their respective lower and upper limits, given in metres: '*Gentiana verna*, minimum 600 m, maximum 3,000 m; . . . *Gentiana nivalis*, 2,000–2,600.' Humboldt had embellished the manuscript with some annotations of his own. Where, for instance, Ramond gave '1,800–3,000' as the limits of the Pyrenean *Silene acaulis*, Humboldt inscribed in the margin: 'on the Mont Blanc, at 3,468 m (1,780 toises), Lamarck's *Silene acaulis*'.[80] An example, among many, of the way Humboldt was persistently trying, in his cabinet as in the field, to compile, compare, and co-ordinate data in search of an integrated and global science.

The ambition of taking hold of the whole natural world by an encyclopaedic and quantifying survey culminated in Humboldt's search for graphic methods that would grasp all the phenomena and make them visible – such as the first isothermal lines map, representing the distribution of heat over the world.

Figure 5.3 Plate first sketched out by Humboldt 'from observations and measurements made in the field between 1799 and 1803', and completed later as the magnificent engraving of his *Géographie des plantes équinoxiales* (1807).

Source: Alexander von Humboldt, *Géographie des plantes près de l'Équateur*, 1803 (Museo Nacional de Colombia, Bogotá).

Remarkably, it was to the image of the mountain as a microcosm that Humboldt returned, as Giraud-Soulavie had before him, in order to represent an all-encompassing view of nature, turning the suggestive vision into a rational and precise scheme. The Chimborazo profile was first sketched out in the field during his expedition (Figure 5.3) and completed later as the magnificent engraving entitled 'Géographie des plantes équinoxiales' (1807). 'I made a curious drawing, which represents . . . a mountain profile, with the precise heights, the accurate longitude positions, and the exact height at which every plant grows.'[81] This physical portrait of the tropics represents a cross-section of the Andes with the stratification of vegetation along the volcano slopes, framed with columns of numbers which summarise a myriad of barometric, thermometric, hygrometric and other measurements. On the fieldwork sketch, Humboldt had the slopes of the mountain carpeted with minute inscriptions that enumerated the names of plants. On the final illustrated engraving, next to the textual inscriptions, was a colourful image of the physiognomy of the landscape representing the various aspects of the vegetal strata, ranging from palm trees to alpine plants. The shimmering profusion of figures and words seems to express the 'co-operation of forces' which are invisible to the naked eye and construct the order and unity of nature, while the coloured painting simultaneously gives way to an emotional appreciation of the beauty of nature displayed to the eyes. In Humboldt's undertaking, the instrumental method, which sought to overcome the variety of nature and unravel its laws, even in most inhospitable sites, was in no way incompatible with the aesthetic contemplation of its unity and harmony. To enumerate a landscape in columns, lines and numbers was both a scientific endeavour and a new dimension given to the human aesthetic experience: an attempt to make science and art a shared encounter with nature.

By the beginning of the nineteenth century, carried along by a curiosity for the links between botany, climate, and geography, practices of measurement and precision had thus made their way into the habits of a number of naturalist-travellers. Still, Humboldt's insistent concern with his view of the Chimborazo as a mountaintop experience of nature, adding an aesthetic dimension to the analytical and quantified approach, raises a final issue: what was gained, what was lost, in the end, through the instrumentation of nature?

The naturalist traveller's stance in approaching nature was indeed quite singular, if compared to other travellers: whereas a navigator, for instance, might consider his instruments – a sextant, a compass or a watch – as means for mastering nature and preserving himself from the danger of the world that surrounds him, the naturalist's objective is, in contrast, to use his instruments to engage more deeply with nature, to experience its vivid quality, to become intimately acquainted with its order. All the travellers who have appeared in this story, whether French or German, Catholic or Protestant, enthusiastic or doubtful about instruments and quantification, shared an awareness of the danger of losing their object – nature – in the process of counting or measuring, as Ramond de Carbonnières nearly did at some point. Humboldt himself, at the end of his life, when science was becoming more and more instrumental and quantified, ventilated his worries and fears:

> It may seem a rash attempt to endeavour to separate, into its different elements, the magic power exercised upon our minds by the physical world, since the character of the landscape, and of every imposing scene in nature, depends so materially upon the mutual relation of the ideas and sentiments simultaneously excited in the mind of the observer.

The repetitive routine of the scientific investigator, he admitted, could not but chill any feeling of empathy and contemplation:

> It is almost with reluctance that I am about to speak of . . . the fear entertained by certain persons that nature may by degrees lose a portion of the charm and magic of her power, as we learn more and more how to unveil her secrets, comprehend the mechanism of the movements of the heavenly bodies, and estimate numerically the intensity of natural forces . . . The astronomer who . . . measures patiently, year after year, the meridian altitude and the relative distance of stars . . . does not feel more excited – and this is the very guarantee of the precision of his labours – than the botanist who counts the divisions of the calyx, or the number of stamens in a flower.

But far from driving observers away from a holistic experience of nature and impoverishing their sense of self, any increase in knowledge of nature would also strengthen its spell over the mind: 'Yet the multiplied angular measurements, on the one hand, and the detail of organic relations, on the other, alike aid in preparing the way for the attainment of higher views of the laws of the universe.'[82] In Humboldt's view, the science the naturalist was aiming at was a picture from multiple angles, each one opening new paths to the discovery of nature's laws, serving the pleasure of both the intellect and the senses. Beyond the diversity of their practices and programmes, all the naturalist-travellers whose itineraries towards precision and quantification were followed throughout this chapter shared such scientific and aesthetic expectations, tinged with apprehension. At the close of the eighteenth century and the beginning of the nineteenth, this distinctive combination of reason and emotion gave the naturalists' use of instruments and quantification its characteristic tone.

Acknowledgements

I am most grateful to Jean-Marc Drouin, Myles W. Jackson, Ilana Löwy, R. Josué Seckel, Skúli Sigurdsson, Fernando Vidal, and to all the contributors of this volume, for reading and discussing with patience and incisiveness the draft versions of this chapter, and to Lorraine J. Daston, Richard Staley and Mary Terrall for both their comments and their help with the English language. The Max Planck Institute for the History of Science in Berlin provided me with the time and resources to conduct this research, discuss it in workshops and conferences, and write its final version. The Alexander-von-Humboldt-Forschungsstelle at the Berlin-Brandenburgische Akademie der Wissenschaften, Berlin, and the Staatsbibliothek zu Berlin-Preussischer Kulturbesitz kindly made it possible for me to work on Humboldt's correspondence and manuscripts. Except when mentioned otherwise, the translations are mine.

Notes

1 A. von Humboldt to Forell, s. l. n. d. (1799), in E.-T. Hamy (ed.) *Lettres américaines d'Alexandre de Humboldt (1798–1807)*, Paris: E. Guilmoto, 1905, p. 8.
2 S.F. Cannon, 'Humboldtian science', in *Science in Culture. The Early Victorian Period*, New York: Dawson and Science History Publications, 1978, pp. 73–110.
3 L.-F. Ramond de Carbonnières, *Observations faites dans les Pyrénées pour servir de suite à des observations sur les Alpes*, Paris: Belin, 1789, p. 116.
4 See Christian Licoppe's contribution to this volume; M.-N. Bourguet and C. Licoppe, 'Voyages, mesures et instruments: une nouvelle expérience du monde au siècle des lumières', *Annales. Histoire, Sciences Sociales* 52 (5), 1997, 1115–51.
5 H. Reboul and J. Vidal, 'Exposition d'un nivellement fait dans les Pyrénées pendant les mois de juillet et août 1787', *Mémoires de l'Académie royale des sciences, inscriptions et belles-lettres de Toulouse*, IV, 1790, pp. 1–13 (p. 4: one of the levels could measure angles with a precision of 15″), and *Annales de chimie*, XIII, 1792, pp. 225–42.
6 There is no evidence that Ramond actually met Goethe when the poet was in Strasbourg in 1770–1 or in 1775, but he liked to suggest he had, evoking his university years as 'l'époque où l'on y a vu Goethe, Lenz, et où j'y étais moi-même'. L.-F. Ramond de Carbonnières, *Au Pic du midi, 1787–1810. Trente-cinq voyages*, Paris: H. Beraldi, 1922, p. 25; C. Girdlestone, *Poésie, politique, Pyrénées. Louis-François Ramond (1755–1827), sa vie, son œuvre littéraire et politique*, Paris: Minard, 1968, pp. 21–8.
7 L.-F. Ramond de Carbonnières (trans.), *Lettres de M. William Coxe à M.W. Melmoth, sur l'état politique, civil et naturel de la Suisse, trad. de l'anglais et augmentées des observations faites dans le même pays par le traducteur*, Paris: Belin, 1781, 2 vol. (Coxe's first English edition, 1779).
8 C.E. Engel, *La littérature alpestre en France et en Angleterre aux XVIIIe et XIXe siècles*, Chambéry: Dardel, 1930; N. Broc, *Les montagnes vues par les géographes et les naturalistes de langue française*, Paris: Bibliothèque nationale, 1969; J.-Cl. Pont and J. Lacki, *Une cordée originale. Histoire des relations entre science et montagne*, Genève: Georg, 2000.
9 Ramond, *Observations*, p. v.
10 L.-F. Ramond de Carbonnières, *Carnets pyrénéens*, Lourdes: Éditions de l'Échauguette, 1931, 4 vol.; Idem, *Herborisations dans les Hautes-Pyrénées, ou essai pour servir à l'histoire naturelle, tant des végétaux qui y croissent spontanément que de ceux qu'une culture habituelle y a naturalisés*, Ibos: Randonnées Pyrénéennes, 1997.
11 Ramond, *Observations*, pp. 285–6. Ramond had already struck that pose at the time of his tour of Switzerland: 'We left Lucerne, my companion and myself, without any guide or servant; our luggage was packed in our pockets; our walking sticks in hand, we began to walk with the determination and lightness of those who are free from the burden of equipment' (Ramond (trans.), *Lettres*, vol. 1, p. 226).
12 L.-F. Ramond de Carbonnières, *Voyages au Mont-Perdu et dans la partie adjacente des Hautes-Pyrénées*, Paris: Belin, an IX–1801, p. 121.
13 Ibid., p. 263. One can find an echo of Ramond's remark in many mountain travel accounts. In his instructions for travellers, Saussure emphasised the point: 'Many times, I thought I had just two or three hundred footsteps to go before reaching a summit, which was actually still more than one *lieue* further' (H.-B. de Saussure, *Agenda du voyageur géologue*, Genève: Impr. de la Bibliothèque britannique, 1796, pp. 46–7).

14 E. Hutchins, *Cognition in the Wild*, Cambridge (Mass.) and London: The MIT Press, 1995.
15 In a manuscript relation of his ascents, written at the end of his life, Ramond listed with some detail the names of the guides and companions of his Pyrenean courses (*Au Pic du midi*, passim).
16 S. Shapin, *A Social History of Truth. Civility and Science in Seventeenth-century England*, Chicago and London: The University of Chicago Press, 1994, ch. 8.
17 A few examples: during the summer of 1792, when seeing a bluish fog down in the valleys, Ramond was told that 'it announced the weather was to be good there. That was right, and we found the weather nicer and nicer as we were going' (Ramond, *Carnets*, vol. 1, pp. 6–7). On another occasion, when looking for the Alpine currant bush, 'Jacou told [him] that [they] will find it at the bottom of a small valley' (Ramond, *Herborisations*, p. 110).
18 Ramond, *Observations*, p. 236; Id. *Voyages au Mont Perdu*, p. 167.
19 J. Rennell, 'On the rate of travelling, as performed by camels; and its application, as a scale, to the purpose of geography', *Philosophical Transactions* 81, 1791, pp. 129–45. On the uses of the human pace as a surveying instrument, see Kapil Raj's contribution to this volume.
20 Ramond, *Carnets*, vol. 1, p. 131. Saussure's notebooks give a detailed account of a series of minute calibrating experiments he made in July 1787 while preparing his ascent of Mont Blanc: 'I can climb 132 toises in 28 seconds and 1,310 footsteps, which gives an average of ten steps per toise' H.-B. de Saussure, *Journal d'un voyage à Chamouni et à la cime du Mont Blanc en juillet et août 1787*: (E. Gaillard and H.-F. Montagnier eds), Lyon: M. Audin & Co., 1926, p. 8.
21 Ramond (trans.), *Lettres*, vol. 1, p. 228; Idem, *Voyages au Mont-Perdu*, p. 66.
22 Ramond, *Observations*, p. 73. He held Reboul and Vidal's levelling expedition in such high esteem that he was proud to reproduce their tables and figures in his own account, as if to give the two Languedocians' achievement a larger audience as well as to get some credit for his own work by linking it to a series of scientific endeavours.
23 Ibid., p. 286.
24 Ibid., p. 287.
25 Ibid., p. 303.
26 A. Leitzmann (ed.) *Wilhelm von Humboldts Tagebücher*, Bd. 1, *1788–1798*, Berlin: Behr's Verlag, 1922, pp. 95–101 (p. 98).
27 J.W. von Goethe, 'Der Versuch als Vermittler von Objekt und Subjekt' (1793), in K. Richter *et al.* (eds) *Sämtliche Werke nach Epochen seines Schaffens* (*Münchner Ausgabe*), München: Carl Hanser Verlag, vol. 12, 1989, pp. 684–93 (p. 686).
28 Ramond, *Observations*, p. 38.
29 Ramond (trans.), *Lettres*, vol. 1, p. 211.
30 M.-A. Claret de Latourrette, *Voyage au Mont-Pilat dans la province du Lyonnais, contenant des observations sur l'histoire naturelle de cette montagne et des lieux circonvoisins, suivies du catalogue raisonné des plantes qui y croissent*, Avignon: Regnault, 1770. P. Jacquet, 'Un botaniste lyonnais méconnu du dix-huitième siècle: Marc-Antoine Claret de La Tourrette (1729–1793)', *Bulletin mensuel de la Société linnéenne de Lyon* 68 (4), 1999, 77–84.
31 Latourrette, op. cit., p. 8. Latourrette's argument closely echoed, here, the *Encyclopédie* of Diderot and d'Alembert on 'Montagne' (vol. 10, 1765, p. 676).
32 Latourrette, op. cit., p. 21.
33 Ramond, *Carnets*, vol. 1, pp. 72–7 (14 September 1792).
34 Ramond, *Observations*, pp. 333–5.
35 Ibid., p. v.
36 Ibid., pp. 330–1.
37. Ibid., p. 335.
38 The report is printed as an appendix at the end of the book (*Observations*, p. 443). The same year, Ramond published a German translation of his book: *Reise nach den höchsten*

französischen und spanischen Pyrenäen, Strasbourg: In der akademischen Buchhandlung, 1789.

39 J.-L. Giraud-Soulavie, *Histoire naturelle de la France méridionale*, Paris: J.-F. Quillau, and Nîmes: C. Belle, 1780–1784. The academicians who reported on Giraud-Soulavie's work to the Académie des sciences were the naturalist Auguste-Denis Fougeroux de Bonderoy and the mineralogist Jean-Étienne Guettard. On Giraud-Soulavie: A. Mazon, *Histoire de Soulavie, naturaliste, diplomate et historien*, Paris: Fischbacher, 1893, 2 vol.; L. Aufrère, *De Thalès à Davis. Le relief et la structure de la terre.* t. IV, *La fin du XVIIIe siècle*, 1. *Soulavie et son secret*, Paris: Hermann, 1952.

40 Giraud-Soulavie, *Histoire naturelle*, Part II, 1: *Les végétaux* (1783), pp. 245–7. Cf. J.-M. Drouin, 'Comprendre et dominer le monde végétal. La vulgarisation de la géographie botanique au XIXe siècle', in P. Acot (ed.) *La maîtrise du milieu*, Paris: Vrin, 1996, p. 51.

41 Giraud-Soulavie, *Histoire naturelle*, I, 1, p. 151; II, 1, pp. 33, 187, 224.

42 Ibid., I, 1, p. 312; II, 1, pp. 259–73.

43 Ibid., II, 1, pp. 222, 217, 248.

44 Ibid., II, 1, pp. 148–50. Michel Adanson, convinced of the necessity to observe the plants in their natural environment, made a series of correlated thermometric, barometric and hygrometric observations while he was in Senegal. Michel Adanson, *Voyage au Sénégal* (D. Reynaud and J. Schmidt, eds), Saint-Étienne: Publications de l'Université de Saint-Étienne, 1996, pp. 111–14; J.-P. Nicolas, 'Adanson, the man', in *Adanson. The Bicentennial of Michel Adanson's 'Familles des Plantes'*, Pittsburgh: The Hunt Botanical Library, 1963–1964, vol. 1, p. 25.

45 Giraud-Soulavie, *Histoire naturelle*, II, 1, p. 228.

46 Ibid., II, 1, p. 253.

47 Ibid., p. 222. On Giraud-Soulavie's plant geography: G. Ramakers, 'Die "Géographie des plantes" des Jean-Louis Giraud-Soulavie (1752–1813). Ein Beitrag zur Problem- und Ideengeschichte der Pflanzengeographie', *Die Erde* 107, 1976, 8–30.

48 C.-M. de La Condamine, *Journal du voyage fait par ordre du roi à l'Équateur, servant d'introduction historique à la mesure des trois degrés de méridien*, Paris: Imprimerie royale, 1749, pp. 48–9.

49 Giraud-Soulavie, *Histoire naturelle*, II, 1, p. 267.

50 Ibid., IV, pp. 300–2. On Giraud-Soulavie's programme to study the history of the earth as a physicist, see Aufrère, op. cit., pp. 39–40, 70.

51 Giraud-Soulavie, *Histoire naturelle*, II, 1, pp. 25–6. He developed that view in his *Œuvres complètes de M. le chevalier Hamilton, . . . commentées par M. l'abbé Giraud-Soulavie*, Paris: Moutard, 1781, in particular pp. 344–7.

52 Giraud-Soulavie, *Histoire naturelle*, II, 1, p. 39.

53 M. Rudwick, 'The emergence of a visual language for geological science, 1760–1840', *History of Science* XIV, 1976, 149–95.

54 Dupain-Triel had engraved some sheets of Lavoisier and Guettard's *Atlas minéralogique*, and published the work of the Languedocian geographer Du Carla, which gave the first full exposition of contour lines. Cf. J. Konvitz, *Cartography in France, 1660–1848. Science, Engineering, and Statecraft*, Chicago: Chicago University Press, 1987, pp. 77–81, 90–1.

55 The map also appeared separately as 'Carte géographique de la nature, ou disposition naturelle des minéraux, végétaux, etc. observés en Vivarais', with the prospectus for the publication of the *Histoire naturelle* (*Géographie de la Nature, ou distribution naturelle des trois règnes sur la surface de la terre*, Paris: Clousier, 1780).

56 Giraud-Soulavie, *Histoire naturelle*, II, 1, pp. 192, 274–6.

57 Ibid., II, 1, pp. 187–9.

58 Ibid., I, 1, p. 3; I, 5, p. 192.

59 [Abbé Augustin Barruel], *Les Helviennes ou Lettres provinciales philosophiques*, Amsterdam and Paris: Laporte, 1781 (Giraud-Soulavie is mocked and attacked, under the name of 'M. Rupicole', in the letter XXVIII, pp. 329–49 (p. 338)).

60 Friedrich Stromeyer, *Commentatio inauguralis sistens historiae vegetabilium geographicae specimen*, Göttingen: H. Dieterich, 1800; A. von Humboldt, *Mineralogische Beobachtungen über einige Basalte am Rhein*, Braunschweig: Schulbuchhandlung, 1790, and 'Sur les lois que l'on observe dans la distribution des formes végétales', *Annales de chimie et de physique* 1, 1816, p. 226. For a comparison between Giraud-Soulavie and Humboldt, see Ramakers, op. cit., pp. 26–30. In early nineteenth-century France, Augustin-Pyramus de Candolle described Giraud-Soulavie as 'the first botanist who acknowledged the true importance' of the relationship of vegetation with geography ('Mémoire sur la géographie des plantes en France, considérée dans ses rapports avec la hauteur absolue', *Mémoires de physique et de chimie de la société d'Arcueil* 3, 1817, pp. 262–322). At that time, however, Humboldt was so anxious to claim for himself the use of precise and accurate data that he wrote to his colleague: 'Giraud-Soulavie qui parle déjà de limites supérieures des oliviers, qui en trace même les lignes par le Vivarais . . . mérite de justes éloges . . . Tout ce que je vous demande se réduit à l'observation que j'ai donné la première carte botanique et le premier ouvrage fondé sur des mesures réelles et observations de température', A. de Humboldt, *Correspondance scientifique et littéraire, recueillie, publiée et précédée d'une notice et d'une introduction par M. de la Roquette* . . . , Paris: E. Ducrocq, 1865, pp. 197–200. (I am grateful to Ruth Scheps and Jean-Marc Drouin for this reference).

61 Giraud-Soulavie, *Histoire naturelle*, II, 1, p. 280. On the early history of geo-botany, and the early links between plant distribution, geography and physics: J. Larson, 'Not without a plan: Geography and Natural History in the late eighteenth century', *Journal of the History of Biology* 19 (3), 1986, 447–88; C.W.J. Withers, 'Geography, natural history and the eighteenth-century Enlightenment: Putting the world in place', *History Workshop Journal* 39, 1995, 136–63.

62 D. Villars, 'Observations de météorologie et de botanique sur quelques montagnes du Dauphiné', *Observations sur la physique, sur l'histoire naturelle et sur les arts*, t. 22, April 1783, pp. 269–79.

63 Ibid., pp. 269, 271, 272.

64 Ramond, letter to the Pyrenean botanist Picot de Lapeyrouse, 26 floréal IV, quoted in Girdlestone, *Poésie*, p. 289. Dangos was, then, most interested in the daily variations of the barometer, a topic that Ramond was going to take on: J.-A. Dangos, 'Sur la variation diurne et périodique du baromètre', *Journal de physique*, XXX, 1787.

65 On Ramond's character: C.-A. Sainte-Beuve, *Causeries du lundi*, t. 10, Paris: Garnier, 1855, pp. 362–410.

66 L.-F. Ramond de Carbonnières, 'Voyage au Mont-Perdu', *Annales du Muséum national d'histoire naturelle*, 3, an XIII (1804), p. 74; 'Voyage au sommet du Mont-Perdu', *Journal des Mines*, XIV, no. 83, thermidor an XI (1803), p. 330.

67 L.-F. Ramond de Carbonnières, 'Sur l'état de la végétation au sommet du pic du midi de Bagnères', *Mémoires du Muséum d'histoire naturelle*, XIII, 1825 (publ. 1826), p. 220.

68 L.-F. Ramond de Carbonnières, *Mémoires sur la formule barométrique de la mécanique céleste et les dispositions de l'atmosphère qui en modifient les propriétés, augmentés d'une instruction élémentaire et pratique, destinée à servir de guide dans l'application du baromètre à la mesure des hauteurs*, Clermont-Ferrand: Landriot, 1811, préface.

69 There is in Humboldt's papers a note by Ramond, giving detailed instructions about the way to calculate and make use of Laplace's formula (Staatsbibliothek zu Berlin-Preussischer Kulturbesitz, Handschriftenabteilung, Nachlass A. von Humboldt, gr K5, Nr 35 ab, 1).

70 Girdlestone, op. cit., pp. 333–9.

71 Letter to Pictet, 30 December (*c.* 1810?), in A. Rilliet (ed.) *Lettres de Humboldt à Marc-Auguste Pictet (1795–1824), Mémoires de la Société de géographie de Genève*, t. VII, 1868 (1869), p. 202.

72 Ramond, *Mémoires sur la formule barométrique*, p. 222 (4th memoir, 1809); p. 73 (3rd memoir, 1808).
73 Ibid., p. 118 (3rd memoir, 1808).
74 Ramond, 'Sur l'état de la végétation', pp. 219–20.
75 Ibid., pp. 241–3. Similarly, in 1827, in a letter to the Pyrenean naturalist Saint-Amans, Ramond complained about the 'grands feseurs' who speculated about plant geography (quoted in Girdlestone, op. cit., p. 420).
76 E. Weigl, *Instrumente der Neuzeit: Die Entdeckung der modernen Wirklichkeit*, Stuttgart: Metzler, 1990, pp. 201–27; M. Dettelbach, 'Humboldtian science', in N. Jardine, J.A. Secord and E. C. Spary (eds) *Cultures of Natural History*, Cambridge: Cambridge University Press, 1996, pp. 287–304; M.-N. Bourguet, 'La république des instruments. Voyage, mesure et science de la nature chez Alexandre de Humboldt', in É. François, M.-C. Hoock-Demarle and M. Werner (eds) *Marianne–Germania. Deutsch-französischer Kulturtransfer im europäischen Kontext (1789–1914)*, Deutsch-Französische Kulturbibliothek 10 (2), Leipzig: Leipziger Universitätsverlag, 1998, pp. 405–36.
77 Humboldt to Lalande, 19 November 1799, in A. von Humboldt, *Briefe aus Amerika, 1799–1804* (U. Moheit, ed.), Berlin: Akademie-Verlag, 1992, p. 68.
78 Humboldt to the professors of the Muséum, 27 frimaire an XIII (18 December 1804), in E.-T. Hamy, 'Alexandre de Humboldt et le Muséum d'histoire naturelle', *Nouvelles archives du Muséum d'histoire naturelle*, 4e sér., t. 8, 1906, pp. 10–11; A. von Humboldt, *Relation historique du voyage aux régions équinoxiales du nouveau continent*, Paris, 1814–1815, repr. Stuttgart: Brockhaus, 1970, t. 1, p. 3. On Humboldt's geo-botany: M. Nicolson, 'Alexander von Humboldt, Humboldtian science and the origins of the study of vegetation', *History of Science* 25, 1987, p. 167; Idem, 'Alexander von Humboldt and the geography of vegetation', in A. Cunningham and N. Jardine (eds) *Romanticism and the Sciences*, Cambridge: Cambridge University Press, 1990, pp. 169–85; Idem, 'Humboldtian plant geography after Humboldt: the link to ecology', *British Journal of History of Science* 29, 1996, 289–310.
79 Humboldt, *Relation*, vol. 1, pp. 4–5. Humboldt's Tagebüch, VII a u. b, fol. 142r, printed in *Reise auf dem Rio Magdalena durch die Anden und Mexico* (M. Faak, ed.), Berlin: Akademie-Verlag, 1986–1990, t. 1, p. 174. Letter to the professors of the Muséum: Hamy, 'Alexandre de Humboldt et le Muséum', p. 11.
80 Staatsbibliothek zu Berlin-Preussischer Kulturbesitz, Handschriftenabteilung, Nachlass A. von Humboldt, gr K1, M2, Nr 11a; gr K6, Nr 82b.
81 Humboldt to Delambre, 29 July 1803, in Humboldt, *Briefe aus Amerika*, p. 245.
82 Alexander von Humboldt, *Cosmos. A Sketch of a Physical Description of the Universe* (English trans. by E.C. Otté), London: H.G. Bohn, vol. 1, 1849, pp. 6, 18. On science and aesthetics in Humboldt's work: E.V. Bunkse, 'Humboldt and an aesthetic tradition in geography', *The Geographical Review* 71 (2), April 1981, 127–46; M. Dettelbach, 'Global physics and aesthetic empire: Humboldt's physical portrait of the tropics', in P.H. Reill and D.P. Miller (eds) *Visions of Empire. Voyages, Botany and Representations of Nature*, Cambridge: Cambridge University Press, 1996, pp. 258–92; idem, 'The face of nature: precise measurement, mapping, and sensibility in the work of Alexander von Humboldt', *Studies in History and Philosophy of Biological and Biomedical Sciences* 90 (4), 1999, 473–504; see also Dettelbach's 'Introduction to the 1997 edition', in A. von Humboldt, *Cosmos*, Baltimore and London: The Johns Hopkins University Press, 1997, vol. 2, pp. xvii–xxviii.

6 Appropriating invention

The reception of the voltaic battery in Europe

Giuliano Pancaldi

Introduction

Hailed by Michael Faraday in the 1830s as a 'magnificent instrument of philosophic research', in several European countries throughout the nineteenth century the voltaic battery enjoyed pride of place in the literature on science and technology produced by both specialists and popular science writers.[1] Within that literature the battery acquired a symbolic role, coming to embody a set of characteristics that science and technology were credited with in the age of progress.

This chapter explores the early reactions to the voltaic battery in several European countries *c.* 1800. Amateur as well as expert opinion will be examined, the former playing an important role in those days, as we shall see, and the attitudes of various local communities will be compared. The success of the battery with expert and lay audiences in the early nineteenth century makes it an interesting and rarely pursued case study, allowing us to assess the status of the physical sciences, chemistry, and their practitioners around 1800 in society at large, and to verify to what extent similar developments were taking place across Europe's cultural frontiers. The early circulation of the voltaic battery also lends itself to a discussion of a number of historiographic issues. These include the interaction between instrument, theory and interpretations at a time that was crucial for the transition from eighteenth-century natural philosophy to nineteenth-century science. The early impact of the voltaic battery offers grounds for testing the comparative merits of historical approaches based either on realist or social constructivist views of scientific instruments and practice, or a combination of both. The apparently easy replicability of the voltaic battery on the one hand, and the circumstance that widely different meanings were attached to the same basic instrument by different practitioners on the other, will be discussed in detail. As will be seen, the fact that, compared with other scientific instruments, the battery travelled comparatively easily across cultural frontiers did not save it from getting involved in a maze of interpretations and unintended developments. These circumstances contributed to the notion, frequently expressed by Volta's contemporaries, that the battery had opened up a new continent for the endeavours of scientific investigators.

This chapter is a contribution to the geography of this new continent, as it emerges from a survey of the battery's early travels throughout Europe; a geography in which the instruments, their interpreters, and the differences and similarities of the cultural contexts involved all receive due attention.

Spreading the news

Volta's battery happened to be invented in Como, Lombardy, during the brief interval – thirteen months, from 28 April 1799 to early June 1800 – when the Austrians, allied to the British, regained temporary control over Lombardy after the French invasion of the Italian peninsula three years before.[2] This, plus Alessandro Volta's well-known fondness for Britain and the recent award of the Copley medal, explain why in March 1800 he chose the Royal Society to announce his invention. On scientific grounds, too, he must have thought that the magnitude of his latest invention was such as to make it best publicised in London, which he regarded as the capital city of the ideal, experts' community he belonged to.[3] As to the language for the announcement, Volta chose French, a language he had mastered completely and that at the time made communication across Europe easier.

The troubled political and military situation of the Continent in spring 1800 played a role in Volta's decision to split the paper announcing the battery into two parts. The first part was a four-page, self-contained, 1,000-word account of the column or pile apparatus, with no diagrams and no discussion of its effects. It was dated Como, 20 March 1800, and sent to the President of the Royal Society, Joseph Banks, by mail.[4] The second part, announced in the first, was a 5,000-word account accompanied with an elegant diagram, describing both the pile and the crown of cups apparatus. It discussed the effects of the new machine on sense organs in detail, and assessed the merits (but above all the shortcomings) of William Nicholson's suggestion on how best to imitate the organs and effects of the torpedo fish by means of an electrical device, similar in some respects to the battery. In Volta's manuscript the text dealing with Nicholson was kept separate from the rest by a dotted line, with the implication that the author left Joseph Banks with the decision whether to include the discussion of Nicholson in the published version or not. Volta dated the second part of his paper 1 April 1800, and put it in the hands of a spice and drug merchant from Como, Pasquale Garovaglio, who was leaving Lombardy for England to try his fortune abroad.[5] The first, short description of Volta's battery arrived in London around the middle of April;[6] the second part weeks later, but apparently before 3 June.[7]

Figure 6.1 visualises the early dissemination process of the news concerning the battery, showing the major directions and timing of the information flows. Figure 6.2 represents column voltaic batteries, contained in boxes, of a type Volta described in 1801 and recommended especially for the travelling natural philosopher, like himself. As we shall see, by the end of April and early May 1800 several 'philosophers' were making experiments with voltaic batteries in London, relying only on Volta's first account which had been circulated by Banks.

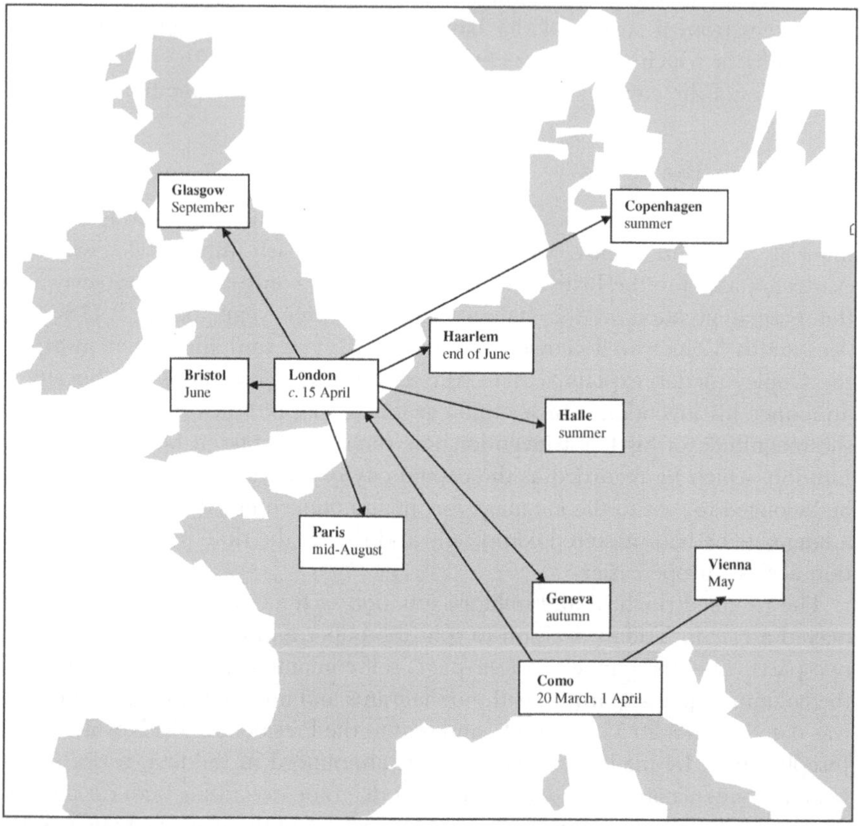

Figure 6.1 The main information flows announcing the voltaic battery, from 20 March to autumn 1800. See text for details.

Meanwhile, Volta began spreading the news of the battery closer to home. During April, several 'cognoscenti' in Milan saw a voltaic battery in operation in the house of the chemist Valentino Brugnatelli, Volta's colleague and friend. On 26 April Brugnatelli reported to Volta that he had observed a number of chemical effects in the battery, though not the decomposition of water performed at the same time by Anthony Carlisle and William Nicholson in London.[8]

Sometime during April, Volta sent another description of the battery (including an account of the crown of cups apparatus) to his main contact at the court of Vienna, who in those days was the physicist, chemist and official Marsilio Landriani. By early May Landriani and his Austrian colleagues in Vienna had built a sort of 'trough' voltaic apparatus, of the kind that became common later in England and the German states. Despite his success in operating with the battery, and improving on its design, Landriani declared to Volta that he had problems in understanding 'the theory of these new, surprising phenomena'.[9]

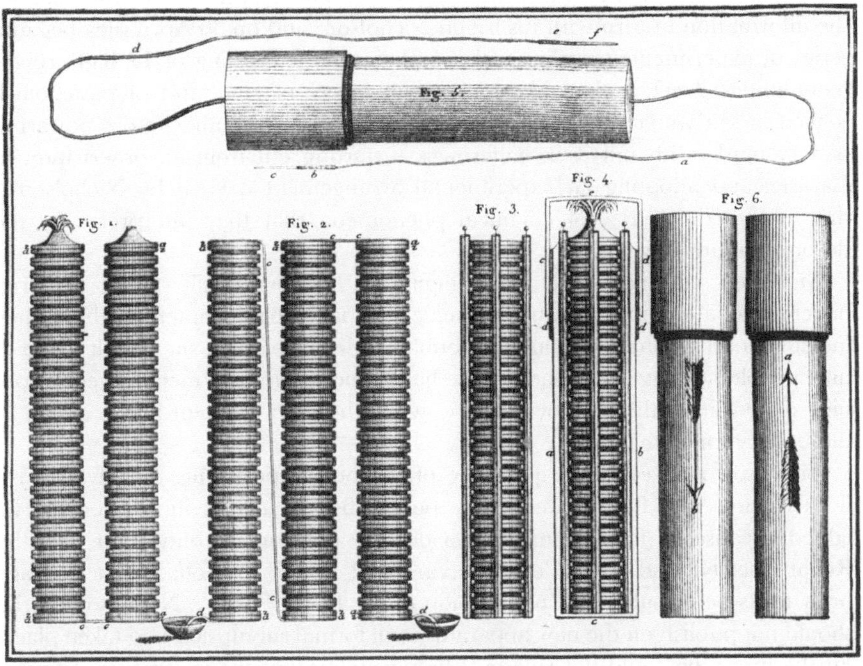

Figure 6.2 Voltaic batteries for the travelling natural philosopher. From A. Volta, 'Description de la pile électrique communiquée par Brugnatelli', *Journal de chimie et de physique* (Van Mons, Bruxelles), 6 November 1801, in VO, vol. 2, pp. 127–35, p. 132.

Source: Alessandro Volta, *Le opere*, Milan, 1918.

This, as we shall see, was to be a common reaction among those who engaged in experiments with the voltaic battery.

Landriani and his friends in Vienna also became an important centre in the informal network of acquaintances that kept Volta informed of the early career of the battery on the European scene. It was from Vienna and via Landriani that Volta received a detailed account of the decomposition of water experiments performed with the battery in London, soon to be repeated in Vienna.[10]

It took a little more time for news of the battery to reach Paris. For all we know, news arrived there via London's newspapers. Before considering this side of the story, however, the early circulation of the voltaic battery in London is worth examining in detail. Here news of the battery became immediately tied up in an intricate story, involving also the French, that brought the announcement of Volta's achievement to an unexpectedly wide audience, while exposing the tensions internal and external to the expert community of natural philosophers in the London daily press.

On receiving Volta's first account of the battery, Banks had passed it on to some of his acquaintances, including Carlisle, a distinguished physician at

Westminster Hospital and a former pupil of William Hunter.[11] Carlisle shared the information in turn with his friend Nicholson, and on 30 April they began a series of experiments together with a voltaic pile 'consisting of 17 half crowns [containing silver], with a like number of pieces of zinc, and of pasteboard, soaked in salt water'. On 2 May, using this as well as another, more powerful, battery made with thirty-six halfcrowns – starting out from an observation of Carlisle, and adopting an experimental arrangement devised by Nicholson – they observed a series of chemical phenomena that they interpreted as the decomposition of water.[12]

Nicholson continued the experiments on his own, built a one hundred halfcrowns battery and obtained some very impressive phenomena with it. They included loud detonations, clouds of bubbles, gleams of light, shocks felt by up to nine people holding each other by the hand, and a ramified, metallic vegetation, nine or ten times the bulk of the wire, when the wire was kept in the circuit of the battery for four hours.

Nicholson realised the importance of the new experiments quickly. He refrained, however, from immediately publishing the results in the journal he edited because he found himself in a delicate position in connection with the Royal Society, and with Volta. Having had access to Volta's first message prior to its being submitted to a session of the Royal Society, Nicholson felt he should not publish on the new apparatus until formal submission had taken place. Furthermore, he could not rule out the possibility that the impressive chemical phenomena he and Carlisle had observed with the battery might be mentioned in the second part of the paper Volta had promised to Banks. Nor could Nicholson anticipate that Volta had quoted him and his intended artificial torpedo in this second part. All these uncertainties kept Nicholson from publishing his results during May; but he must have talked to many people, if we are to judge from the events that took place in London towards the end of that same month.

Acquaintance with the new apparatus and the decomposition of water was such in London during May 1800 – that is, within weeks after the arrival of Volta's first description – that the recently appointed lecturer of the newly established Royal Institution, Thomas Garnett, decided to present Volta's apparatus and the decomposition of water experiment to the Institution's fashionable public. Garnett chose for the purpose his Wednesday evening lecture, 28 May 1800, using a voltaic apparatus lent by Edward Howard. The lecture was a great success as far as the battery and the public were concerned, though it also became a source of trouble for the lecturer and his patron, Rumford.[13]

Prompted by the success of the Wednesday lecture at the Royal Institution, and clearly inspired by Garnett himself, the London *Morning Chronicle* published a detailed account of the battery and the decomposition of water experiment performed at the Royal Institution on the following Friday, 30 May 1800.[14] The account – the first ever published of the voltaic battery – duly mentioned Volta, the president of the Royal Society, Nicholson, and Carlisle, but it conveyed several misunderstandings about their respective merits, as well as about some of the scientific issues at stake.

The newspaper's report led readers to believe that Volta had conceived both the new apparatus and the decomposition of water experiment, and that Nicholson and Carlisle had merely repeated the experiment rather than devised it in London. This was of course unjust to Nicholson, who by then must have read the second part of Volta's letter to Banks, where no mention was made of the decomposition of water, and where Nicholson's earlier suggestion on how to imitate the electric fish was discussed. The newspaper's inaccuracy, plus Nicholson's already mentioned delicate position in connection with the Royal Society and Volta – and now also as editor of a journal that found itself unexpectedly in competition with the daily press, as well as with the venerable but slow *Philosophical Transactions* – prompted Nicholson to a spirited reply.

Published in the *Morning Chronicle* the following Tuesday, Nicholson's letter to the editor reveals the mixed feelings generated among experts by the enthusiasm for scientific matters the daily press had been displaying on the occasion of the announcement of the voltaic battery.[15]

Nicholson perceived that changes were under way in the informal rules controlling the circulation of scientific news, as well as merit assessment, because of the role played by the daily press and by new social arenas like those provided by the Royal Institution. These were not, however, the only tensions that mingled with the early news of Volta's achievement in London. Other tensions were caused by the clash of national feelings exacerbated by the Napoleonic wars.

On introducing the battery to the public attending his 28 May lecture, Garnett – who very likely knew that the first account of the battery circulating in London was written in French, and maybe also knew that Volta, the author, was from Lombardy, and that Lombardy had been under the French since 1796; but perhaps did not know that the Austrians had temporarily regained control over the region, and certainly could not predict that they would lose it again to the French within days – attributed the discovery of the battery 'to the French'.[16] News of Garnett's statement quickly reached the President of the Royal Society, who immediately wrote to Rumford, patron of the Royal Institution, pressing for public emendation. The statement of rectification agreed between Banks and Rumford, and read by Garnett at his next public lecture, retracted the attribution of the battery to the French. It emphasised, instead, that the first news of the discovery had been addressed to the Royal Society, and acknowledged the merits of the discoverer, 'professor Volta of Milan', as well as those of Nicholson, Carlisle, and Howard.[17] The issue of nationality was clearly a sensitive one, and national rivalries, embittered by war, added to the usual individual and corporate jealousies.

The impression made by the battery and its chemical effects on London's 'philosophical men' (to use Nicholson's phrase) and their public was remarkable. Two months later, another newspaper thought it appropriate to report on Garnett's public lecture again. The *Courier de Londres*, a journal printed in London in French for a continental audience, published a translation of the *Morning Chronicle*'s piece on the battery and the decomposition of water on 8 August 1800.[18] The French translation was seized upon in Paris by the *Moniteur universel*, and it was published on 17 August 1800.[19]

The *Moniteur* was no ordinary newspaper: it was General Bonaparte's mouthpiece, and it styled itself as France's 'only official journal' – a definition that, in the wake of the battle of Marengo, had widespread implications for the rest of Europe. It was apparently through the *Moniteur* that news of the battery first arrived in Paris via London, and it was through the French newspaper that Volta first learnt about the reception of his new device in London.[20]

It was not unusual for the *Moniteur* to report on scientific matters. Spring 1800 issues of the journal contained frequent reports on the metrical system, emphasising the involvement in the project of scientists belonging to countries recently included in France's sphere of influence: General Bonaparte and his men knew how to use the popularity of science, and its practitioners among the ruling classes, to convey a reassuring image of Parisian ambitions. Volta, who in the summer of 1800 found himself again in a region within French control, was soon to be involved in the French campaign aimed at the European intelligentsia. In Paris, meanwhile, the battery won the attention of electricians from a variety of backgrounds, as it had in London.

To judge from his promptness in publishing a report on Volta's new device, the person who was most quickly won over to Volta's invention was Étienne-Gaspard Robertson.[21] He was then a popular figure in Paris because of a peculiar show he ran each night near the Opera. It was called the 'Fantasmagorie de Robertson', and it was a fashionable entertainment Parisians flocked to see. The key attraction was apparitions: Robertson produced ghosts using specially adapted magic lanterns mounted on wheels.[22] In addition, he included electrical experiments that gave his public very special sensations: a big crank friction machine, used in the electrical part of his show, was probably the most widely known electrical device in Paris before the advent of the voltaic battery.[23]

Relying only on the information published by the *Courier de Londres*, Robertson quickly had some voltaic batteries built for him, and ran a series of experiments in which he declared he had involved 'more than fifty people'.[24] He read a report of his experiments to the Institut National de France on 30 August 1800; a mere three weeks after the first arrival of news of the battery in Paris. A long-time adept of galvanism, he was intrigued by the new sensations that the steady current of the battery produced on the human body; sensations that could be turned into interesting additions to the repertoire of his 'Fantasmagorie'. Robertson's amateur results were later published in the experts' *Annales de chimie*, a not unusual step in those days. The paper he had read at the Institut National thus became the first article-length report on the voltaic battery to be published in a French scientific journal. By September, experiments had also been carried out at the École de Médecine, and soon afterwards several groups of experts and amateurs engaged in experiments with the battery in Paris.[25]

Through the daily press, in Paris as in London, the battery and its chemical effects were made known to a public that went well beyond that of expert electricians. An analysis of the press reports, however, helps clarify what exactly was regarded as news in connection with the battery among lay audiences in the early months following Volta's announcement.

What made the news in the spring and summer of 1800 in several European countries was not so much the battery *per se*, as we would now expect from our own, technological perspective, as much as the decomposition of water *plus* the instrument that had achieved it. What really made the news was the notion that water, still regarded in many quarters as a simple element (despite decisive experiments to the contrary by Lavoisier and his followers), had been shown to be compound. This surprising result had been achieved by means of an intriguingly simple instrument of a kind familiar in popular experiments on so-called animal electricity, the notions of animal electricity and galvanism being linked in turn with speculations on the nature of life, matter, and the nervous fluid. The voltaic battery, in other words, was being presented to lay audiences as closely if mysteriously linked with very broad issues debated within the domain of natural philosophy – a circumstance that enabled every cultivated person to form an idea, however vague, of the nature of the novelty announced, and of Volta's own achievement in building the new instrument.

If this was the message conveyed to lay audiences by the daily press in London and Paris, the experts had their own networks for spreading the news: personal contacts, correspondence, and the scientific journals. In the early stages of the case under examination these channels seem to have supplemented the information circulated by the press.

Joseph Banks, for example, described Volta's new device to Martinus Van Marum, of Haarlem, in a letter dated 14 June 1800.[26] Van Marum, however, joined the front line of research on the battery only a year later, when Volta sent Christoph Heinrich Pfaff, whom Volta had met in Paris, to invite his Dutch colleague to carry out experiments aimed at reinforcing Volta's own interpretation of the battery against the galvanic interpretations.[27] Compared to the kind of correspondence exchanged by Banks as president of the Royal Society with colleagues on the Continent, Nicholson's monthly journal was more effective in spreading an interest in the battery.[28] In the months following the experiments on the decomposition of water carried out in London, Nicholson published a long series of articles on the subject, written both by him and a number of British authors who had joined the rush to experiment, speculate and publish on Volta's new device and its performance. Several of the articles published in Nicholson's journal were quickly translated and reprinted on the continent.

The *Bibliothèque Britannique, Sciences et arts* of Geneva, having direct connections with circles in London, and especially the Royal Institution, first announced Volta's new 'electrical or galvanic apparatus' in an autumn issue of 1800.[29] It also announced that a Dr Marcet, who had witnessed the experiments carried out in London with the new apparatus, had repeated them in Geneva, where the Société de Physique et d'Histoire Naturelle was to devote considerable attention to the matter. The next issue of the *Bibliothèque*, for spring and summer 1801, opened with no less than four papers devoted to the voltaic battery and its chemical effects, all taken from Nicholson's journal; the authors were Nicholson himself, William Cruickshank and William Henry. A number of additional contributions on the apparatus by British authors, including Humphry Davy, were published in subsequent issues.

From autumn 1800, the same papers taken from Nicholson's journal, and published by the *Bibliothèque* in French translation in Geneva, were also published in German in the *Annalen der Physik*, edited by Ludwig Wilhelm Gilbert in Halle. During the following three years the 'Voltaische Säule' attracted the attention of the majority of the authors that contributed to Gilbert's journal, a few issues being devoted almost exclusively to the topic. The journal's interest in the voltaic apparatus culminated in the autumn of 1801, when it published reports and a letter written in Paris by Pfaff – who had meanwhile been won over to Volta's interpretation of the battery – describing the experiments carried out by the Italian there.[30]

If we accept the testimony of young Hans Christian Ørsted, news of the voltaic battery arrived in Copenhagen from London.[31] By the late summer and early autumn of 1800, through the daily press, the specialist journals and the experts' informal networks, information on the voltaic apparatus had reached a large number of both expert and amateur electricians from a variety of backgrounds across Europe.

From London, the news was communicated quickly to the British provinces. Within weeks after the first battery had been built in London, voltaic batteries were built in Bristol and Glasgow. In the summer of 1800 Thomas Beddoes, the patron of the celebrated Pneumatic Institute in Bristol – the institutional birthplace of Humphry Davy – added a voltaic battery of 110 metallic pairs to the paraphernalia of the place.[32] Early in the autumn John Robison had a battery of seventy-two pairs ready for experiments in Glasgow, and expected to learn more about its functioning from James Watt and an unnamed young chemistry student who had promised to assist him.[33]

Volta's choice of London was the right one: the mature, well-travelled natural philosopher knew how to ensure the appropriate circulation of his latest invention. Given the simplicity of Volta's apparatus – quickly remarked upon by amateurs and semi-amateurs like Robertson and the editors of the *Bibliothèque Britannique* – the wide European circulation of accounts describing the voltaic battery implied that, very likely, a correspondingly high number of batteries were built in the first few months after Volta's announcement. However, the passage from the availability of accounts describing the battery to the actual building of the instrument deserves to be analysed with care.

Replicating the instrument

The anonymous journalist of the *Morning Chronicle*, who first gave the news of the battery to the public, managed to describe Volta's device in just 131 words. These are as good now as they were then for anybody wanting to build a working voltaic battery:

> A number of pieces of zinc, each of the size of a half-crown, were prepared, and an equal number of pieces of card cut in the same form; a piece of zinc was then laid upon the table, and upon it a half-crown; upon this was placed a piece of card moistened with water; upon the card was laid another piece

> of zinc, upon that another half-crown, then a wet card, and so alternately till more than forty pieces of each had been placed upon each other; a person then, having his hands well wetted, touched the piece of zinc at the bottom with one hand, and the half crown at the top with the other: he felt a strong shock, which was repeated as often as the contact was renewed.[34]

Such straightforward descriptions of the battery, as we have seen, were easily translated from French into English, and vice versa. Nicholson and his colleagues in London had to rely on Volta's original manuscripts, written in French. Volta himself found the *Moniteur*'s description of his device – taken from the *Courier de Londres*, that had translated it from the English of the *Morning Chronicle* – 'passable'.[35] Judging from the early success of the battery in the German states and the Low Countries, further translations into German, Dutch and Danish caused no problems.

Such exercises in word counting and translation seem to show that the voltaic battery *was* a simple apparatus, at least from the point of view of the instructions needed to describe it. What about building the instrument itself, and making experiments with it?

Another easy exercise, this time in chronology, shows that building the battery and replicating some of its effects was also a relatively simple matter. As we already know, Volta's first letter describing the battery was in the hands of Joseph Banks around the middle of April 1800. From that moment it took only two weeks to have several voltaic batteries built in London, and another two weeks to have the electrical decomposition of water, or electrolysis, observed. In another three weeks, the decomposition of water and the battery were being displayed to fashionable audiences in London: within six weeks of the arrival in London of Volta's first letter, the battery, the decomposition of water, and their popular representations in the lecture hall of the Royal Institution were making headline news in London newspapers. Building the battery, at this early stage of its history, was not a business reserved to a well-defined circle of experts.

Difficult though it is to estimate the number of voltaic batteries actually built in the first few months after the circulation of news describing the instrument in different European countries, there is enough indirect evidence to maintain that there must have been several dozen, since dozens of published reports described experiments performed with the battery. Easy replicability and the common availability of the materials needed to build a battery, on the other hand, also meant easy disposability: that explains why only a few voltaic batteries from the early nineteenth century are preserved in museums today.[36]

The fact that the voltaic battery was easy to replicate, and that some of its effects were comparatively easy to reproduce, did not prevent a great variety of interpretations of the functioning and meaning of the instrument being put forward. In the case of the battery, easy replication did not carry uniformity of interpretation with it. Even assuming that the different interpreters in several European countries around 1800 were all facing the same basic instrument, the multiplicity of interpretations of the battery *vis-à-vis* the new machine *is* a challenge for historians.

Appropriating the battery

Indeed, the early history of the battery is just as challenging in a realistic perspective as it is in a social constructivist perspective.[37] In analysing issues of this kind in the following paragraphs, our aim is to explore how instruments, interpreters, and the differences as well as similarities in cultural contexts interacted in the assessment of a device that was almost instantly recognised as bringing about a momentous breakthrough in science. In order to cover a significant spectrum of interpretations and contexts, three major interpreters belonging to three different milieux will be discussed: Alessandro Volta in Lombardy, William Nicholson in London, and Jean-Baptiste Biot in Paris.

Volta's battery, 1

Volta's paper announcing the battery, finally published in the *Philosophical Transactions* late in 1800,[38] was the result of the combination in London of the two letters already mentioned dated Como 20 March and 1 April. For the sake of convenience we will call the presentation of the battery contained in these documents 'Volta's battery, 1'.

Volta's presentation strategy embodied all the know-how he had acquired in dealing with electricity *and* a variety of expert communities of electricians around Europe in the previous thirty-seven years of his career.[39] Volta's 1800 presentation focused on the following issues, in order of decreasing importance: first, the instrument itself; second, its performance; third, its relevance to the controversy over galvanism; fourth, and last, Volta's own explanation of what happened inside the instrument.

In adopting these priorities in such an order Volta was taking advantage of lessons he had learnt on previous occasions within the international community of electricians to which he belonged. When presenting his other major achievements – like the electrophorus (1775), and the *condensatore* (1780) – Volta had learnt that his instruments tended to be more easily accepted, and adopted by fellow electricians, than his reasons explaining how the instruments worked, or how he had come to build them.[40] The battery was no exception, and Volta, now fifty-five, was even less willing to take risks than on previous occasions: by 1800 he was well aware that electrical instruments enjoyed a life of their own (up to a point) within the mixed community he belonged to.

Accordingly, the paper announcing the battery to the Royal Society of London began with a short, straightforward description of the instrument itself, offering a hint of its marvellous performance.[41] The performance of the instrument was conveyed easily by saying that it behaved like a battery of Leyden jars that did *not* need to be charged and charged again to produce its effects. Volta's allusion to a 'perpetual impulse', 'action' or 'motion' of the electric fluid achieved by the battery added a sensational touch to his description of the device,[42] while the fact that the instrument in the construction he recommended could be built using just a few dozen discs, made of three easily available materials, added the appeal of simplicity.

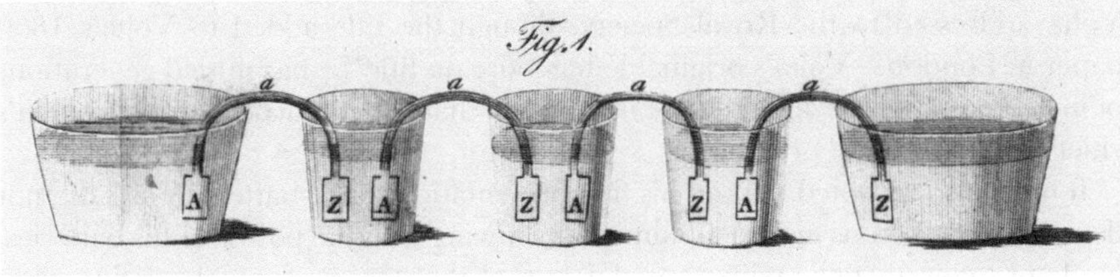

Figure 6.3 Volta's 'crown of cups' battery. From Volta, 'On the electricity excited by the mere contact of conducting substances of different kinds' (1800), in VO, vol. 1, pp. 563–82, p. 570.
Source: Alessandro Volta, *Le opere*, Milan, 1918.

The instrument, its performance and the description of a repertoire of experiments to be carried out with it occupied more than two-thirds of Volta's first presentation of the battery. The remaining third, to the surprise of subsequent generations of physicists until recently, was devoted to an analysis of the effects of the battery on sense organs, and to a comparison between the battery and the electric organ of the torpedo fish.[43]

The two latter issues were of course connected with the controversy over galvanism in which Volta had been involved over the previous eight years.[44] Yet, when first presenting the battery, Volta did not think it appropriate to present the new instrument as just another step in the controversy, nor as a winning move in his fight against galvanism. Popular as these interpretations became in the later literature, they were not encouraged by Volta when he first presented the battery.

The instrument itself, in Volta's presentation, came in two different formats: the column or pile battery (see Figure 6.2), and the crown of cups (see Figure 6.3). In his presentation, Volta claimed to regard the two formats as perfectly equivalent, though crediting each one with special strengths and limitations. According to Volta, the pile was more handy and compact, and it could be easily built in pocket or travelling versions: an important feature for a natural philosopher who liked travelling and showing his expertise in electrical matters to varied audiences throughout Europe.[45] The crown of cups battery, on the other hand, was more suitable for demonstrations, as the cups and the bimetallic arcs could be added or subtracted easily in a way that, Volta emphasised, 'spoke directly to the eyes'.[46] Volta also mentioned that he had developed the same basic device in other formats as well, but he thought this was not an issue.[47]

Thus, already in its author's first presentation the battery was a *family* of instruments rather than a single device. There is no evidence, however, that either Volta or the other early experimenters with the battery wished to emphasise the multiplicity of formats. Volta certainly called his colleagues' attention to the instrument being regarded as the same basic device, and to its extraordinary performance, rather than to its different formats.

The fact that, in his first presentation of the battery, he did not emphasise theoretical and interpretative issues was probably part of the same strategy. This is quite clear from the content and the line of argument developed in the texts Volta addressed to the Royal Society, though the title added to Volta's 1800 paper in London – Volta's original letters bore no title[48] – has misled generations of interpreters, conveying the false impression that the contact theory was Volta's chief concern on the occasion.

It must also be noted that, in his first presentation of the battery, Volta did not develop any rigorous argument aimed at showing that the power of his batteries, made of many metallic pairs, was the sum of the power of the electricity set in motion by each metallic pair – a point which, as we shall see, he insisted was a crucial one in his later presentations of the battery, especially the one addressed to the Institut National in Paris one year later.[49]

All these circumstances are worth remembering. Volta's strategy when first presenting the battery was in keeping with the steps that had led him to the battery a few months earlier. According to my reconstruction,[50] no new, decisive conceptual turns had taken place in his investigation in the three years preceding the building of the battery. The event that precipitated the building of the new instrument – his reading of an article by William Nicholson, which proposed building a device imitating the electric organ of the torpedo fish – took place at a quite special level of expertise. The kind of expertise De Solla Price has described as requiring, on the part of the experimenter, the ability to operate with one's 'brain in the fingertips': a level of expertise, that is, in which ingenuity and manipulative ability were just as important as theory and natural philosophy. Volta's chosen strategy when he first presented the battery, focusing on the instrument itself, reflected this background, as well as his awareness that instruments tend to be less controversial than the theories explaining them.

Instruments also tend to have a life of their own, however, and Volta was well aware of that too: when dealing with the battery soon after spring 1800 he had to adjust his strategy to the needs deriving from the interpretations that other experts were suggesting for his own creature. This leads us back to William Nicholson.

Nicholson's battery

Nicholson was one of only two authors (the other was Tiberius Cavallo) that Volta mentioned in his first description of the new instrument. Although Volta did not acknowledge the role that Nicholson's article on the torpedo fish had played in leading him to the battery, he did discuss Nicholson's paper at length.[51] Thus Nicholson had several reasons for taking note of the battery, including some personal ones. As is well known, while operating in the company of Carlisle with one of the first voltaic batteries built in London, he soon noticed something chemical going on inside the instrument; something Volta had *not* mentioned in his account of spring 1800.

Having devoted about two months to experiments with the battery, Nicholson was ready to acknowledge that Volta's latest invention was a 'most curious and important combination'.[52] He also acknowledged that the discovery 'must for ever remove the doubt whether galvanism be an electrical phenomenon'.[53] Nicholson, however, showed amazement at finding that Volta had entirely neglected the chemical phenomena associated with the battery. As he put it in July 1800:

> But I cannot here look back without some surprise, and observe that the chemical phenomena of galvanism, which had been so much insisted on by Fabroni [*sic*], more especially the rapid oxidation of the zinc, should constitute no part of his [Volta's] numerous observations.[54]

The allusion was to a paper by Giovanni Fabbroni published in Nicholson's journal some time earlier.[55] Nicholson's emphasis on the chemical phenomena of the battery, which had been ignored by Volta, may have been prompted by his desire to adopt an original stance *vis-à-vis* Volta's contact explanation of the battery, which as we have seen he found less than perspicuous. Some jealousy generated in Nicholson by Volta's critical remarks about his earlier suggestions on how to imitate the torpedo, voiced in the same paper in which Volta announced his new, brilliant apparatus, should also be taken into account. But Nicholson's enthusiasm for the powerful, chemical performance of the battery was genuine. Having combined Carlisle's battery and his own to make an apparatus totalling sixty-eight metallic pairs, Nicholson reported the following impressive observations:

> A cloud of gas arose from each wire, but most from the silver, or minus side. Bubbles were extricated from all parts of the water, and adhered to the whole internal surface of the vessels. The process was continued for thirteen hours, after which the wires were disengaged, and the gases decanted into separate bottles. On measuring the quantities, which was done by weighing the bottles, it was found, that the quantities of water displaced by the gases, were respectively, 72 grains by the gas from the zinc side, and 142 grains by the gas from the silver side; so that the whole volume of gas was 1.17 cubic inches, or near an inch and a quarter. These are nearly the proportions in bulk, of what are stated to be the component parts of water . . . From the smallness of the quantity no attempt was made to detonate the air from the zinc side, but a portion of that from the silver side, being mixed with one third of atmospheric air, gave a loud detonation.[56]

Reflecting on the power of the battery to set the electric fluid in motion, in Lombardy Professor Volta had evoked the old, mechanical notion of a 'perpetuum mobile'. Viewing the same instrument from the standpoint of an independent philosopher, fully immersed in the scientific and technological ferments of Britain's first industrial revolution, Nicholson perceived the battery instead as a powerful chemical machine; a machine to be compared with, and tested like, other chemical apparatus or, perhaps, like the steam engines of his times. As he put it:

> We are in want of a measure of the intensity of the action of these machines. Will this be derived from the quantities of water decomposed, or of gas extricated under like circumstances in given times? Or from any change of temperature? Or what other commensurate incident? – Mr. Carlisle has not found that the water in the tube, while under this agency, did produce the slightest effect on a very small and delicate thermometer.[57]

The impressive chemical phenomena Carlisle and Nicholson had observed with the battery made their apparatus a quite different kind of instrument from the one presented by Volta a few weeks earlier in his letter to Banks. The apparatus built in Como and in London was indeed basically the same, but the uses described by Nicholson and Carlisle had several distinct features.

In a manner wholly unanticipated by Volta, in London (and, via London, in several other places in Europe) the battery was being perceived as closely linked to long-debated chemical issues, and especially to discussions on the nature of water. Volta's work on the battery subsequent to spring 1800 was devoted to coming to terms with this unexpected turn of events, and to rescuing the battery and his own contact theory from the swamping effect produced by the chemical interpretations, by the discussions on the nature of water, and by the reiterated, public assertions of faith in galvanism on the part of many philosophers interested in the battery.

Volta's battery, 2

News of the reception of the battery and the decomposition of water experiments carried out in several European countries reached Volta in Como after delays, and in a patchy manner. We know that he first learnt of the experiments of Nicholson and Carlisle of May 1800 from the Parisian *Moniteur*, and again via his friends in Vienna, late in August;[58] further, as a subscriber, sometime in the autumn of 1800 or early in 1801, he must have seen the long series of papers devoted to experiments with the battery that was printed in Nicholson's *Journal*.

Volta was of course pleased by the big impact his latest invention was making on expert and lay audiences. Nevertheless, he was definitely worried to see the little attention the battery itself was receiving, as well as his contact theory as an appropriate explanation for it, compared to the emphasis given to the chemical phenomena he had failed to mention when first announcing the battery. As a consequence, in his correspondence (and few publications) of 1800–1, Volta adopted a guarded stance towards the chemical phenomena of the battery; a stance he would preserve for the rest of his life.

In addition, when reading declarations like the one published by Humphry Davy in the autumn of 1800 to the effect that 'the oxydation of the zinc in the pile, and the chemical changes connected with it are *somehow* the cause of the electrical effects it produces',[59] Volta must have realised that a prudent stance on the chemical effects of the battery was not enough if he wanted to keep a grip on his invention, to assert his own interpretation, and to obtain the rewards he expected from it.

The need to adopt a new strategy in dealing with the battery (compared to the strategy pursued in the spring 1800 letters to the Royal Society, described above as *Volta's battery, 1*) merged with the new needs urged on Volta by the political and personal situation in which he found himself during the summer of 1800. After the battle of Marengo, the French had regained control over Lombardy, and the Austrians, who had closed the University of Pavia and suspended Volta's stipend as professor of physics there, left. In the autumn of 1800, with the recent discovery of his brilliant apparatus in his favour, and the University of Pavia reopened by the French, Volta began to seek the kind of recognition that would place him above the uncertainties of the times, or at least to offer him protection. Under the new circumstances, Volta realised, rewards were to be sought in Paris, not in London.

On 28 September 1800 Volta wrote to General Guillaume Brune, head of the French troops in Italy. He had already been introduced to the general, and had informed him about the invention of the battery. A few weeks earlier, we know, Bonaparte's newspaper had announced the discovery of the battery in appreciative terms. Volta asked the general for two things: first, for leave of absence to go to Paris to thank Bonaparte formally for the restoration of the University of Pavia; second, for protection for his relatives and a list of friends during his absence.[60]

Volta still doubted, however, whether the times were right for a mission to Paris: the sudden changes in fortunes on the European military and political scene advised caution.[61] But on 9 February 1801 the Lunéville peace treaty was signed. Only Britain remained at war with France, and that too soon changed when peace negotiations, begun in London, were concluded in October that same year. In May 1801 Volta resumed contacts with Gaspard Monge and Claude-Louis Berthollet in Paris: both of them had attended Volta's experiments on galvanism in Lombardy at the time of their mission on behalf of the French government in Italy in December 1796.[62] Volta also sent a description of the battery to chemist and minister Jean Antoine Chaptal in Paris.[63]

Meanwhile, pro-French administrators in Milan realised that Volta's trip to Paris might benefit them too. When Volta – still personally uncertain about the advisability of the trip – asked for a substantial sum to cover travel expenses, it was granted. The letter written by the Minister of the Interior of the Cisalpine Republic in Milan to finally arrange for Volta's trip to Paris, spoke of two aims of the mission. One was 'cementing an alliance of talents and knowledge' between the Cisalpine and the French Republics. The other was to ensure 'the progress of scientific pursuits', that 'benefits immensely from the fast communication of enlightenment'.[64] The rhetoric of the new Minister was more emphatic than that used before the French Revolution, but the substance of the message was the same as Volta's old Austrian patrons had had recourse to on similar occasions during the twenty-five years in which Volta had worked for them.

Volta left for Paris on 1 September 1801, and arrived there three weeks later, having stopped in Geneva to meet his several friends there, and to win them over to his interpretation of the battery. Volta brought with him some fine 'pocket' batteries, and drafts of a paper he had written over the previous twelve months with the purpose of establishing his own explanation of the instrument, opposing the galvanic and the chemical interpretations, and re-appropriating his own invention.[65]

As on previous occasions, Volta's new arguments were the result of close interaction with a number of other researchers. The more so this time because he had decided to involve the Paris natural philosophers in his new campaign for rewards, and he had had some problems in his relations with them since his previous visit there in 1782.[66] As on the previous occasions when he had acted to defend his claims to the invention of a new instrument, he developed an elaborate set of theoretical arguments and empirical proofs to secure his rights.[67]

Volta's key step in the new strategy, worked out in 1801 with Paris (no longer London) in mind, consisted in rooting the 'demonstration' of the battery on a fundamental experiment, in which a single metallic pair without any humid conductor gave tangible, if extremely weak, signs of electricity. The experiment was clearly conceived to exclude any role for chemical or galvanic phenomena. However, whereas the instrument by then popularly known as Volta's battery, combining many metallic pairs and humid bullets, was easy to replicate even by amateurs, and produced surprising effects to the satisfaction of lay and expert audiences alike, Volta's fundamental experiment using a single metallic pair was extremely difficult to replicate, and its effects controversial.[68]

In order to obtain signs of electricity from a single pair, and to show that the '*force électromotrice*' generating the current of the battery resided in the mere contact of two different metals, Volta needed his '*condensatore*', a device that, like other 'doublers' or 'multipliers' of electricity at the time, could be suspected of generating, by unintended friction, the electricity it was supposed to detect. The elaborate set of practical rules that Volta kept repeating in written and oral presentations of his fundamental experiment in 1801 – ostensibly with the purpose also of reasserting the merits of his earlier inventions like the *condensatore* and the straw electrometer – was not such as to dispel all doubts. In fact, Volta admitted that this experimental setting made only the electricity of the silver disc detectable. In order to detect the electricity of the zinc disc, a piece of wet cardboard had to be interposed between the metal and the plate of the *condensatore*.[69]

There could be no sharper contrast between the easy replication of the voltaic battery, and its by then popular effects, and the extremely difficult replication of the experiment Volta wanted to use to demonstrate the principle on which the battery worked in the autumn of 1801.

As a consequence of all these difficulties it took repeated demonstrations, and all the personal qualities of Volta (including his well-known proselytising ability), to convince the '*commissaires*' of the Institut National on the occasion of his frequent meetings with them. He managed it, however, through an impressive *tour de force*.

He first won over to his cause the amateur Robertson, whom he visited as often as he visited the elite members of the Institut. Second, after stern lobbying among the First Consul's entourage he convinced another, more powerful amateur: Bonaparte himself, who was also a member of the Institut. Finally, he had his own explanation of the battery adopted in the final report of the Institut's experts' commission, written by Biot. Volta's impressive endeavours in Paris in 1801 are worth following in detail.

They began on 26 September, when the *Journal de Paris* announced his arrival in the French capital, accompanied by the chemist Brugnatelli, his colleague and friend, who kept a diary of Volta's achievements.[70] By 30 September Volta was a guest in Arcueil and had already met Claude Louis Berthollet, Antoine François de Fourcroy, Georges Cuvier, René-Just Haüy, Biot and Robertson. On 2 October some of Volta's experiments were repeated at Fourcroy's after a lunch. The next day Volta and Brugnatelli were made members of the commission on galvanism previously appointed by the Institut. On the 4th and 6th he met and won over to his cause young Christoph Heinrich Pfaff, whom he soon sent to Van Marum in Haarlem with the purpose of having the performance of the big Teyler machine compared with the battery in the decomposition of water. By 8 October, as already mentioned, Pfaff was advertising Volta's achievements in Paris to the German states in letters addressed to German journals.

On 6 October Volta had lunch at the Ministry of War and met there, among others, General Brune (who had been instrumental in bringing Volta to Paris) as well as Monge and Berthollet. On the 7th he met Minister Chaptal, and was his guest for lunch on the 17th. Soirées were mostly spent attending Robertson's 'Fantasmagories': on 23 October Bonaparte's journal, the *Moniteur*, announced that Robertson had carried out public experiments confirming Volta's interpretation of the battery, and disproving the galvanic interpretations previously endorsed by Robertson himself. By 3 November the list of Volta's new or renewed acquaintances in Paris included also Pierre-Simon de Laplace, Jean Nicolas Pierre Hachette, Charles-Bernard Désormes, Jean-Henri Hassenfratz, Louis Bernard Guyton de Morveau, Louis Joseph Dumotiez and Benjamin Thompson (Count Rumford).

On 6 November Ferdinando Marescalchi, the representative in Paris of the Cisalpine Republic (of which Volta was now a citizen), introduced Volta to Bonaparte, with whom he would also have eaten lunch, had it not been for the delayed delivery of the ticket inviting him. The next day Volta, who had already attended three previous sessions of the Institut's commission on galvanism, began reading a memoir in French that he had meanwhile polished up with the help of two other acquaintances in Paris: the physicians Biron and Tourdes.

On that same day, 7 November 1801, Bonaparte was able to spend more than an hour at the Institut attending Volta's presentation. The length of the First Consul's stay was carefully noted down by methodical Rumford, whose report of the event conveyed a mixture of pride and jealousy at the attention Volta was being paid by Bonaparte.[71] The First Consul's interest in electricity, on the other hand, was genuine, if generic. Like many amateurs in those days he regarded electricity and galvanism as among the major ingredients of a broad, vague, but compelling naturalistic world-view.[72]

At the end of the 7 November session at the Institut, the First Consul invited his colleagues to confer a gold medal on Volta. The next day Volta let Bonaparte know that he was ready to repeat his electrical experiments for him whenever he pleased.

The members of the Institut complied with Bonaparte's request to reward Volta with some hesitation. Many of them of course remembered the uneasy relationships Volta had had with Parisian natural philosophers since 1782, and they were aware of his negative reactions to Coulomb's memoirs on electrostatics.[73] They realised, however, that in his latest initiative Bonaparte's amateur interests in science combined with his usual political goals. These were indeed momentous days for France, and for Europe. Two days after Volta's first presentation at the Institut, Paris celebrated the (temporarily) restored peace, favourable to the French, with a great fanfare. Volta, an Italian whose name was known by then to the cultured elites of most European capitals, who had served Austria for twenty-five years and who had announced the battery in London (alas!), though writing in French, was now conferring with his peers in Paris: Volta embodied all the symbolic characteristics needed to convey a reassuring, 'scientific' image of French ambitions in Europe.

Bonaparte also attended Volta's second presentation at the Institut, which took place the day after the big peace celebrations. Volta concluded his presentation on 13 November. The next day he was chosen to represent the Cisalpine Republic at the congress of the Italian regions under French influence that was to convene shortly in Lyon.[74]

Biot dated his final report to the Institut on Volta's apparatus 2 December. On the 4th Volta left Paris for Lyon; Bonaparte's *Moniteur* and other Parisian newspapers, however, continued to keep their readers informed about Volta's achievements. On 9 December the *Moniteur* reported that Van Marum and Pfaff in Holland had confirmed the common electrical (not galvanic) nature of the battery's effects. On the 20th, the journal once again published instructions on how to build a voltaic battery. Over the same period the popular *Journal de Paris* hosted an exchange of letters on the battery, in which Robertson defended Volta's interpretation. As advertised every day in the same newspaper, by then Robertson had made of 'physique expérimentale' a permanent attraction of his evening performance in Cour des Capucines. On Christmas Eve, 1801, the *Moniteur* published an abstract of Biot's report on Volta's experiments to the Institut (on which more later).[75] Meanwhile, on 8 December, the Consuls of the French Republic had decreed a gratuity of 6,000 francs to Volta.[76]

By the end of December 1801 Volta and Bonaparte would certainly have agreed that they had achieved their respective goals in dealing with the battery in front of the savants of the Institut, the Paris daily press, and, through them, Europe's cultivated elites.

Biot's battery

Young Biot's first reactions to the voltaic battery had been in tune with the generic galvanic framework within which the voltaic battery was first reviewed in Paris.[77] Thus, the uncertain connections between the galvanic and the electric fluids, and the chemical effects of the battery, claimed Biot's main attention in the first experiments he carried out with the new apparatus.[78]

Inspired by Laplace, however, Biot soon introduced speculations on the speed of the electric fluid in metallic and wet conductors of different kinds, and in the summer of 1801 he sketched an interpretative model of the battery that was partly based on mechanical assumptions.[79] As to the chemical effects of the battery, Biot wanted to keep them within the framework of the same electrical interpretation.[80]

Though many important issues placed Volta and Biot worlds apart in scientific matters,[81] in the interpretation of the battery they shared two basic goals: to keep galvanism within the domain of electricity and to prevent the chemical phenomena of the battery from undermining the efforts being made at a physical interpretation of electrical phenomena.

A hint in the direction of a mechanical and mathematical treatment of the battery had been offered to Biot by Laplace himself, who had observed that attractions and repulsions could be detected at the two ends of the instrument.[82] Further hints in that same direction derived from Biot's own speculations on the supposed speed the electric fluid acquired inside the battery, depending on the size and shape of the metallic discs being used. The measuring techniques necessary to substantiate this physical approach to the battery, however, simply did not exist. In the end, Biot's strategy when dealing with the battery in the report he finally submitted to the Institut was a dual one, and somewhat inconclusive.[83] In the first part he asserted the priority, in principle, of Coulomb's notions over the entire science of electricity, while in the second part he introduced a mathematical model of the battery that established no link with Coulomb's law.

Biot's quantitative approach to Volta's contact theory was much more sophisticated than Volta's. However, Biot's mathematisation of the battery did not exert an impact on his contemporaries comparable to the one exerted by Cavendish's memoirs on a previous generation of electricians, nor to that played by Poisson's treatment of electrostatics on scientists of the same generation as Biot. Apparently, there was more to the battery than the mathematical physics of Biot could convey.

After 1801, in any case, the majority of the researchers interested in the battery focused their attention on its effects and how to improve them, rather than on a mathematical treatment of what happened inside it. As to Volta, despite the considerable satisfaction he had got in Paris, and the ample role Biot had granted to his contact theory, he never mentioned Biot in his subsequent writings, either published or private.

A name for the instrument

Having described the attempts made to appropriate the battery to various individuals, scientific traditions and cultural contexts, it may be instructive to trace the history of the names that were attached to the instrument being appropriated.

As it turns out, the names that circulated most extensively were those that did *not* identify the instrument with a single interpretation, interpreter or group of interpreters. It was as if the experts agreed, amidst much controversy, that it was reasonable after all to adopt a name for the instrument that allowed it to enjoy a certain amount of independence from the clash of different interpretations. The most popular names adopted to designate the voltaic battery were in fact those that focused on its most obvious and least controversial features.

When first presenting the battery, indeed, Volta himself had tried a different strategy, dictated by the need to be recognised as its inventor. He was well aware of the role the names attached to new instruments played within expert and amateur communities; he had learnt that from the time of his electrophorus, when the name imposed on the instrument had helped him to vindicate his merits in his subsequent, controversial claims for originality. As he declared on presenting the battery to the Royal Society, 'il faut donner des nouveaux noms à des instruments nouveaux, non seulement par la forme, mais aussi par les effets, ou par le principe d'où ils dépendent'.[84] Accordingly, Volta had recommended two possible names for the battery. The first alluded to the organ of the torpedo fish that had inspired him, and to Volta's own claims against galvanism. It was the name of '*organe électric artificiel*', the organ of the fish being the '*organe électric* naturel'.[85] The second one alluded instead to the principle by which the battery worked according to its inventor: '*appareil électromoteur*'.[86] This was of course an elaboration of Volta's own notion – developed well before the invention of the battery – according to which different conductors, especially metals, when in contact, had the power to set the electric fluid in motion.

Volta's own recommended names, however, were not to be adopted widely within expert or amateur circles in Europe. These apparently needed a term that did not imply Volta's own appropriation of the instrument.

When presenting the battery Volta had used a couple of other names in order to distinguish between the two forms of the instrument he described on the occasion. He used the phrase '*appareil à colonne*', or column apparatus, for the first battery he had built, the one described in his 20 March letter; and '*appareil . . . à couronne de tasses*', or crown of cups apparatus, for the one illustrated in his 1 April letter. Early interpreters – who had built Volta's apparatus and replicated its basic effects, but were unwilling to commit themselves to Volta's interpretation of it – preferred similar names or phrases describing the instrument, rather than its controversial working principles.

Thus Nicholson, who in his July 1800 title cautiously mentioned 'the new *Electrical or Galvanic* apparatus of Sig. Alex. Volta', in that same article called the instrument 'the pile', or 'the electric pile'.[87] Landriani, in a letter of August 1800 already mentioned, called it the 'apparato a colonna', or the column apparatus.[88] In those same months the battery was called 'l'appareil électrique ou galvanique' in Paris and Geneva, while the Germans called it 'Volta's Säule', or Volta's column. In the course of 1801, the French and English 'pile', the Italian 'pila' (occasionally also 'piliere'), and the German 'Säule' prevailed, and Volta accepted it for convenience's sake,[89] while continuing to use his own term of 'elettromotore' whenever possible.

Due to well-known interpretative uncertainties, the adjectives 'galvanic' or 'voltaic' (or 'Volta's'), or both, continued to be attached to the 'pile' or 'column' for several years, with Volta of course discreetly campaigning in favour of the latter. Meanwhile, another name emerged: battery.

Volta and all the early experimenters with the new apparatus noticed that the electricity produced by it resembled the electricity delivered by a battery of Leyden jars; that is, by a condenser made by connecting a number of Leyden jars together. The batteries of Leyden jars were known to be able to retain and discharge large quantities of electricity at a weak charge; or, to use Volta's terminology, their electricity displayed a high capacity and a low tension, just like Volta's new apparatus. Since Franklin's studies in the 1740s, batteries of Leyden jars were known to any amateur and expert electrician, and they were described in all manuals and treatises devoted to the subject. In 1800 Volta was proud to remind his readers of those earlier batteries, while emphasising that, contrary to the batteries of Leyden jars, his new device did not need to be charged again and again with an external machine after discharge.[90]

Because of these earlier electrical batteries – and perhaps also because of the common usage of the word 'battery', with its military origins, to denote a set of similar pieces used for combined action[91] – experimenters using Volta's device made of many identical metallic pairs combined together began to call it 'a battery'. Humphry Davy used the expression 'galvanic battery' in entries of his manuscript *Physical Journal* from November 1800.[92] By 1803, 'galvanic batteries', 'Volta's batteries', and 'voltaic batteries' obtained regular entries in English electricity textbooks.[93] Later names, like the English 'trough',[94] the German 'Trog', and, later still, the English 'cell',[95] indicate that the forms of the instrument continued to be regarded as a safer ground for attaching a name to the apparatus than the controversial interpretations accompanying it.

The story of the names attached to the voltaic battery in different countries at the beginning of the nineteenth century points to the fact that at a basic, but important, communication level an instrumentalist attitude was shared by the many who were interested in the new device within expert and amateur circles. By providing some common, relatively unproblematic terms, an instrumentalist attitude made translations and travel easier for the new device, as well as for its many textual and visual representations. Indirectly – but in an important respect – that same instrumentalist attitude favoured the circulation of the conflicting interpretations of the new apparatus that were being put forward in several European countries. Some common terms were indeed prerequisites for the different appropriations of the battery to take place, and their respective merits to be assessed.

Conclusion

The early reception of the voltaic battery in several European countries has been surveyed in the present chapter with the purpose of exploring how instruments, experts, amateurs, and cultural contexts interacted in the assessment of a device that brought about a momentous breakthrough in the history of science.

The events associated with the early reception of the voltaic battery point to two distinct sets of circumstances. On the one hand, the rapid diffusion of the battery in expert and amateur circles, approaching the instrument from widely different backgrounds, indicates that the battery and its basic effects were comparatively easy to replicate and, in some obvious sense at least, the battery was regarded as being the same, basic instrument in London, Paris, Geneva, Vienna, Halle, Copenhagen, and Como, where it had first been conceived and built.

It is also clear, however, that – side by side with the rapid, relatively unproblematic diffusion of the instrument – a wide range of different, sometimes contradictory, interpretations of the battery were put forward in various quarters, occasionally by the inventor himself. As already pointed out, in the case of the battery easy replicability did not carry uniformity of interpretation with it.

How are we to assess this dual, apparently contradictory, set of circumstances?

An easy way out of the difficulty, of course, would be to explain away one of the two contradicting sets of evidence. We should refrain, however, from underplaying either easy replication or disagreement over interpretation: both played a part in the early history of the voltaic battery, and we seem to need an interpretative framework compatible with both. The following concluding remarks are aimed at sketching such an interpretative framework.

To begin with, it is useful to recall how the battery was introduced to the scene in the first place.[96] According to our reconstruction, the battery enjoyed a certain amount of independence from the possible interpretations provided by natural philosophy in the investigative experience and presentation strategies of its own inventor. As previously pointed out (under *Volta's battery, 1*), this relative independence of the instruments from the interpretations provided by natural philosophy was something Volta was aware of, because he had experienced it when introducing his earlier inventions. On the other hand, Volta also knew that within the mixed community he belonged to of expert and amateur electricians, natural philosophers and instrument makers, and considering his position as a university professor, his claims to be the inventor rested mainly on his ability to provide evidence that he *possessed the theory explaining the instrument.*

As has been argued, these circumstances may well explain why, when first presenting the battery, Volta emphasised the instrument and its effects rather than the theory he thought could explain it; and why he could adjust his presentation strategies to his peers' reactions and expectations, as he did shifting from *Volta's battery, 1* to *Volta's battery, 2*, at the same time as his quest for rewards shifted from London to Paris.

Other circumstances favoured the relative independence of the instrument from the author's interpretation of it, and, *a fortiori*, from the interpretations of others. Within the mixed community Volta belonged to, recognition and reward for the invention depended on well-administered publicity for the instrument, its construction, its effects and its implications for natural philosophy. That publicity had to be addressed to a broad group of expert and amateur electricians belonging to different cultural, social and national contexts. This meant providing a description of the instrument that appealed to many different audiences, rather than to a closely knit group of experts; and that was what Volta pursued with his letters to the Royal Society, and with his propaganda trip to Paris and other European centres in the autumn and winter of 1801.

The news leaks about the device and its performance in London in spring 1800 – from the circles inside and around the Royal Society to those of the Royal Institution, and hence to the London and Paris daily press – further encouraged the public career of the battery independently of the interpretations recommended by its author, or by any self-selected group of experts. As a result of those several news leaks – and because of other, more cogent circumstances – for many months the battery was built, and its functioning and effects assessed in many circles around Europe, independently of the presentation the author had thought fit to accompany it with. Under the circumstances, the fact that entirely different interpretations were soon attached to the battery can be regarded as an obvious consequence.

The degree of independent life the voltaic battery enjoyed during its earliest appearances on the public scene was no doubt enhanced by its simplicity, and by the easy availability of the ingredients needed to build it.

In subsequent years, with the introduction of different formats, greater size, and above all chemical improvements on the voltaic battery, it no doubt became a highly sophisticated device, and its use a complex enterprise on the part of accomplished natural philosophers like Humphry Davy and Antoine-César Becquerel, and technologists like John Frederic Daniel and William Robert Grove, who were often supported by institutions that, in several countries, invested considerable amounts of money in building big and improved batteries. The fact, however, that only Frederick De Moleyns, in England, in 1841, thought it fit to patent improvements in battery construction, and regarded all earlier batteries as 'voltaic batteries' that he considered to be 'in general use',[97] points again to the circumstance that the basic identity of the voltaic battery had been agreed upon up to that time, despite the considerable technical developments in battery construction in the four decades following its first introduction.

As pointed out, however, the agreed-upon replicability of the voltaic battery went hand in hand with an impressive array of diverse appropriations of the instrument from many different quarters. These appropriations, as we have seen, responded to different needs, depending on the intellectual and social roles of the persons or groups interested in the new instrument.

For Volta himself, because no controversy developed over his claim to be regarded as the inventor, appropriating the battery was above all a matter of intellectual pleasure, as well as part of the social strategy aimed at obtaining rewards for his new contribution to knowledge. As all natural philosophers like Volta knew, the chief strategy for appropriating a new instrument was to show mastery of the theory explaining it. Those who read Volta's second presentation of the battery, or saw him performing with the battery during his trip of 1801, could nurture no doubts about his main intellectual ambition in connection with the instrument: Volta regarded the battery as a decisive blow in the battle to show that there was no special, animal or galvanic electricity, but only the same plain, physical electricity everywhere in nature. This broad, natural philosophical, physicalist tenet was, according to Volta, what conferred general meaning to the battery, and gave an appropriate measure of his achievement in building it.

The many who disagreed with Volta over his interpretation of the instrument agreed in any case on the kind of knowledge, and the broad intellectual implications, that the battery was perceived to contribute to – however confusedly. Nicholson, for example, when he emphasised the chemical phenomena accompanying the battery's operations that had gone unnoticed by Volta, was at the same time appropriating part of the battery (its chemical effects), claiming originality and hence recognition, and vindicating the role his model of the torpedo fish had had in leading Volta to the battery – a role the Italian had not acknowledged in his letters to the Royal Society. Nicholson in any case implicitly agreed that what made the battery special (and his own contribution to it important) were the broad implications the instrument was supposed to have for the natural philosophies involved.

Because there was a general, if vague, understanding of the unexplored but momentous philosophical implications of the battery, widespread disagreement about which particular philosophy explained it better did not detract from what was universally a generous appreciation of Volta's achievement in inventing it, and of the merits of Nicholson and Carlisle in discovering its chemical effects.

As we have seen considering the remarkable degree of popularity the battery enjoyed beyond expert circles, some form of appropriation of the battery appealed also to a wide variety of non-expert cultivators of the sciences. If appropriating the battery within expert circles meant to contribute original explanations and/or new evidence and technical improvements, and to get recognition for them, among lay audiences it was linked to other, arguably not too dissimilar intellectual and social pleasures. Judging from the early press reports describing the battery discussed above, and from the fascination journalists and lay audiences alike felt for the water, electricity and galvanism business in London, Paris, Geneva or Vienna, what made the battery popular was not so much the instrument *per se* but rather its vague and yet intriguing connections with broad speculations concerning the nature of life, matter, and human beings; speculations that apparently appealed to many cultivated people scattered across the borders of Europe *c.*1800. To them, the easy replication of the battery added wonder to an already sensational subject.

Judging again from the early reception of the battery among lay audiences, an understanding of the broad implications the instrument had for natural philosophy and an adhesion to the cultural and social rituals involved in the recognition of original contributions to expert knowledge were relatively widespread commodities in Europe around 1800.

It was the significant amount of shared, if uncertain, understanding of the instrument and its implications, and the similar social rituals associated with the assessment of knowledge production in various European countries, that allowed Volta to search for and obtain recognition as inventor of the battery outside his own country, allowing him to spend in Lombardy the rewards he had obtained in London, Paris and Vienna. That same amount of shared understanding, and the not too dissimilar rituals associated with the reward of intellectual achievement across Europe's troubled borders, allowed Bonaparte to use the prestige of Volta and his battery to please the cultured elites of several European countries at a time, leaving a lasting impression on the imagination of subsequent generations of scientists and lay audiences alike of the intriguing connections between science, power and public culture.

Yes, the history of the voltaic battery does not seem to fit easily either into a traditional realist interpretation of scientific instruments or into a constructivist view of science conceived as the product of strictly local cultures: a combination of both seems to be needed if we want to understand the geography of the new continent opened up by the invention of the battery, and by its early travels across Europe's cultural borders.

Notes

Abbreviations: DSB = *Dictionary of Scientific Biography*, 18 vols. (New York, 1970–90); OED = *The Oxford English Dictionary*, 12 vols. and Supplement, Oxford, Clarendon Press, 1933, reprint 1961; VE = Alessandro Volta, *Epistolario* (F. Massardi ed.), 5 vols. (Bologna, 1949–55); VMS = Volta Papers at the Istituto Lombardo, Milan; VO = Alessandro Volta, *Le opere*, 7 vols. (Milan, 1918; reprint: The Sources of Science, No. 70, New York and London, 1968).

1 M. Faraday, *Experimental Researches in Electricity*, 3 vols., London, Taylor, 1839–55, vol. 1, p. 300. On the battery see J.L. Heilbron's magisterial *Electricity in the 17th and 18th Centuries. A Study in Early Modern Physics*, Mineola, N.Y. Dover Publications, 1999 (1st edition 1979), pp. 491–4, and H.O. Sibum, 'Die Mechanisierung der Lebensvorgänge – der Weg zum elektrischen Strom', in J. Meya and H.O. Sibum, *Das fünfte Element. Wirkungen und Deutungen der Elektrizität*, Reinbeck: Rowohlt, 1987, pp. 117–41. On the invention of the battery: G. Pancaldi, 'Electricity and life: Volta's path to the battery', *Historical Studies in the Physical and Biological Sciences* 21, part 1, 1990, pp. 123–60. On the impact of the battery: R.H. Schallemberg, *Bottled Energy. Electrical Engineering and the Evolution of Chemical Energy Storage*, American Philosophical Society, *Memoirs*, vol. 148, Philadelphia, 1982; and G. Ramunni, 'L'obstacle de la vérité: la pile, objet permanent d'investigation', *La revue, Musée des arts et métiers*, no. 31, December 2000, pp. 43–50. For a recent reassessment of eighteenth-century science: *The Sciences in Enlightened Europe*, edited by W. Clark, J. Golinski and S. Schaffer, Chicago and London: The University of Chicago Press, 1999.

2 On Napoleon and Europe: S.J. Woolf, *Napoleon's Integration of Europe*, London and New York, Routledge, 1991; M. Broers, *Europe under Napoleon, 1799–1815*, London: Arnold, 1996.

3 See G. Pancaldi, *Volta: Science and Culture in the Age of Enlightenment*, Princeton: Princeton University Press, forthcoming chapter 5.

4 Royal Society Archives, London, Letters and Papers, XI, No. 137.

5 Royal Society Archives, London, Letters and Papers, XI, No. 133, fol. 2v. For the identification of Pasquale Garovaglio: Archivio di Stato, Como, Notarile, cartella 5221, n. 676.

6 'The latter end of last April', according to W. Nicholson, 'Account of the new electrical or Galvanic apparatus of Sig. Alex. Volta, and experiments performed with

the same', *A Journal of Natural Philosophy, Chemistry, and the Arts*, July 1800, pp. 179–87, p. 179. 'In the beginning of last Month [i.e. April]', according to a public statement agreed between Garnett, Rumford and Banks, to be read at the Royal Institution on the evening of 30 May 1800: Rumford to Banks, 30 May 1800, Dartmouth College Library, Hanover, N.H.

7 As evinced from W. Nicholson's letter to the editor in *The Morning Chronicle*, 3 June 1800, p. 3. Here Nicholson vindicates his own and Carlisle's merits in the discovery of the decomposition of water as if knowing that Volta had not treated the issue in the second part of the paper describing the battery announced in the first part, which had been in Nicholson's hands since April.

8 VE, vol. 4, pp. 1–2.

9 VE, vol. 4, p. 6.

10 VO, vol. 2, p. 3.

11 On Carlisle: R.J. Cole, 'Sir Anthony Carlisle, FRS (1768–1840)', *Annals of Science* 8 (3), 1952, pp. 255–70.

12 W. Nicholson, 'Account of the new electrical or Galvanic apparatus', p. 181.

13 Rumford to Banks, 29 May and 30 May 1800, Dartmouth College Library, Hanover, N.H. On the Royal Institution: Morris Berman, *Social Change and Scientific Organization. The Royal Institution 1799–1844*, London: Heinemann, 1978. On Garnett: S.G.E. Lythe, *Thomas Garnett (1766–1802)*, Glasgow: Polpress, 1984.

14 *Morning Chronicle*, 30 May 1800, p. 3.

15 W. Nicholson, Letter to the Editor, *The Morning Chronicle*, 3 June 1800, p. 3.

16 As reported by Rumford in Rumford to Banks, 29 May 1800, Dartmouth College Library, Hanover, N.H.

17 Rumford to Banks, 30 May 1800, Dartmouth College Library, Hanover, N.H.

18 *Courier de Londres*, 8 August 1800, p. 96.

19 *Gazette nationale ou Le moniteur universel*, 29 thermidor an 8 (17 August 1800), No. 329, p. 1.

20 VO, vol. 2, p. 7.

21 E.-G. Robertson, 'Expériences nouvelles sur le fluide galvanique, . . . lues à l'Institut National de France, le 11 fructidor an 8', *Annales de chimie*, vol. 37, December 1800, pp. 132–50. On Robertson's career: E.-G. Robertson, *Mémoires récréatifs, scientifiques et anecdotiques du physicien aéronaute E.-G. Robertson*, Paris, 1831–1834. On the *Annales de chimie*: M. Crosland, *In the Shadow of Lavoisier: the Annales de Chimie and the Establishment of a New Science*. London: British Society for the History of Science, 1994.

22 J. Remise, P. Remise and R. van de Walle, *Magie lumineuse. Du théâtre d'ombres à la lanterne magique*, [Paris], Balland, 1979, pp. 39–61.

23 Years later, when Robertson won new popularity as an aeronaut, a big, friction electrical machine was represented on the illustrations advertising his balloon, called 'La Minerve', together with other scientific and technological devices that he supposedly carried on board: Bibliothèque Nationale, Paris, Estampes, Collection Hennin, vol. 164, No. 14356.

24 E.-G. Robertson, 'Expériences nouvelles sur le fluide galvanique', p. 134.

25 C. Blondel, 'Animal electricity in Paris: from initial support, to its discredit and eventual rehabilitation', in *Luigi Galvani International Workshop. Proceedings* (edited by M. Bresadola and G. Pancaldi), *Bologna Studies in History of Science*, vol. 7 (CIS, University of Bologna), 1999, pp. 187–209.

26 Banks to Van Marum, 14 June 1800, British Library, W.R. Dawson MSS, 68 (I), 28–9, copy.

27 See W.D. Hackmann, 'Electrical researches', in R.J. Forbes (ed.) *Martinus Van Marum. Life and Work*. Haarlem, Willink, 1969–76, 6 vols., vol. 3, pp. 329–78, esp. pp. 359–61.

28 See S. Lilley, 'Nicholson's Journal (1797–1813)', *Annals of Science* 6, 1948–50, pp. 78–101.

29 Vol. 14, pp. 398–9.
30 *Annalen der Physik*, vol. 6, 1800, pp. 340–75; vols 8–11, 1801–1802, passim.
31 Hans Christian Ørsted, *Selected Scientific Works* (translated and edited by K. Jelved, A.D. Jackson and O. Knudsen, with an introduction by A.D. Wilson), Princeton, N.J.: Princeton University Press, 1998, p. 101.
32 H. Davy, 'An account of some experiments made with the galvanic apparatus of Signor Volta', *A Journal of Natural Philosophy, Chemistry, and the Arts*, September 1800, p. 275.
33 Robison to Watt [October 1800], published in *Partners in Science. Letters of James Watt and Joseph Black* (edited, with introductions and notes, by E. Robinson and D. McKie), Cambridge, Mass.: Harvard University Press, 1970, p. 358.
34 *Morning Chronicle*, 30 May 1800, p. 3.
35 VO, vol. 2, p. 7.
36 A few early voltaic batteries preserved in Lombardy were destroyed by fire during an exhibition in 1899: Cencio Poggi, 'Il salone dei cimelii', in Società Storica Comense, *Raccolta Voltiana*, 1899, pp. 12–15. The Musée d'Histoire des Sciences de Genève preserves a battery that is said to have been donated by Volta to Nicholas-Théodore de Saussure in 1801; see Association pour le Musée d'Histoire des Sciences de Genève (J. Trembley (ed.)) *Les savants genevois dans l'Europe intellectuelle du XVIIe au milieu du XIXe siècle*, Editions du Journal de Genève, 1987, p. 134. The Royal Institution, London, preserves a battery which is said to have been donated by Volta to Faraday, when the latter visited Volta on the occasion of a trip to Italy with Davy in 1814.
37 For an in-depth review of the literature on constructivism: J. Golinski, *Making Natural Knowledge. Constructivism and the History of Science*, Cambridge: Cambridge University Press, 1998.
38 Also, in English translation, in *The Philosophical Journal*, vol. 7, 1800, pp. 289–311.
39 Volta's first recorded letters to expert electricians (Beccaria and Nollet) were dated 1763; see Pancaldi, *Volta*, ch. 1.
40 Ibid., chs 3 and 4.
41 Volta, 'On the electricity excited by the mere contact of conducting substances of different kinds', in VO, vol. 1, pp. 563–82, esp. pp. 565–7.
42 Ibid., pp. 565, 566, 576.
43 Ibid., pp. 573–4, 576–81.
44 Pancaldi, 'Electricity and life'.
45 VO, vol. 1, p. 575.
46 Ibid., p. 572.
47 Ibid., p. 569.
48 Royal Society of London, Archives, Letters and Papers, Nos. 133, 137.
49 See *Volta's battery, 2*, below (pp.140–4).
50 Pancaldi, 'Electricity and life'.
51 VO, vol. 1, pp. 581–2.
52 *A Journal of Natural Philosophy*, July 1800, p. 179.
53 Ibid., p. 181.
54 Ibid.
55 Giovanni Fabbroni, 'On the chemical action of different metals on each other at the common temperature of the atmosphere', *A Journal of Natural Philosophy, Chemistry and the Arts*, 1799, pp. 300–10; and 1800, pp. 120–7. On Fabbroni, Nicholson, and the battery: Ferdinando Abbri, 'Il misterioso "spiritus salis". Le ricerche di elettrochimica nella Toscana napoleonica', *Nuncius*, vol. 2, 1987, pp. 55–88, esp. pp. 67–8.
56 *A Journal of Natural Philosophy*, July 1800, p. 186.
57 Ibid., p. 187.
58 VO, vol. 2, p. 7.
59 *A Journal of Natural Philosophy*, November 1800, p. 341; Davy's emphasis.
60 VE, vol. 4, p. 17.

61 Ibid., p. 24.
62 VO, vol. 1, p. 525; VE, vol. 2, p. 32n.
63 VE, vol. 4, p. 37.
64 Ibid., pp. 52–3.
65 VO, vol. 2, p. 16.
66 Pancaldi, *Volta*, ch. 4.
67 Ibid., ch. 3.
68 A. Volta, 'De l'électricité dite galvanique', *Annales de chimie*, 30 Frimaire an 10 (December 1801), pp. 225–56. The phrase 'expériences fondamentales', or 'fait principal', to designate Volta's experiments with a single metallic pair without humid conductors, was first introduced by Biot in his report of Volta's presentation at the Institute (dated 2 December 1801): Biot, 'Rapport', note 83, below, p. 196. Later, to designate these same experiments, Volta himself occasionally adopted the phrase 'sperienza fondamentale' (VO, vol. 2, p. 52; from a paper originally published in 1802).
69 A. Volta, 'De l'électricité dite galvanique', op. cit., 235.
70 The information reported in this and the following paragraphs is taken from Brugnatelli's diary (published in VE, vol. 4, pp. 461–533), and from the newspapers mentioned in the text.
71 Rumford to Banks, 11 November 1801, Massachusetts Historical Society.
72 See J. Fischer, *Napoleon und die Naturwissenschaften*, Stuttgart, Steiner, 1988, p. 283.
73 G. Pancaldi, op. cit., ch. 4.
74 VE, vol. 4, p. 124.
75 *Le Moniteur*, 2 Nivose an 10 (24 December 1801), pp. 369–70.
76 Bonaparte to Volta, 17 Frimaire an 10 (December 1801), VE, vol. 4, p. 123.
77 On galvanism in Paris see Blondel, 'Animal electricity in Paris'. On Biot: Eugène Frankel, 'Jean Baptiste Biot: the career of a physicist in nineteenth-century France' (Ph.D. dissertation, Princeton University, 1972); Ivor Grattan-Guinness, *Convolutions in French Mathematics, 1800–1840*, Basel, Boston, Berlin: Birkäuser, 1990, 3 vols, vol. 1, pp. 187–9.
78 J.-B. Biot and F. Cuvier, 'Sur quelques propriétés de l'appareil galvanique', *Annales de chimie*, 30 Messidor an 9 (July 1801), pp. 242–50. See also *Bulletin des sciences, par la Société Philomatique*, an 9 (1801), p. 40.
79 J.-B. Biot, 'Sur le mouvement du fluide galvanique', *Bulletin des Sciences par la Société Philomatique*, Thermidor an 9 (July 1801), pp. 45–8 (p. 48).
80 Ibid.
81 See J.-B. Biot, 'Volta (Alexandre)', *Biographie universelle*, Paris, 45 vols., 1880, vol. 44, pp. 77–81.
82 See Biot, 'Sur le mouvement', op. cit., p. 45.
83 J.-B. Biot, 'Rapport sur les expériences du citoyen Volta', *Mémoires de l'Institut National des Sciences et Arts. Sciences mathématiques et physiques*, Tome V, Fructidor an 12 (August 1804), pp. 195–222.
84 VO, vol. 2, p. 576.
85 Ibid., p. 566.
86 Ibid.
87 Nicholson, 'Account of the new electrical or Galvanic apparatus', op. cit., pp. 179 (emphasis added), pp. 182–4.
88 Landriani to Volta, VO, vol. 2, p. 3.
89 VO, vol. 2, p. 169.
90 VO, vol. 1, p. 566.
91 OED.
92 H. Davy, Royal Institution, London, MSS Notebook 22B, fol. 9ff.
93 T. Cavallo, *The Elements of Natural or Experimental Philosophy*, 4 vols, London, Cadel & Davies, 1803.

94 From 1806, according to the OED.
95 From 1828, according to the OED.
96 G. Pancaldi, 'Electricity and life', op. cit.
97 *Specification of Frederick De Moleyns, Production or Development of Electricity, and its Application to Illuminating and Motive Purposes*, London Patent Office, 1841, No. 9053: London, Spottiswoode, 1855, pp. 2–3.

7 When human travellers become instruments

The Indo-British exploration of Central Asia in the nineteenth century

Kapil Raj

> . . . the manners and sentiments of the eastern nations will be perfectly known; and the limits of our knowledge no less extended than the bounds of our empire.
>
> Sir William Jones, *A Grammar of the Persian Language*

> 'Yes, and thou must learn how to make pictures of roads and mountains and rivers – to carry these pictures in thine eye till a suitable time comes to set them upon paper. Perhaps some day, when thou art a chain-man, I may say to thee when we are working together: "Go across those hills and see what lies beyond." '
>
> . . . But as it was occasionally inexpedient to carry about measuring-chains a boy would do well to know the precise length of his own foot-pace, so that when he was deprived of . . . 'adventitious aids' he might still tread his distances.
>
> Rudyard Kipling, *Kim*

Kashmir 1863

On a warm spring day in May 1863, two men, a South Asian and a European, could be seen walking endlessly up and down a sinuous mountain path in the Himalayan heights somewhere to the east of Srinagar, the capital of Kashmir. A pilgrim's staff in hand, the former leads the way, the latter closely follows meticulously counting and measuring his companion's paces: yes, his 2,000 steps do indeed make a mile. Suddenly they stop, light a small oil stove and place a copper vessel upon it – to make some tea, one might think. But as soon as the water begins to boil, the native produces a small boiling-point thermometer from the depths of his long woollen dress and plunges it into the pan. He watches it for a while and then whispers something to his fellow traveller. They start walking afresh. A few moments later, they stop again. This time, the former lays down his staff and pulls out from under his dress a smoked glass and produces, almost from nowhere – very like a magician – a tiny sextant which he holds above the glass and begins to observe the sun. An instant later, he mumbles a number to his companion who, in spite of his forbidding look, lets slip from behind his large beard the slightest hint of a satisfied smile. Then the Indian examines a compass that materialises mysteriously onto his palm. Next, seeking

Figure 7.1 Captain Thomas George Montgomerie.
Source: Royal Geographical Society.

the shade of a rock, he pulls out another thermometer, which he exposes for some minutes before squinting at it and muttering something to the European. As the sun sets, the two men make their way to their camp tent but, as soon as night falls, they come out again and begin to observe the stars through the sextant. Day after day, night after night, they repeat the same gestures. On 12 June, the European returns to Srinagar, while his native companion joins a dark-blue-uniformed troop column making its way towards Leh, the capital of Ladakh.

Contrary to what one might think, these two men were not rehearsing for some kind of theatrical performance. The European, Captain Thomas George Montgomerie of the Royal Engineers (Figure 7.1), was at the time in charge of the mapping of the kingdom of Kashmir, an area of more than 160,000 square miles.[1] In 1845, at the age of 15, this Scotsman from Ayrshire joined the East India Company's Military Academy at Addiscombe to train as an engineer for the Company's domains in India. There, for four years, he was drilled and disciplined into the rigour of military marching of course, but also into that of integral and differential calculus, the theory of equations, spherical trigonometry and astronomy, winning the Pollock Medal awarded to the best student of the year.[2] He was commissioned second lieutenant in 1849, arrived in India in 1851, and soon opted for the Great Trigonometrical Survey of India, one of the most prestigious departments of the Company's overseas services. The success of the triangulation operation of Kashmir, which he completed in 1864, earned him the Founder's Gold Medal of the Royal Geographical Society the following year.[3] This region is one of the most hostile in the world. In his speech Sir Roderick Impey Murchison, then President of the Royal Geographical Society, said:

> When we reflected upon the remarkable facts that you had passed from the hot plains of Hindustan to the loftiest region on the face of the globe and that there, amidst enormous glaciers, you had made accurate scientific observations at stations, one of which was 5,000 feet higher than the summit of Mont Blanc, we could not fail to applaud and reward such noble feats, displaying, as they did, the great abilities and energy with which you conducted so arduous a survey.[4]

Indeed, so arduous was the Kashmir operation that it permanently affected Montgomerie's health: forced to retire and return to England in 1873, he died in Bath on 31 January 1878 at the age of 47. Author of a dozen articles in the *Journal of the Royal Geographical Society*, elected to the Royal Society in 1872 and promoted to the rank of Colonel in 1876, he represented the Royal Geographical Society and the British and Indian governments at the *Congrès international de géographie* held at Paris in 1875, where he was decorated '*Officier de l'Université de Paris et de l'Instruction Publique*'.

In May 1863, in the Himalayan heights somewhere to the east of Srinagar, Montgomerie was quite simply disciplining his companion, Mahomed-i-Hameed, in order to transform him into an intelligent instrument of measure. Thus metamorphosed, this under civil servant of one of the local administrations was to participate in a survey of the Central Asian regions north of the Himalayas, a vast *terra incognita* encompassing an area of almost 1,400,000 square miles which, on paper at least, was part of the Chinese empire.[5]

However, China, extremely weakened in the light of the Opium Wars against the British, was pulling out of its most distant imperial possessions, allowing the ever-advancing shadow of Russia to spread to these regions, a shadow which had already spread across the Ottoman Empire and northern Persia. Ever since Napoleon's defeat in 1812, the British had begun to consider Russia as their main rival in the race to corner trade in Central Asia. Some even went as far as to see in Russia's acts a grand design to invade India and wrest it out of British control. Russophobia reached its peak in the years following the 1857 rebellion in northern India, as Himalayan hill-stations, like Simla and Darjeeling, began acquiring strategic importance as political and military headquarters of British India and/or of the provinces. From these highland resorts the British sought to dominate the subcontinent, while at the same time isolating themselves from the mass of South Asian society.[6] In this context the problem of securing the Empire's northern frontiers took on a special urgency for, contrary to our received ideas, the Himalayas were not a naturally impregnable frontier: the Tibetans had been invading the lower Himalayas and the Indo-Gangetic plain since the sixth century, even ruling over Bengal from the seventh century to the tenth, and the Sikhs had attacked western Tibet and briefly occupied the region around Lake Manasrovar as recently as 1841. Consequently, the Transhimalaya had, by the mid-nineteenth century, become the scene of the Great Game, a tournament of shadows played out between British and Russian spies for the political control of High Asia and immortalised years later by Kipling in his novel *Kim* (published in 1901). Thus, to ensure the stability of their prize possession, the British felt they had at all costs to map out and stabilise this vast no-man's-land that lay between them and the Tsarist Empire. And in order to achieve this, what to Victorian eyes could surpass geography, that queen of imperial sciences, which in times of peace was the continuation of politics by other means?

Between 1863 and 1885 no less than fifteen native Indians – almost all small-time functionaries of the Great Trigonometrical Survey of India – were to follow in Mahomed-i-Hameed's steps in order to map this region meticulously – Hindus and Buddhists to cover Tibet, Muslims to criss-cross what was then known as Eastern Turkistan. Two of them were murdered, one sold into slavery by his Chinese colleague; yet another, suspected of spying, spent seven months in a Mongolian prison. Almost all of them had near fatal encounters with bandits, but succeeded nevertheless in accomplishing their geographical mission. They were code-named the 'Pundits', and became celebrities of the period, the subject of many an article in the British press and recipients of a number of awards from European scientific societies.

One could of course ask why Montgomerie, himself an international celebrity precisely for his survey of mountainous regions, could not go out and map these Transhimalayan lands. After all, during the century since they had first conquered any sizeable territory in South Asia, the British had succeeded in taking the measure of the whole Indian subcontinent – even those parts which were not yet under their direct control. But it was not the same for the Transhimalaya.

In spite of the weakening of their empire, the Chinese were nonetheless still present in the region and looked on the British with a jaundiced eye. The Tibetans, extremely jealous of their autonomy and identity, were even more suspicious of Europeans. Besides, a number of British emissaries to the few Central Asian khanates that were still outside the ambit of Tsarist power had met with a gruesome end, either hanged or beheaded in public or else murdered in their sleep (Figure 7.2). Ironically, the inventor of the phrase 'the Great Game', Captain Arthur Conolly, was himself one of the first to pay this deadly price: he was beheaded in Bukhara in June 1842. Geography being a continuation of politics by other means, the political situation in turn demanded the continuation of geography by other means.

Figure 7.2 Khirgiz tribesmen from the Oxus region holding the severed head of a European intruder. From J.A. Mac Gahan, *Campaigning on the Oxus* (London: Sampson, Low, Marston, Low & Searle, 1874).

Eastern Turkistan, 1863–4

Thus it was that Montgomerie's instructions during his mission to Kashmir explicitly mentioned that he 'obtain the means of rectifying our imperfect geographical knowledge of the regions beyond British influence', but warned him that 'it will neither answer to risk the safety of the party nor to entangle Government in political complications'.[7] While patiently triangulating his way through this remote mountainous state, Montgomerie kept these desiderata in mind. On 2 April 1862, in the course of a talk on the geography of Turkistan that he delivered to the Asiatic Society of Bengal, of which he was a member, he made the first public mention of 'the advisability of employing native agency' for the purpose of surveying the hostile Transhimalayan regions.[8] Having noticed how Himalayan merchants were allowed to pass quite freely into Tibet and Chinese Turkistan, he reasoned that thus disguised his collaborators might also be able to slip into Transhimalayan Asia. However, he also realised that even if he succeeded in enrolling a few natives into his scheme, they would not be able to use conventional surveying techniques (Figure 7.3). His idea was that they should measure distances by counting their paces. Each of his collaborators was

Figure 7.3 As this drawing of the measurement of the Calcutta baseline in 1831 illustrates, conventional surveying techniques mobilised a host of men and instruments: (1) sheltered from the sun's rays, verifying the length of the baseline using a micrometer; (2) surveying chains; (3) various tripods; (4) transit apparatus; (5) observation tower. Adapted from a sketch by J. Prinsep in the *Journal of the Asiatic Society of Bengal*, 1 (1832): 71.

to be disciplined to regulate the length of his stride whatever the terrain or the incline. They were to count 2,000 paces to the mile; that is, to maintain a step of $31^1/_2$ inches. He perfected his idea in a letter to the Asiatic Society dated 21 July 1862. The latter backed his project and forwarded it to the Viceroy and the Council of India. The plan was most warmly approved and the Government of India 'agreed to support it liberally'.[9]

The moment he received the approval, Montgomerie began to look for his first collaborator whom he found in Mahomed-i-Hameed, better known by his code name, 'the Moonshee'. Immediately after the month-long intensive training I mentioned at the outset, Montgomerie sought to test the feasibility of his plan by mounting a full-scale experiment.

Hameed reached Leh on 4 July 1863 having mapped his route from Kashmir. Reassured of the reliability of his new 'instrument', Montgomerie directed him to continue. Accordingly, on 24 August the Moonshee, disguised as a merchant and accompanied by two servants and a pony-load of merchandise, joined a caravan heading for Yarkand. This destination was not chosen by chance. Yarkand was an oasis on the ancient silk route, and was thus a major crossroads for trade between China, India and Central Asia. Although its position had been determined by French Jesuits at the beginning of the eighteenth century, and again in the 1850s by the Schlagintweit brothers, Montgomerie – along with all his other British colleagues – believed their calculations to be false. Moreover, Yarkand was only fifteen days' march from the frontier of British India and, according to Montgomerie's estimation, lay almost due north of Leh. This last factor would greatly simplify things, as longitude measurements at the time were a tricky affair. By way of equipment, Hameed carried a pocket sextant, a smoked-glass artificial horizon, a pocket compass, a prismatic compass, a pilgrim's staff with its knob flattened so as to provide a stable base for compass readings, two thermometers, two silver watches, a copper jug and an oil stove to boil the thermometers, a lantern to help him read his instruments at night and two pocket notebooks to record his observations.

The party arrived at its destination on 30 September. The Moonshee had meticulously traced his whole route, carefully noting all that he had observed on his way, especially the vegetation and human dwellings. He seems not to have had any difficulty in settling in Yarkand for the winter. The winter was harsh, the temperature rarely rising above −20 °C during the months of January and February. Undeterred, Hameed assiduously wrote his thermometric and hygrometric observations into his notebooks: it snowed for only two days, 19 and 20 January. He drew a map of the city and even of the region, carefully identifying its major cities (Kashgar, Khotan, Sirikul, and others), and plotted Tsarist encroachments into the territory as well as the positions of their fortifications. He even had time to transcribe the history of the region as told by its own people. He described the local political structure, the nature of Chinese presence and the size and organisation of their troops. Sure enough, he took great care to hide his notebooks, along with all his other instruments, in the folds of his great robe. And, of course, he did not forget to take regular astronomical observations – eleven for Yarkand alone – the last dated 27 March 1864. On that day, the Governor of Yarkand, whom Hameed had befriended, told him that the Chinese were becoming suspicious of his activities. Hameed promptly packed his belongings and joined another caravan returning to Leh, accompanied by a cousin he had met during his stay in Yarkand. They crossed the most dangerous section of the route without great difficulty and reached British-controlled Indian territory towards the end of April. Unfortunately, just before arriving at Leh the Moonshee and his cousin died, either as a result of eating poisonous rhubarb (according to his companions), or as victims of murder by Chinese or Russian agents (if one were to believe the suspicions of Montgomerie's assistant, William Johnson, who was the first to reach the scene of the tragedy). Hameed's fellow travellers had meanwhile made off with his most saleable belongings. Fortunately for Montgomerie, no one had touched his notes and instruments nor, curiously enough, his silver watches. All these were forwarded to Montgomerie who transmitted the papers concerning the political situation in Yarkand and Eastern Turkistan to the Punjab government. Taking the rest of the Moonshee's notes and instruments with him, Montgomerie sailed for England on 20 February 1865 after thirteen years in India, using his leisure to draw a map of the route between Leh and Yarkand (Figure 7.4) and correct the co-ordinates of the latter city: the French Jesuits had declared its latitude to be 38° 19′, Hameed's observations put it at 38° 20′; its longitude was 76° 16′ according to the Jesuits,

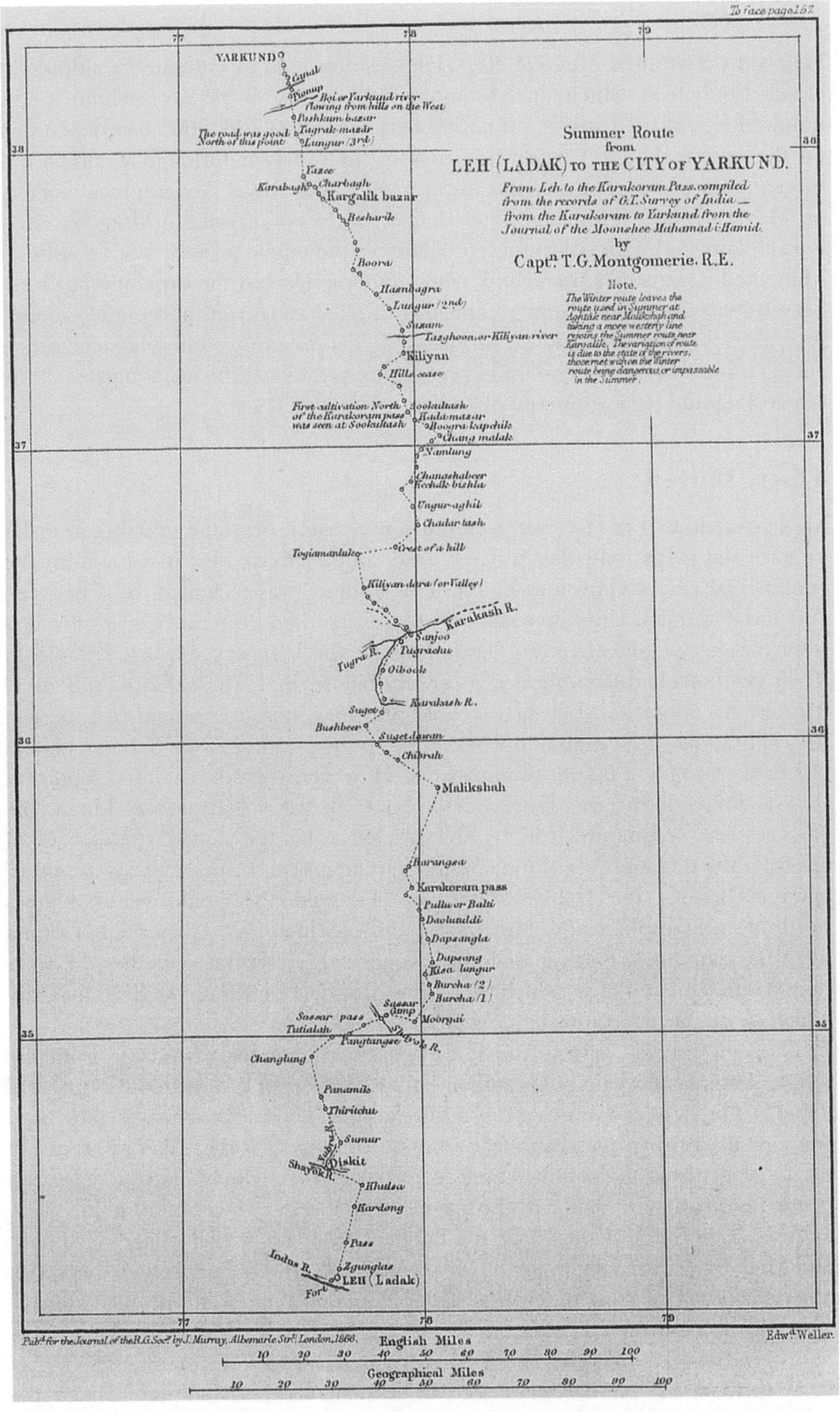

Figure 7.4 Map of the summer route between Leh and Yarkand compiled by Montgomerie from the journal of the Moonshee, Mahomed-i-Hameed. From *The Journal of the Royal Geographical Society* 36 (1866): facing p. 157.

Hameed determined it at 77° 30′.[10] He was also able to estimate its altitude at about 4,000 feet, which, in Montgomerie's words, 'is perhaps within a few hundred feet of the truth'[11] – it varied between 2,000 and 5,000 feet depending on the source. He also made the most of his two-year furlough to prepare a presentation of the results of this experiment to the Royal Geographical Society on 14 May 1866.[12] This step, as we shall see later, was crucial to Montgomerie's future plans and, indeed, to the credibility of the whole project. For, in spite of Mahomed-i-Hameed's tragic end, Montgomerie deemed the expedition to have been sufficiently successful to go ahead with plans to recruit more native agents. And already before leaving on furlough, he had begun to refine his techniques and to train a whole generation of prospective pundits at the headquarters of the Great Trigonometrical Survey of India.

Tibet, 1864–6

Montgomerie had in fact even sent out a new team, this time to Tibet in order to trace the great road that traverses the country from west to east, from the commercial city of Gartok to Lhasa. This distance was reckoned to be between 700 and 800 miles. Until then the only places that had been more or less located were the city of Shigatze, and Tashilhumpo, the lamasery which overlooks it. Their positions had been plotted a century before, in 1783, by Captain Samuel Turner. A comparison between the old and new measures would help perfect the calibration of the human probes. Besides, the position of the Tibetan capital of Lhasa was only a matter of guesswork. However, there were more important reasons for choosing this itinerary: the route between Gartok and Lhasa, the *Junglam*, was commonly held by the British to be the main artery of Tibet ensuring the flow of most of the country's trade and civil and military communications. Besides, the *Junglam* follows the Tsangpo, the great river that flows north of and parallel to the Himalayas, and of which neither source nor outlet were then known. Charting its course would help determine whether the river flowed into the Irrawaddy, which runs across Burma, or else into the Brahmaputra, which shares an enormous delta with the Ganges.

This new mission was entrusted to two *Bhotiya* cousins who came from the remote Himalayan region of Kumaon, on the borders of British India, Nepal and Tibet.[13] The *Bhotiyas* were reputed to be familiar with Tibet because they used traditionally to earn part of their livelihood trading with the Tibetans, crossing the border during the summer months to barter rice, wheat and manufactured goods mainly against salt but also against borax, horses, wool and gold.[14] The cousins, Nain Singh (Figure 7.5), a schoolteacher in his native village of Milum, and Mani Singh, the village *patwari*, or record-keeper, were not likely to meet with any difficulty in crossing the frontier from Kumaon. Besides this decided advantage, they had already taken part in topographical surveys with the Schlagintweits and were thus familiar with geodesic and astronomical instruments.[15]

Apart from the same miniaturised instruments as the Moonshee used,[16] the new pundits carried a prayer wheel and a rosary – perfectly normal adjuncts,

Figure 7.5 Pundit Nain Singh or 'Number One'.
Source: Royal Geographical Society.

one would think, for two lamas on a pilgrimage to the holy city of Lhasa (which in Tibetan means 'God's abode'). In fact, these two outwardly innocuous objects constituted instruments of a very special nature for their owners. The prayer wheel (or *mani-chakra*) consists of a hollow cylindrical box to which is attached a metal chain weighted at the other end. The whole device revolves around a spindle, one end of which forms the handle (Figure 7.6). A slight movement of the wrist makes the box rotate and each revolution represents one incantation of the sacred verse of Tibetan Buddhists – *Om Mani padme hum* ('Hail! O, jewel in the lotus') – which is either engraved on the exterior of the wheel or else written on a scroll kept inside it. For Tibetan Buddhists the prayer wheel is the very emblem of a precision instrument: it always ensures an error-free incantation on condition, of course, that the prayer be correctly inscribed. In oral recitation, on the other hand, the slightest slip of the tongue constitutes not

Figure 7.6 Tibetan prayer wheel, as drawn by Pundit Sarat Chandra Das.
Source: Royal Geographical Society.

just a fault in terms of the ritual but a genuine catastrophe. Besides, owing to its sacred status, the prayer wheel could conceal small objects or inscriptions from curious frontier officials ever suspicious of the British and their native employees.

As for the rosary, it helped count paces: one bead for every hundred paces. The Buddhist rosary has 108 coral beads; those of the cousins only had a hundred, every tenth bead being much larger than the others made of the dark corrugated seed of the *rudraksh*. A complete round represented exactly 10,000 paces or five miles.

Besides, Nain Singh and Mani Singh carried quicksilver in an empty coconut shell and a reserve in cowrie shells closed with wax. When poured out into the wooden bowl that the *Bhotiyas* always carry with them as a recipient for food and drink, the glistening surface of the mercury served as a far better artificial horizon than the late Moonshee's smoked glass. In order to complete their panoply of instruments, one must mention a six-inch Elliot sextant with a precision of ten seconds of arc. Too cumbersome to be hidden under their clothes, it was ingeniously secreted in a false compartment of their large wooden travelling chest. In spite of the risk and inconvenience of pulling it out and using it unseen, Nain Singh was able to use it ninety-nine times.

After months of rigorous drilling in the use of the sextant, compass and other survey instruments and techniques, Nain Singh and Mani Singh left for Tibet in March 1864 and returned to Dehra Dun on 27 October 1866. Turned back at the Tibeto-Kumaoni border, they tried subsequently to enter Tibet through Nepal, all the while charting the route they followed. Despite the advantage of their origin, they had to use the most devious ploys in order to succeed in entering into Tibetan territory: at times trying to pass off as merchants in search of Tibetan horses; on another occasion as the legal representatives of fellow *Bhotiyas* travelling to Lhasa to plead their case before the Dalai Lama; ingratiating themselves with Tibetans by readily lending them money, curing their families of minor ailments; adopting different disguises including a pig-tail, and so on.

It took more than a year of the utmost perseverance for Nain Singh to succeed in entering Tibet, alone, and on condition, under penalty of death, that he keep away from Lhasa. However, disguised as a pious tradesman, the resourceful Pundit managed to join a Ladakhi merchant's caravan on its way to Lhasa carrying presents from the Maharaja of Kashmir to the Dalai Lama. He passed through Shigatze, the second most important city of the region, and paid his respects to the eleven-year-old Panchen (or Teshoo) Lama, head of the great lamasery at Tashilhumpo. He even managed to gain access to the forbidden city of Lhasa and have an audience with the Dalai Lama in person. Short of money at various times, he was forced to earn his living teaching Hindu mathematics and accounting to local merchants. He was once even reduced to pawning his watch and servant so as to borrow money from Tibetan traders. On another occasion, he was forced to sell his bull's-eye lantern to curious Tibetan monks and had to content himself with an oil wick in this windy region to read the nocturnal measurements from his sextant and thermometers.

During his two-and-a-half-year expedition, Nain Singh had walked 1,200 miles at a steadily maintained pace. He had counted 2.5 million paces on his rosary. He successfully measured the position of Lhasa and a number of other important Tibetan cities, thirty-one stations in all. He had also mapped the upper course of the Tsangpo from its source to Lhasa, a distance of about 700 miles, and reported that according to his Tibetan interlocutors the Tsangpo and the Brahmaputra were the same river. Nain Singh, however, did not confine his mission to the collection of geographical data. He also gained useful insights into Tibetan culture. He gave a detailed description of Tibet's general climatic conditions, its population, ethnology, cities and monasteries, agricultural production, economy and trade, especially with China and Kashmir, the state of its roads, transport and communications, and of the Tibetan army.[17] The prayer wheel and the rosary served as a perfect alibi, giving the Pundit a sufficiently pious look for him not to be bothered by his fellow travellers when he took his astronomical and meteorological readings.

Not having succeeded in entering Tibet, Mani Singh spent his time exploring western Nepal, which was also out of bounds to Europeans. But when Nain Singh returned to India, exhausted after having spent eighteen months at an average height of about 15,000 feet, Mani succeeded at last in entering Tibet. He bought back Nain Singh's watch and servant and completed the last 100-odd miles of the *Junglam* that his cousin had had to abandon during the last stage of his odyssey.

High Asia, 1868–82

The success of this mission prompted Montgomerie to request further financial support for the purpose of organising other Transhimalayan expeditions; the governmental authorities

> cordially approve[d] of the proposal to employ Asiatics in the exploration of the Trans-Himalayan Regions, and direct[ed] that it be intimated to each explorer who may be sent out by the Trigonometrical Survey that any important political intelligence he may bring back [would] receive a separate pecuniary acknowledgement, according to its value, from the Foreign Secretary.[18]

The way was now open for the large-scale extension of corporeal techniques to map all of Central Asia.

Between 1868 and 1875 three further expeditions were sent out to map the regions to the north-east of Afghanistan: the Hindu-Kush range, the Pamir and Chinese Turkistan. Using the rosary, but abandoning the prayer wheel (one must remember we are in Islamic lands) and working in extreme conditions, the 'Mirza' (from 1868 to 1869), the 'Havildar' (in 1870) and the 'Mullah' (from 1873 to 1874, then from 1875 to 1876 and in 1878) – such were their code names – accurately determined the hydrology of High Asia, mapped the course of the Upper Indus and identified the sources and course of the Amu-Darya (the Oxus of the Greeks) which, following the Anglo-Russian agreement of 1873, was to constitute the frontier between Russia and Afghanistan, a river of which hardly anything was known at the time. They also charted the route between Kabul, Kashgar and Yarkand, of which more than 300 miles were then unknown to Europeans, and determined the positions of various important geographical sites on their route. And, of course, they brought back crucial political information concerning the Central Asian khanates and of Tsarist activity in the region.[19] All returned safely to British India, except the Mirza who met a tragic end on his second expedition (1872–3), murdered in his sleep while on his way to Bukhara.

Nain Singh (code named 'the Pundit') himself returned to Tibet three times between 1867 and 1877 in order to complete mapping the *Junglam* right up to Leh and thus link the Tibetan survey with Montgomerie's triangulation of Kashmir, to visit the Tibetan gold mines, reckon its gold resources and finally to explore the regions north of the Tsangpo valley. His ardour and skill won him a pension, a grant of land, the Companionship of the Star of India, international fame, and the Royal Geographical Society's gold watch in 1868 and gold medal in 1875. At the presentation ceremony it was said that 'his observations have added a larger amount of important knowledge to the map of Asia than those of any other living man'.[20] The system of corporeal techniques was by now well oiled and continued after Montgomerie's death in early 1878. His most brilliant 'instrument', the Pundit, or simply 'Number One' as he was sometimes called, became an instrument-reproducing instrument, instructing a new generation of Indian explorer-geographers in these novel surveying techniques.

But no account of this Transhimalayan survey would be complete if it did not mention Sarat Chandra Das (Figure 7.7), the best known of all the Pundits for having written two books on his travels, a history of traditional Indian explorers to Tibet and a Tibetan–English dictionary, but above all as the inspiration for *Kim*'s Babu Hurree Chunder Mookerjee.[21] Nor would any account be complete without a mention of Pundit Kishen Singh (Figure 7.8), a cousin of Nain Singh who travelled across the Tibetan plain and the Kunlun mountains between 1878 and 1882. In Lhasa, Kishen Singh, alias A.K., was delayed for over a year waiting for his caravan to leave. During his sojourn he spent his time usefully, drawing a detailed map of the forbidden city. When his caravan did leave, he met with a succession of impediments. He was twice robbed on the road and imprisoned by the Mongolians for seven months on suspicion of being a spy. Finally freed by a Tibetan lama-merchant who bought his release on condition that A.K. work for him, he was forced by the former to ride a horse in order to

Figure 7.7 Self-portrait of Pundit Sarat Chandra Das returning on a yak from Tibet in 1879.
Source: Royal Geographical Society.

Figure 7.8 Pundit Kishen Singh who transferred the pacing technique to animals by successfully calibrating the pace of his horse.
Source: Royal Geographical Society.

survey the caravan, thereby preventing him from counting his paces. Kishen Singh got round the problem by measuring the length of the horse's pace and counting the number of times the right foreleg hit the ground. When he arrived back in British India in 1882, he had travelled 2,800 miles and counted 5.5 million paces.[22]

Tibet, 1904

All the data collected by the Pundits were published over the years in learned journals, especially in *The Journal of the Royal Geographical Society*. It was also computed into a detailed map of the Transhimalaya (Figure 7.9). Thanks to this

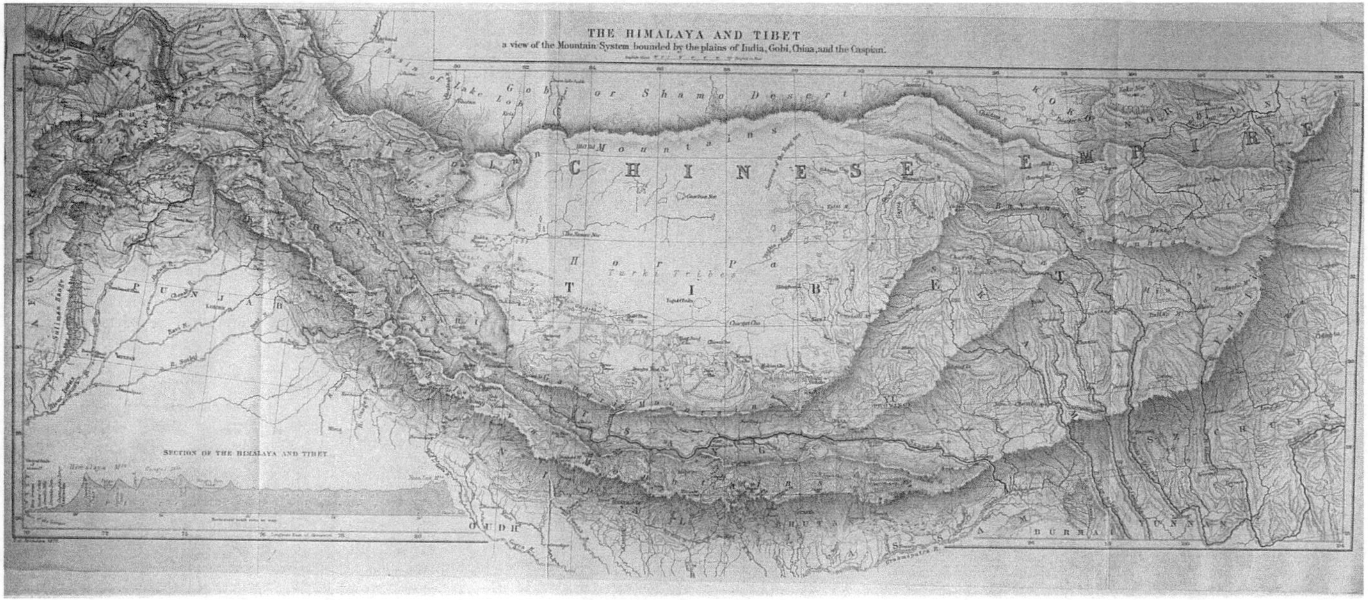

Figure 7.9 One of the first of a series of maps of Tibet based on the information culled by the Pundits. From C.R. Markham, *A Memoir of the Indian Surveys* (London: Her Majesty's Secretary of State for India in Council, 2nd edn, 1878).

Promethean work, the British felt better armed to contain the Tsarist spectre whose *Drang nach Osten* was reckoned to be advancing at the rate of fifty-five square miles a day.[23] In December 1903, convinced that the Russians and Chinese had concluded a secret agreement to hand over Tibet to the former, Lord Curzon, then Viceroy of India and long-time advocate of an armed confrontation – or forward policy, as it was euphemistically termed – with Russia, decided to send a pre-emptive force to invade Tibet. Under the command of Captain Francis Younghusband, the expeditionary force, comprising 1,000 soldiers armed with sophisticated artillery, 10,000 porters, 7,000 mules and 4,000 yaks, entered Tibet early in 1904 and occupied Lhasa in July of the same year, having killed almost 5,000 Tibetan soldiers and civilians on the way. Much ado about nothing: the Russians, bogged down in a protracted war with the Japanese, did not have the means to retaliate. Two months later, the British packed their belongings and returned to India, the expedition having been deemed a fiasco.

Instruments, travel and science

Although the saga of the Pundits has been saved from oblivion by a handful of chroniclers it has usually been told within the restrictive framework of international espionage. Some authors have of course recognised it as a contribution to the history of geography, but have sought to place the narrative within the tradition of mid-Victorian heroic exploration as exemplified by British explorers of Central Africa in the 1860s and 1870s.[24] It must be said that the parallel is tempting to draw. Apart from the coevality of the exploratory thrust into Central Asia and the Lake regions of Central Africa, there is a fascinating similarity in the hydrographical aims of the two projects: the search for the sources of the Tsangpo on the one hand, and the Nile on the other. Indeed, many contemporaries at the Royal Geographical Society quite overtly compared the Pundits and the explorers of the Central African lake regions: Sir Henry Yule, for instance, wrote of Nain Singh that 'to have made two such journeys adding so enormously to accurate knowledge . . . is what no European but the first rank of travellers like Livingstone or Grant have done'.[25]

However, the Pundits' travels cannot be assimilated to the paradigm of Victorian Central African exploration. Quite to the contrary, they contrast very sharply with the latter canon in some quite fundamental ways, shaping the very nature of the knowledge produced. The differences pertain as much to the mode of publication of their work as to their conditions of production and the nature and function of their respective audiences. But the Transhimalayan surveys differ from all other contemporary surveys – including those being conducted in South Asia itself – in the nature of the instruments and techniques used and the specific civility (ethnic and social origins and culture) of the explorers as compared with other explorers. Although I shall take these differences up individually, it will become clear that they are inextricably connected and have to be taken together to understand truly each context and type of scientific activity implied: taken together they point to the different institutional contexts in which the two sets of explorations were organised and which provide the very conditions for the possibility of each mode. Through this contrast I do not seek to show the exceptional nature of the Transhimalayan project with respect to a Victorian expeditionary paradigm, rather that there were at least two canons which had completely different ends. Although the exploration of Central Asia was in a sense exceptional and peripheral, it was certainly not peripheral to the exigencies of mainstream Victorian science and technology. Indeed, I hope through the discussion that follows to show how it mobilised some of the principal actors of Victorian science and, more broadly, how, its peripheral origins notwithstanding, it can help us understand some of the strategies and investments required for general methods and procedures in science and technology to be transferred from one site to another and replicated (blurring *en passant* the traditional distinction between 'centre' and 'periphery'). And because the Transhimalayan surveys were so different from other geographical surveys they required certain crucial problems to be solved and some specific investments to be made in order for the knowledge produced to count as such.

The first difference lies in the means by which these works were rendered public. The Pundits' accounts were reported exclusively in official, and often confidential, reports of the Survey of India and in learned journals, most notably the *Journal of the Royal Geographical Society*. The purpose of these voyages was primarily to delimit and stabilise British India in the aftermath of the 1857 mutiny by mapping and stabilising its surrounding territories. This, as mentioned earlier, was perceived by the British to require the resolution of the geography of the Transhimalayan regions in order to bring them into the ambit of British trade, even if this implied military intervention. Although the European newspapers covered their exploits, the primary audience of the core literature produced was the community of surveyors, professional geographers, the army, the Government of India and the India Office. And in mid-Victorian Britain no better forum could assemble all these different groups than the Royal Geographical Society, the office-bearers of which were largely drawn from the Survey of India or from the Council of India. Terrestrial measurement, the description of the trade routes, trade flows, the state and preparedness of the various Central Asian armies, the sociology of the towns, cities and monasteries of the region are what one finds on reading the different reports. The Pundits' accounts typically form part of a genre that Mary Louise Pratt has called 'autoethnography', by which term she means

> to refer to instances in which colonized subjects undertake to represent themselves in ways that engage with the colonizer's own terms . . . Autoethnographic texts are not . . . what are usually thought of as 'authentic' or autochthonous forms of self-representation . . . Rather autoethnography involves partial collaboration with and appropriation of the idioms of the conqueror.[26]

The Victorian explorers of Central Africa, on the other hand, published their narratives either in the form of books or articles in popular journals like *Blackwood's Magazine*, or, in the case of the Anglo-American John Rowlands, alias Henry Morton Stanley, in leading English and American daily newspapers. In these intensely personal works aimed at a wide readership, the author pleaded his special cause, often with a note of religious and passionate conviction; he reached out, as it were, to his readers, stirring up their sympathy and indignation. And since these appeals were interlarded with themes of bravery and high adventure the response was enormous.[27] While it is true that the Royal Geographical Society also played a major role in the diffusion and legitimisation of these accounts, it did so more as the key sponsor of the African expeditions. The purpose of these accounts was quite overtly to resolve African geography in order to convert the endless stretches of untilled land into useful farms, to open up commerce and to proselytise the natives out of their perceived animism and savagery – in short to possess the land and its resources rather than have control over trade routes.[28]

The opposition between the two genres, their publics and their legitimating strategies went hand in hand with two completely different ways of organising its production. For while the Royal Geographical and other scientific societies seconded the British Foreign Office in sponsoring and funding most of the exploration of *terrae incognitae* and providing them with logistical support, the exploration itself was conducted by private individuals of high social standing who worked mainly in order to outdo their rivals in order to realise their personal ambitions.[29] Controversy was of the essence in order to ensure the dynamic of the geographical enterprise, each explorer trying through his popular writings to muster up as much public support as possible in his ruthless quest for publicity, personal glory and, eventually, royal patronage: 'If he was attacked by jealous rivals,' writes Alan Moorehead in his popular history of the exploration of the White Nile, 'people sprang to his defence; and in the speculative and highly charged arena of African exploration there was a great deal of jealousy . . . It was as patriotic and as partisan as war.'[30] This is most dramatically illustrated by the tragic dispute between Burton and Speke over the source of the Nile we shall speak of below. It was an acute awareness of this all-important role of controversy that led Stanley, the most successful of the Central African explorers, to carefully choose his European assistants for his second expedition (1874–7) from *working-class* families for their toughness and sense of discipline, precisely in order to preclude the possibility of their writing independent accounts of their adventures or to dispute his views in the Royal Geographical Society. However, in the case of the Indo-British survey of the Transhimalaya, there are few signs of controversy. Quite to the contrary, as the foregoing narrative suggests, we are here confronted with a textbook account of consensual, cumulative knowledge formation, the success – and indeed the very possibility – of which required the foreclosure of all dispute and a degree of control over the material and social world which was beyond the means of individuals. As Svante Lindqvist has convincingly shown, this sort of activity could only be carried out within the framework of centralised institutions, with their imperatives of teamwork and a strong sense of hierarchy, which had the requisite authority and competence to exercise such control.[31]

No less striking is the difference in the nature of instruments used. The use of humans as instruments seems at first sight to be a local, rather primitive, *ad hoc* arrangement concocted mainly because of the impossibility of transporting the adequate, but cumbersome, surveying equipment – theodolites, heliotropes, barometers, etc. However, it is important to remember that measurement in surveying had been the epitome of metrological precision for over a century,[32] and since the end of the eighteenth century, explorers, and above all scientific travellers, had been vying with each other in the number and sophistication of the measuring instruments they used in their expeditions. During his American expedition between 1797 and 1804, Alexander von Humboldt, the scientific traveller *par excellence*, carried chronometers, telescopes, sextants of all sizes, compasses, magnetic needles, a repeating circle, a theodolite, an artificial horizon, a pendulum, a magnetometer, barometers, thermometers, hygrometers, electrometers, eudiometers, aerometers, a graphometer and a quadrant, all crafted by the best instrument makers in Europe.[33] And although Humboldt did admit the possibility of using corporeal techniques for measuring, on condition, of course, that the measurement had a determinable limit of error, these techniques were valid according to him only for the naturalist because of his need for a much lesser degree of precision in surveying than the astronomer or geodesist.[34] Indeed, by the second half of the nineteenth century, metrology was at the centre of a ruthless battle between the British and the French to determine which was to be the universal linear unit, the imperial yard or the 'natural' metre.[35] According to the historian of Victorian science, Susan Cannon, the scientific spirit of the period, which she qualifies as Humboldtian, is characterised by:

1. A new insistence on accuracy, not for just a few fixed instruments, but for all instruments and all observations.
2. A new mental sophistication, expressed as contempt for the easy theories of the past, or as taking lightly the theoretical mechanisms and entities of the past.
3. A new set of conceptual tools: isomaps, graphs, theory of errors.
4. An application of these not to laboratory isolates but to the immense variety of real phenomena, so as to produce laws dealing with the very complex interrelationships of the physical, the biological, and even the human.[36]

Indeed, the *raison d'être* of the Great Trigonometrical Survey of India was the securing of extreme accuracy in its measurements. It was for this very reason that its staff was kept distinct, at least on paper, from that of the Topographical and Revenue Surveys of India. In his memoir on the Indian Surveys, their first historian, Clements Markham, himself an old GTS man and future president of the Royal Geographical Society, remarked that the

Figure 7.10 Survey perambulators of the Survey of India: the Madras-pattern perambulator, above, and the Everest perambulator, below. Adapted by the author from F. Smyth and H.L. Thuillier, compilers, *A Manual of Surveying for India* (Calcutta: W. Thacker & Co., 1851).

> mathematical attainments [of the trigonometrical surveyor] must be of the highest order, and he would have neither the time nor, as a rule, a special aptitude for the collection of topographical details for filling up the exact skeleton furnished by his scientific labours. The two kinds of work are distinct, and require a different training in many respects and a different turn of mind.[37]

A major part of the training entailed the cultivation of extreme patience and a perfect mastery over the use, care and daily recalibration of the great three-foot theodolite, zenith sectors, sextants, thermometers and watches, astronomical clocks and telescopes and perambulators. Professional-cultural pressures and traditions apart, it is also not difficult to imagine the havoc even the smallest error in determining the co-ordinates of Lhasa or Shigatze would have caused for the logistics of the Younghusband expedition. How then did this instrumental 'innovation' square with these exigencies? In order to answer this question we first have to understand what this innovation actually consisted of and how it was made to fit in with other map-making techniques, thereby anticipating tensions between the protagonists of different metrological instruments.

Outmoded as it might seem, the technique of pacing is in fact an adroit transfer of some of the functions of the perambulator (or 'waywiser'), the principal instrument of measure for long-distance traverse surveying, the chain being the standard instrument for estate and revenue surveys ever since the start of the British surveys of the Indian subcontinent. Although inspired from the traditional European perambulator (which was found to be unsuitable for India conditions: 'flimsy, bad in principle, and incapable of working except on a smooth road or bowling green'[38]), the Indian instrument was from the start designed for South Asian conditions. Thus the 'Madras-pattern 8-mile perambulator', designed in the mid-1780s by the Madras surveyor John Pringle and his autochthonous assistants, consisted of an iron wheel about seven feet in diameter, fitted with differential brass axle plates worked by an endless screw. The dial recorded revolutions in terms of miles, furlongs, yards, feet and inches (Figure 7.10). It was propelled by two men, with handles passing through the axis about breast high, but was not suited to rough ground.[39] Many modifications were to be introduced into this instrument in the decades that followed before it was replaced by the 'Everest 6-mile perambulator', devised between 1832 and 1836. The latter had a wheel of diameter just under three feet, with differential dials reading to miles and decimals of miles. It needed only one person to propel it, the handle being inspired by the South Asian plough; the wheel was made strong and handy for work over rough ground, and specially heavy so that it would be easier to wheel on the ground than carry idle on the head or shoulder as was the case with the Pringle instrument.[40] Local gestures and customs were thus continuously inscribed into the changes wrought to the instrument. Montgomerie's Pundits were a reconfiguration of some of the principal functions of the perambulator: the circular motion of the wheel was translated into the paces of the operator (who was also the surveyor in that he himself measured his paces), the dials to his rosary.

This said, Montgomerie still had to standardise his human 'instrument' in order to render it replicable. When he received the late Moonshee's possessions, we might recall that Montgomerie retained the latter's notes and instruments. In the talk he gave to the Royal Geographical Society on 14 May 1866, he meticulously described the way he re-tested all of Mahomed-i-Hameed's instruments – his thermometers, sextants, watches, compasses, etc. – in a variety of conditions. He also ascertained the reliability of his human instrument. The Moonshee's performance did not vary between the simulated campaign in Kashmir and the real one in Chinese Turkistan – his pace was a constant $31^1/_2$ inches.

> I . . . desired he should simply record the bearing and direction of the road as far as he could see along it at one time, and with his watch note the time he marched in that direction. By this means I trusted that a very fair route-survey might be obtained, if the value or rate per hour could be satisfactorily established. This rate I was able to determine from the average of his marches before and after reaching Leh. And again, beyond Leh, the regular survey had been carried three marches down the Turkistan side of the Karakoram Pass. The latter enabled me to follow the moonshee's work for 18 out of the 30 marches; and by this means it was easy to form an opinion on the amount of reliance that ought to be placed on his route-survey of the other 12 marches.
>
> The observations for latitude formed a further very valuable check, and prevented any great accumulation of error. A great number of observations were taken at Yarkand, and by combining the route-survey with the latitude, I think it may be concluded that the latitude and longitude of Yarkand have been determined within narrow limits.[41]

The Moonshee's gestures were indeed what Otto Sibum has referred to as 'gestures of accuracy'.[42] The demonstration was received with much enthusiasm by his colleagues at the Royal Geographical Society – Sir Andrew Scott Waugh, ex-Surveyor General of India, corroborated Montgomerie's interpretation of the Moonshee's observations; for Murchison this represented an exploit 'of the greatest importance for geographers'.[43] With this approval by the community of surveyors and geographers, his 'instrument' was now officially calibrated.

This step was crucial for it is only through calibration that instruments or techniques developed in a given place can be compared with those developed in another place, that the results obtained through the one set can be compared with those obtained through the other set – and thus that contemporary science can claim to be universal. It was thus that Montgomerie could legitimately extend his techniques and instruments to other places. All instruments and methods have to be calibrated: a watch's dial or a thermometer's stem are marked with values which correspond to the movements of the hands or of the liquid for a known time or temperature. Scientists assume that any identical time lapse or temperature identical with that of an initial, or standard, source will produce an identical response in all other watches or thermometers. Assured of the calibration of his human instrument, Montgomerie could now legitimately link two distinct surveying techniques on the same map – classical triangulation and the geodetically and astronomically controlled route surveys of the Pundits. He could also envisage the extension of his corporeal techniques to map other places.[44]

The operation of calibration is crucial in survey practices, the more so in the Indian subcontinent. All surveyors are almost obsessed by it. Chains have to be recalibrated before, after and during each use, and the same can be said for compensating bars, thermometers, sextants and theodolites. In his *Account of the Measurement of an Arc of the Meridian between the Parallels of 18° 3′ & 24° 7′*, published in 1830, for instance, George Everest, FRS, Surveyor-General of India from 1830 to 1843, devotes nearly a dozen pages to a detailed description of the recalibration procedures for the great theodolite after its disastrous fall in 1808 from the top of the tower of the Tanjore temple. Montgomerie himself gave detailed accounts of his calibration activities in his narrative reports of the Kashmir survey operations.

Finally, of course, lies the problem of the very special civility of the explorers and the trust accorded to them. For, while almost all explorers, like all men of science of the time, were white and of high social standing, with the time and resources to engage in such risky pursuits, those engaged in this Transhimalayan project were, with one notable exception (of whom more later), all South Asians. It could, of course, be argued that the enrolment of colonised people to execute what their colonial masters were not themselves in a position to carry out is explicable through the rationality of imperialism, and the story of this exploration could at first sight lend itself to a narrative of the domestication of colonial bodies through the enforcement of discipline. However, this explanation alone cannot respond to a series of questions which arise in the light of the Indo-Victorian intellectual and social context in which this enterprise is embedded. This confidence in natives to record and narrate facts so vital for the survival of the British Empire is especially surprising when one remembers that geography in the nineteenth century consisted not only in taking topographical readings but also in collecting cultural, ethnological, political and commercial information.[45] Although men of science had since the seventeenth century been conducting experiments on humans which required the subjects to relate their reactions, the subjects they chose for this type of experiment were very special – if not themselves, they were carefully chosen from amongst their (European) colleagues: 'for the two grand requisites of a witness,' wrote Robert Boyle, the great theoretician of the experimental form of life, '[are] the knowledge he has of the things he delivers, and his faithfulness in truly delivering what he knows'.[46] Indeed, at least until the mid-nineteenth century, women, servants, children, the sick and the insane were not considered to be reliable witnesses.[47] It might also be useful to remember that although native informants are crucial in the culling of much knowledge, they are hardly ever acknowledged in most narratives, and that at precisely the same moment that Montgomerie was putting together his scheme John Hanning Speke was embroiled in a fatal conflict with his arch rival and former colleague in the Survey of India, Richard Francis Burton, over the discovery of the source of the White Nile. At the heart of the controversy lay Burton's intense contempt for sub-Saharan Africans and thus for Speke's key informants. At one point in the duel Burton was able to turn the tables on his rival by reversing the direction of a river that had been reported to flow into Lake Tanganyika and thus identify this lake (instead of Lake Victoria) as the source of the Nile. He argued quite simply that the originally supposed direction had been culled from local, therefore unreliable, sources. How then could any credibility be accorded to the testimony of a few South Asian underlings? What legitimised the Pundits as reliable producers of knowledge?

Ever since their arrival in the Indian subcontinent and for any continued presence there, all Europeans had had to rely on a close collaboration between them and native intermediaries, commonly called *banians* (bankers) and *moonshees* or *munshis* (interpreter-secretaries), as well as a close interaction with *karigars* (artisans: mechanics, weavers, shipbuilders, artificers).[48] With territorial conquest in the subcontinent in the mid-eighteenth century, this collaboration spread to tax collection and the administration of justice, then to education. For this, the British retained the traditional administrative structures and most of the under civil servants of the South Asian kingdoms and empires they took over. Thus, *tehsildars* (prefects), *kazis* (magistrates), *kotwals* (police officers), *pargana sarrishtadars* or *kanungos* (rural registrars), *patwaris* (record keepers), *amils* (tax collectors), *katibs* (scribes), *maulvis* and *pundits* (teachers), inherited from the Mughal and other princely administrations, served as intermediaries between the British and the local populations.[49] This dependence on local intermediaries is perhaps nowhere more overtly exhibited than in the different surveys undertaken in the subcontinent. The first detailed English map of the Indian subcontinent was published in 1783 by Major James Rennell (1742–1830), first Surveyor-General of Bengal. In order to make it Rennell had recourse to indigenous gazetteers, like the *Ain-i Akbari*, as well as autochthonous route surveys.[50] In the frontispiece to the first edition of this map of the subcontinent, one can see as an open acknowledgement of the co-operation between Indian and British elites, a Brahman giving sacred manuscripts (Shastras) to Britannia while other Brahmans, each loaded with other manuscripts, patiently await their turn (Figure 7.11).[51] This collaboration is also celebrated in contemporary portraiture. For instance, the portrait of Colonel Colin Mackenzie (1754–1821), Surveyor-General of India, painted by Thomas Hickey in 1816, is a good example (Figure 7.12). One sees the Scotsman surrounded by a Jain priest with a palm-leaf manuscript in his hand, his peon Kistnaji holding his telescope and, behind him, his pundit, Kavali Venkata Lakshmaiah, who has only his – immaterial – knowledge to offer.[52]

Caste appellations and titles soon incorporated this new sense of specialist intermediaries in the emerging Anglo-Indian language and by the mid-nineteenth century had wound their way into current English usage.[53] Now, one of the two legitimising strategies deployed by Montgomerie consisted in carefully choosing Anglo-Indian code names that related to intellectual activities in traditional South Asian societies: *pundit* after all is the Sanskrit for a learned man, in Anglo-Indian it denoted an adviser to British judges on matters of Hindu law; *Munshi* is the

Figure 7.11 Frontispiece to James Rennell, *Memoir of a Map of Hindoostan* (London: for the author, 1783).
Source: British Library, Oriental and India Office Collections.

Persian for a scribe or a writer, and Anglo-Indian for a native teacher of Indian languages; and *mirza* and *mullah*, the Persian for the learned and the cleric respectively.[54] In Western traditions, the credibility of a report normally lies in the nominal identity (and thus the credibility) of the reporter.[55] However, in the present case, these identities had to be kept secret. And although this could have been achieved by using banal code names or by simply changing the names of the native explorers, their reports could not have been legitimised. It was thus

Figure 7.12 Portrait of Colin Mackenzie by Thomas Hickey (1816).
Source: British Library, Oriental and India Office Collections.

precisely through the use of Anglo-Indian appellations with intellectual connotations as code names that this credibility could be conferred.

The other strategy consisted in showing how, when compared with Europeans who had in some cases taken geodesic measurements of the same places as the Pundits, the latter were at least as accurate. In the article he published on Nain Singh's first mission to Tibet in *The Journal of the Royal Geographical Society* in 1868,

Montgomerie writes: 'There is no doubt that the Pundit is an excellent and reliable observer.' Then he goes on to devote more than five pages in order to rigorously establish the credibility he accords to his instrument. He compares Nain Singh's mapping of known territories with those made previously by British observers and shows that they are at least as accurate. For Tibet, Captain Samuel Turner, who had been sent by the East India Company as an emissary to that country in 1783, had left a series of observations as well as a map of his route from Bhutan to Lhasa via Shigatze. Montgomerie scrutinises these and concludes that they lack precision:

> Turner observed the latitude at Tashilumbo (Shigátze), and made it 29° 4′ 20″, the Pundit makes it 29° 16′ 32″. Turner's latitude of Chumulárí is 28° 5′, the Great Trigonometrical Survey latitude is 27° 50′. Turner very possibly was not accustomed to take latitudes, and as the Surveyor (Lieutenant S. Davis) sent with him was not allowed to go beyond Tassísudon, it is not to be wondered that there are differences in his latitudes. The comparison of several latitudes now well-known, tend to show that the semi-diameter of the sun may well have been omitted by Turner, as his observations were to the sun only.
>
> The Pundit's observations at Shigátze extend over many days, and include thirteen observations to the sun and a variety of southern stars, as well as to the pole star. The latitudes derived from these observations agree capitally *inter se.* The Pundit was thoroughly practised in the method of taking latitudes, and as his determinations of many well-known points, such as Bareilly, Moradabad, &c., have proved to be correct with only a pair of observations, there can be no doubt about accepting his latitude of Shigátze, where he took so many. The Pundit followed the same river as Turner for 50 miles between Gyangze and Shigátze. They agree in making the bearing between those places 62° west of north. The bends of the river as given by them agree in a general way, but the distance by Turner is 39 miles, and by the Pundit 46 miles. As the former appears to have only estimated his distances by guess, while the latter paced them carefully, the result by the Pundit has been adopted as the most correct.[56]

With this, Montgomerie manages to elevate his collaborator from the rank of a simple instrument, a 'docile body', to that of a peer, comparable to, if not better than, his fellow countryman Samuel Turner. The Royal Geographical Society awarded Nain Singh a gold watch in 1868 and the Victoria gold medal in 1875. The following year he won the French *Société de géographie's* gold watch. Indeed, these awards can be taken as much to be signs of recognition of the individuals to whom they were awarded as a reinforcement of the legitimation of the knowledges and techniques mobilised to acquire them.

Thanks to these, albeit massive, investments in calibration and legitimation then, the knowledge produced by the 'pundit-instruments' was validated and travelled far out of the sites of production in the form of maps, anthropology and linguistic texts, texts on Central Asian statecraft, and so on. But what was the fate of the human 'instruments' and their incorporated gestures? The itinerary of their displacement and appropriation is not a simple one to trace. However, two episodes might help throw some light on their dissemination.

The first relates to an attempted appropriation of corporeal surveying techniques by another imperial power – Russia. In large measure owing to the internationalism of the community of Victorian geographers, almost all the reports and technical publications relating to the Transhimalayan expeditions, along with the prayer wheel and rosary, were transmitted to their colleagues at the Imperial Russian Geographical Society at St Petersburg. However, in spite of at least seven attempts the Russians could not use this knowledge, nor the instruments, for what was missing were the human instruments and the trust and complicity on which the whole system functioned – the latter conditions had, as we have seen, been co-produced with the Anglo-Indian Empire and could not be transmitted to other empires.[57] If Montgomerie's scheme fed on the dual umbilical cord of the replicability of experiments and on investments of confidence and credibility in native agents, this was only conceivable within institutional traditions built through the preceding centuries.

The second example is that of Pundit Kintup from Darjeeling, who, because he was illiterate, was sent on his mission in 1880 under the command of a Chinese lama to determine the course of the Tsangpo downstream of Lhasa. The latter decamped after having sold Kintup to a Tibetan monastery. Kintup, however, bought his freedom by working for the abbot of the monastery and continued his mission – which he completed successfully.[58] Obviously, the British had not been able to establish the required trust and complicity with the Chinese that they had with Kintup and the other Pundits.

Notes

1 C.R. Markham, *A Memoir on the Indian Surveys*, London: Her Majesty's Secretary of State for India in Council, 2nd edition, 1878: 113, 132.

2 Some scant details of the early life and education of T.G. Montgomerie can be gleaned from his application to the military academy at Addiscombe (British Library, Oriental and India Office Collections, hereafter referred to as OIOC, L/MIL/9/215 folios 238–46) and from the monthly reports of the academy, February 1836 to May 1851 (OIOC: L/MIL/9/341). See also R.H. Phillimore, *Historical Records of the Survey of India*, 5 volumes, Dehra Dun: Survey of India, 1945–68: V, 513.

3 'Montgomerie, Thomas George', in L. Stephen and S. Lee (eds) *Dictionary of National Biography from the Earliest Times to 1900*, 22 volumes, Oxford: Oxford University Press, 1917: XIII, 758–60.

4 Quoted in Markham, *A Memoir on the Indian Surveys*: 428. Readers might note that Mont Blanc always functions as the standard against which any other scientific climb is compared, after Horace Bénédict de Saussure's expedition.

5 I have, unfortunately, not been able to find any information about Hameed's life or career.

6 See D. Kennedy, *The Magic Mountains. Hill Stations and the British Raj*, Berkeley, Los Angeles, London: University of California Press, 1997.

7 Letter dated 9 June 1853 from the Surveyor-General to T.G. Montgomerie, National Archives of India, Survey of India Records, Dehra Dun Volumes, Decade 1850–60, 641 (393), quoted in Phillimore, *Historical Records*: V, 480.
8 *Proceedings of the Asiatic Society of Bengal* 2, 1862: 209–13.
9 *Journal of the Royal Geographical Society* 38, 1868: 129.
10 The most recent edition of the *Times Atlas* places Yarkand at 77° 16′ longitude and 38° 27′ latitude.
11 *Journal of the Royal Geographical Society*, 38, 1868: 164.
12 Extracts of the presentation with an account of the discussion that followed were published in *Proceedings of the Royal Geographical Society*, X, 1865–6: 162–5. The complete text, T.G. Montgomerie, 'On the geographical position of Yarkand, and some other places in Central Asia', appeared in *Journal of the Royal Geographical Society* 36, 1866: 157–72.
13 *Bhotiya* is a generic term generally designating the Tibetan-speaking and mainly Buddhist peoples of the Himalayan regions.
14 See C. von Fürer-Haimendorf, *Himalayan Traders*, New York: St Martin's Press, 1975: ch. 7.
15 H. Schlagintweit, A. Schlagintweit and R. Schlagintweit, *Results of a Scientific Mission to India and High Asia Undertaken between the Years MDCCCLIV and MDCCCLVIII, by Order of the Court of Directors of the Honourable East India Company*, 4 volumes, Leipzig, 1861–6: I, 38.
16 These consisted of two miniature sextants, pocket compasses, prismatic compasses, a pocket chronometer, air and boiling-point thermometers and a watch.
17 'Extracts from a diary kept by Pundit ——, during his journey from Nepal to Lhasa, and from Lhasa through the upper valley of the Brahmaputra to the Source of that River near the Mansarowar Lake', *Journal of the Royal Geographical Society*, 38, 1868: 154–79.
18 Letter dated 23 August 1867 from Foreign Department, Foreign Department Proceedings, March 1868, National Archives of India, New Delhi.
19 For a detailed account of these missions, see T.G. Montgomerie, 'Report of "the Mirza's" exploration from Caubul to Kashgar', *Journal of the Royal Geographical Society* 41, 1871: 132–93; idem, 'A Havildar's journey through Chitral to Faizabad, in 1870', *Journal of the Royal Geographical Society*, 42, 1872: 180–201; Anon., *Narrative of Surveys made, during 1876, by 'The Mullah', in connexion with the operations of the Great Trigonometrical Survey of India*, Simla: Government of India, 1877.
20 *Journal of the Royal Geographical Society* 47, 1877: cxxvi.
21 S.C. Das, *Journey to Lhasa and Central Tibet*, London: John Murray, 1902; idem, *Autobiography: Narratives of the Incidents of My Early Life*, Calcutta: K.L. Mukhopadhyay, 1969; idem, *Indian Pandits in the Land of Snow*, Calcutta: Baptist Mission Press, 1893; idem, *A Tibetan–English Dictionary*, Calcutta: Government of Bengal, 1902.
22 For a complete account of this mission, see J.B.N. Hennesey, 'Report on Pandit Kishen Singh's Explorations in Great Tibet and Mongolia', in S.G. Burrard (ed.) *Exploration in Tibet and Neighbouring Regions, Part 1, 1865–1879; Part 2, 1879–1892.* Records of the Survey of India, Volume VIII, 2 parts, Dehra Dun: Survey of India Press, 1915: Part 2, 215–324.
23 P. Fleming, *Bayonets to Tibet*, London: Rupert Hart-Davis, 1961: 23.
24 See, for instance, T.H. Holdich, 'Tibet', in *Encyclopaedia Britannica*, 11th edition, Vol. 26, 1911: 916–28; J.N.L. Baker, *A History of Geographical Discovery and Exploration*, London: George G. Harrap, 1931; I.S. Rawat, *Indian Explorers of the 19th Century*, New Delhi: Government of India – Ministry of Information and Broadcasting, 1973; K. Mason, *Abode of Snow. A History of Himalayan Exploration and Mountaineering from Earliest Times to the Ascent of Everest*, London: Diadem Books, 1987; D.J. Waller, *The Pundits. British Exploration of Tibet and Central Asia*, Lexington: University Press of Kentucky,

1990; J. Pleydel-Bouverie, 'On a wheel and a prayer', *Geographical Magazine* 64, 5, 1992: 12–15.

25 See for instance the letter of Henry Yule to Rutherford Alcock, dated 18 July 1876, in the Royal Geographical Society Archives: London.

26 M.L. Pratt, *Imperial Eyes. Travel Writing and Transculturation*, London: Routledge, 1992: 7.

27 A. Moorehead, *The White Nile*, London: Penguin, 1973: 65.

28 The following quote from Paul Du Chaillu's *Explorations and Adventures in Equatorial Africa*, New York, 1861, superbly parodies this hegemonic goal: 'The murmur of the rapids below filled my ears, and as I strained my eyes toward those distant mountains which I hoped to reach, I began to think how this wilderness would look if only the light of Christian civilization could once be fairly introduced among the black children of Africa. I dreamed of forests giving way to plantations of coffee, cotton, spices; of peaceful negroes going to their contented daily tasks; of farming and manufactures; of churches and schools; and luckily, raising my eyes heavenward at this stage of my thoughts, saw pendent from the branch of a tree beneath which I was sitting an immense serpent, evidently preparing to gobble up this dreaming intruder on his domains' (p. 83, quoted in Pratt, *Imperial Eyes*: 209).

29 H.M. Stanley was the only exception in that he managed to get himself sponsored jointly by two newspapers, the *New York Herald* and the London *Daily Telegraph.*

30 Moorehead, *White Nile*: 65.

31 S. Lindqvist, 'Labs in the Woods: the Quantification of Technology During the Late Enlightenment', in T. Frängsmyr, J.L. Heilbron and R. Rider (eds) *The Quantifying Spirit in the 18th Century*, Berkeley, Los Angeles, London: University of California Press, 1990: 291–314.

32 See S. Widmalm, 'Accuracy, Rhetoric, and Technology: the Paris–Greenwich Triangulation, 1784–88', in Fränsgmyr, Heilbron and Rider (eds) *The Quantifying Spirit*: 179–206. See also, S. Schaffer, 'Accurate measurement is an English science', in M.N. Wise (ed.) *The Values of Precision*, Princeton, N.J.: Princeton University Press, 1995: 135–72.

33 S.F. Cannon, *Science in Culture. The Early Victorian Period*, New York: Science History Publications, 1978: 75–6.

34 'Cette assurance devient beaucoup plus grande encore, on peut mesurer jusqu'à 40t. de hauteur, en se servant plus près des objets de grandes bases, que l'on mesure très rapidement par ses pas ou par une longue perche que l'on jette successivement devant soi. De l'égalité du pas de l'homme. A quelle exactitude l'habitude peut porter. Mesures par le tems sans compter les pas. Mesures des angles de hauteur sans horizon avec une perche de 5 pieds de hauteur, qui sert de signal à la hauteur de l'oeil. J'ai trouvé l'horizon artificiel et ce signal souvent en harmonie à 5′–6′ près. Tout ce détail paraît très minutieux. On peut s'inventer 10 méthodes pour une pour s'assurer à peu près de la grandeur des Objets. Mais ces à peu près ont leurs limites d'erreur, qu'il faut connaître pour l'expérience. On peut par telle ou telle méthode se tromper de 100 mais non de 2 ou 300 toises. Les besoins du Naturaliste (je le répète) ne sont pas ceux de l'Astronome. Le grand problème de la vie est de produire beaucoup en peu de temps et il est certain que si des méthodes de mesurer avec des moyens très simples fussent répandu dans le public, si l'attention des hommes qui vivent dans les champs, les forêts et les montagnes, était plus fixée sur la grandeur des objets et leur distance, qu'après tant de voyages et de recherches faites dans les 2 hémisphères nos idées géologiques (la partie la plus belle, la plus intéressante des connaissances humaines) seraient triplement avancées' (Alexander von Humboldt, Tagebüch 1, Bl. 81r, Deutsche Staatsbibliothek, Berlin). I am very grateful to Michael Dettelbach for providing me with this reference, and also for explaining the context in which Humboldt thinks pacing is a valid measuring technique.

35 See S. Schaffer, 'Metrology, Metrication and Victorian Values', in B. Lightman (ed.) *Victorian Science in Context*, Chicago: University of Chicago Press, 1997: 438–74.

36 S.F. Cannon, *Science in Culture*: 75–6.
37 Markham, *A Memoir on the Indian Surveys*: 64.
38 R. Smyth and H.L. Thuillier, compilers, *A Manual of Surveying for India, Detailing the Mode of Operations on the Revenue Surveys in Bengal and the North-Western Provinces*, Calcutta: Thacker, Spink & Co., 1851.
39 Phillimore, *Historical Records*: I, 199.
40 Smyth and Thuillier, *A Manual of Surveying*: 106–8.
41 Montgomerie, 'On the Geographical Position of Yarkand': 163.
42 H.O. Sibum, 'Reworking the mechanical value of heat: instruments of precision and gestures of accuracy in early Victorian England', *Studies in History and Philosophy of Science* 26, 1, 1995: 73–106.
43 *Proceedings of the Royal Geographical Society* 10, 1865–6: 162–5.
44 See H.M. Collins, *Changing Order: Replication and Induction in Scientific Practice*, London: Sage, 1985: 100–6.
45 It is useful to remember that Section E of the British Association for the Advancement of Science, established in 1851, was entitled 'Geography and Ethnology'.
46 R. Boyle, 'The Christian Virtuoso', in T. Birch (ed.) *The Works of the Honourable Robert Boyle*, 6 volumes, London: J. and F. Rivington, 1772: 5, 529, quoted in S. Shapin and S. Schaffer, *Leviathan and the Air-Pump*, Princeton, N.J.: Princeton University Press, 1985: 58.
47 See S. Schaffer, *From Physics to Anthropology – and Back Again*, Cambridge: Prickly Pear Press, 1994.
48 Visible mainly in commercial interactions – the main reason for British presence in the Orient – this complicity is also found in a number of other domains. This is wonderfully illustrated through the case of Edward Bulkley (*c.* 1651–1714), East India Company surgeon at Madras, who used to send parcels of dried plants to London. The plants carried their descriptions written in Tamil on bamboo strips by indigenous collaborators and attached to each stalk. See I.H. Burkill, 'Chapters on the history of botany in India – 1. From the beginning to the middle of Wallich's service', *Journal of the Bombay Natural History Society* 51, 4, 1953: 852.
49 Few authors have examined the question of native intermediaries during the early years of the British Empire. This lacuna has recently been filled by C.A. Bayly, *Empire and Information. Intelligence Gathering and Social Communication in India, 1780–1870*, Cambridge: Cambridge University Press, 1997.
50 On South Asian cartographic traditions, see J.E. Schwartzberg, 'South Asian Cartography', in J.B. Harley and D. Woodward (eds) *The History of Cartography. Cartography in the Traditional Islamic and South Asian Societies*, Volume 2, Book 1, Chicago and London: University of Chicago Press, 1992: 293–509.
51 J. Rennell, *Memoir of a Map of Hindoostan; or the Moghul's Empire*, London: M. Brown for the author, 1783. The explanation of this emblematic image is to be found on page xii of the first edition. At a number of places Rennell acknowledges the use for drawing his own map of indigenous measures and maps; see, for example, pp. ii, iii and vi. See also Phillimore, *Historical Records*: I, 42, and A.H. Anquetil-Duperron, *Zend-Avesta*, 3 volumes, Paris: N.M. Tilliard, 1771: I, ccccxxxviij and dxlj.
52 Phillimore, *Historical Records*: II, Plate 22 facing p. 427. See also M. Vicziany, 'Imperialism, botany and statistics in early nineteenth-century India: the surveys of Francis Buchanan (1762–1829)', *Modern Asian Studies* 20, 4, 1986: 623–60; N.B. Dirks, 'Colonial histories and native informants: biography of an archive', in C.A. Breckenridge and P. van der Veer (eds) *Orientalism and the Postcolonial Predicament*, Philadelphia: University of Pennsylvania Press, 1993: 279–313.
53 Of course, this mediation in the construction of knowledge had to be legitimised and the British were indeed conscious of this need. One of the main functions of British Orientalism was precisely this. See K. Raj, 'Du commerce à la linguistique', *La Recherche*, 300, July–August 1997: 46–9.

54 For the evolution of the meanings of these terms in English, see *The Oxford English Dictionary* and H. Yule and A.C. Burnell, *Hobson-Jobson. A Glossary of Colloquial Anglo-Indian Words and Phrases*, London: John Murray, 1903; new edition: London: Routledge & Kegan Paul, 1985.

55 See S. Shapin, *A Social History of Truth*, Chicago: University of Chicago Press, 1994.

56 T.G. Montgomerie, 'Report of a route-survey made by Pundit ——, from Nepal to Lhasa, and thence through the upper valley of the Brahmaputra to its source: Extracts', *Proceedings of the Royal Geographical Society* xii, 1867–8: 144.

57 On this subject see the correspondence between Frédéric d'Osten Sacken, secretary of the Imperial Russian Geographical Society, and General J.T. Walker, Superintendent of the Great Trigonometrical Survey of India, archives of the Royal Geographical Society, London.

58 For Kintup's story, see H.C.B. Tanner, 'Kintup's Narrative of a Journey from Darjeeling to Gyala Sindong (Gyala and Sengdam), Tsari and the Lower Tsang-po, 1880–84', in S.G. Burrard (ed.) *Exploration in Tibet and Neighbouring Regions, Part 1, 1865–1879; Part 2, 1879–1892*. Records of the Survey of India, Volume VIII, 2 parts, Dehra Dun: Survey of India Press, 1915: 2, 329–38.

8 The manufacture of species

Kew Gardens, the Empire, and the standardisation of taxonomic practices in late nineteenth-century botany

Christophe Bonneuil

This chapter is about the creation of universals and the maintenance of order in the field of systematic botany during the age of Empire. More precisely, it explores how the practice of broadly circumscribing species, or, in other words, the broad and variable species concept, became dominant in the practice of taxonomy in the second half of the nineteenth century, and how such broad species were established and given authority.

By the mid-nineteenth century it seemed that no consensus could ever be reached between botanists concerning the delimitation of species and how this should be done, not to mention more ontological issues opposing evolutionists and fixists. A tremendous controversy arose between those who adopted a broad species concept ('splitters') and those who preferred narrowly delimited species ('lumpers'). At the methodological level, moreover, the right criterion to distinguish a 'true species' from a mere variety was hotly debated by those who valued the study of living plants and those who valued the comparison of dried forms. By the 1880s and 1890s, professional botanists working in huge herbaria (especially at Kew Royal Botanic Gardens near London) had largely imposed their views:

- the broadly circumscribed species (and genera) became the standard for the delimitation of species (and genera);
- the herbarium was recognised as being the essential tool of the taxonomist (rather than field observations or garden cultivation), and the comparison of dried specimens became the major criterion for deciding the circumscription of species;
- a particular nomenclatural practice known as the 'Kew-rule', where the original specific epithet is not necessarily conserved when the plant is transferred to another genus, was widely followed.

This chapter examines how these three norms for species naming and delimitation came to rule taxonomic practices in the second half of the nineteenth century, and why they appeared to such leading professional systematic botanists as Joseph Dalton Hooker (1817–1911), George Bentham (1800–84), Alphonse de Candolle (1806–93) and Asa Gray (1810–88) as the only way to maintain order in the field of botanical knowledge, to ensure its progress and epistemic steadiness, and restore its declining status in the hierarchy of disciplines. The first two sections sketch the situation of taxonomic botany in the middle of the nineteenth century, and the debates on species delimitation. The third section underlines the role of Joseph Dalton Hooker in leading a crusade against the narrow delimitation of species. The fourth and fifth sections examine how professional botanists at the head of huge public herbaria enriched by expanding empires turned their herbaria into metrological institutions that imposed the broad species concept, by controlling the production of new names and setting standards in the practices of classification. They edited colonial floras, monographs and other unprecedented compilatory works. These huge enterprises prepared the grounds for a division of labour among botanists, and were the reasons for and the tools to stabilise names and to standardise practices. The final section reflects on botany's and the British Empire's need for standards.

The proliferation of practitioners and species: a problem of order and authority in nineteenth-century systematics

In the eighteenth century, systematic botany, the activity of naming and describing plants and classifying them into groups according to degrees of difference and similitude, was deeply related to the encyclopaedic project. It appeared as one of the most successful branches among the sciences, the nearest to its final goal.[1] It had set principles which were adopted not only in the other branches of natural history (zoology, mineralogy) but also in other domains in science, especially chemistry and medicine. Botany was considered a most highly philosophical pursuit. Linnaeus's (as well as Adanson's and Jussieu's) insistence on explicit rules, predictivity, abstraction, numeration, clarity and neutrality of language resonated very strongly with the geometrical, quantifying and rationalising spirit of the Enlightenment.[2] Systematics could still by 1750 believe in completeness, a goal which appeared unachievable in the following decades, due to the rapid increase of specimens brought to herbaria by European expansion in the eighteenth and nineteenth centuries. Linnaeus's *Species plantarum* of 1753 described 5,890 species in 1,097 genera.[3] One century later, the Candolles' *Prodromus* described 58,975 species in 5,134 genera (dicotyledons only). In 1883, Bentham and Hooker's *Genera plantarum* estimated that there were 93,605 species (phanerogams only).[4] This led, in the early nineteenth century, to a descriptionist shift that made systematic botanists appear as quibblers in the early nineteenth century.[5] In 1833, a phrenological study from Hewett Cottrell Watson (1804–81) – who was also a botanist – illustrated the declining status of botany. It stated that, relative to other scientists, the heads of leading botanists were smaller in areas where the higher intellectual activities were carried out.[6] Though it remained appealing for colonial actors and helped the professionalisation of botany, the creed of a general inventory of natural productions of the world had, by the mid-nineteenth century, lost its philosophical prestige.

More deeply, this massive specimen influx also constituted a dramatic epistemic challenge to eighteenth-century classification systems. Most new specimens came from extra-European areas, whereas systems like Linnaeus's and Jussieu's were largely the product and the rationalisation of European folk classifications. There was a relatively limited number of tropical species known in Europe before the end of the eighteenth century. Bringing extra-European and extra-temperate forms into the Eurocentric frame of an existing classification represented a huge challenge. This tension is today revealed by statistical data on the distribution of species into genera. First, a tendency to create more genera (revealed by lower numbers of species per genus) is evident in families important in classic and European folk economy and culture (such as Gramineae), as compared to tropical families or to temperate families with little economical and cultural relevance (such as Cyperaceae). Also evident is the typical 'hollow curve' distribution of species into genera in large cosmopolitan families: a few genera (crafted in the time of Tournefort, Linnaeus and Jussieu) are overcrowded with species, whereas numerous genera (crafted in the nineteenth century) contain no more than one or two species.[7] This clearly hints at the conservatism of leading nineteenth-century botanists (with access to the resources of huge collections at the centre of expanding empires), as they tried to fit new exotic forms into the old genera inherited from the great seventeenth- and eighteenth-century botanists, which supposed, as we shall see, to impede local botanists from the periphery (sometimes better placed to avoid this Eurocentric bias) from creating 'too many' new species and new genera.[8] The enormous number of exotic specimens arriving in Europe was therefore not only a quantitative, but also a qualitative, challenge to an established Eurocentric taxonomic order.

Not only new species and specimens were proliferating, but also botany practitioners. In the time of Linnaeus and the Jussieus, botanising in European countries was done mainly by gentlemen of science or by students herborising with their professor. The botanical exploration of remote regions of the world was done mostly by *protégés* indebted to their scientific patrons.[9] The old regime of botany was threatened in the nineteenth century when natural history became very popular, not only in the upper and middle classes but also in the upper working class. Local societies (and local botanical journals) flourished. Botany was now practised by people of diverse social and educational backgrounds, motivations and interests and practices, etc.[10] European expansion enabled thousands of people who practised botany to do so overseas in remote settings. These agents of empire (Scottish doctors, adventurers, missionaries, soldiers, settlers, etc.) had their own agenda. Few of them were ready to use the Latin and were easy to bend to the discipline of a division of labour between collectors and armchair professional systematists.[11]

This endless accumulation of newly discovered living forms due to European expansion and to the multiplication of practitioners made the eighteenth-century dream of a completion, bit by bit, of the inventory of nature a problematic enterprise.[12] The anarchic increase of ('good' or 'bad') species challenged the way in which taxonomic statements had been until then validated and communicated. New actors and new specimens were not readily disciplined into botany's *ancien régime*. The Linnaean canon, for instance, by which a twelve-word diagnosis could characterise a species and distinguish it from all others, was now 'becoming quite inadequate to the requirements of the science' because it would constantly have to be refined and reframed so as to take into account all the new species discovered since Linnaeus and Willdenow.[13] Furthermore, in the middle third of the century, botanical journals became so numerous that it was impossible, even in the richest libraries of Kew or the Paris Muséum, to gather all the information necessary to identify quickly and correctly name a specimen, or to write a monograph of a genus or a family.[14] The cumulative nature of botany was at stake. Bringing order in the field of botanical statements and publications had become a central problem. In the eighteenth century naturalists viewed themselves as revealing *the* order of nature. By contrast, mid-nineteenth-century monographers seemed concerned rather with bringing *some* order in the chaos of previously published literature. Throughout the latter half of the nineteenth century, flora-workers and monographers introduced their work in similar ways: '[the botany of India] presents a perfect chaos of new names for well-known plants, and inaccurate and incomplete descriptions of new ones' or 'Es schien mir . . . eine lohnende Aufgabe zu sein, die bis jetzt bekannten Arten dieser für die Technik so überaus wichtig gewordenen Pflanzengattung möglichst vollständig zu vereinigen und mit Hilfe dieses Materials in das noch bestehende Chaos Ordnung und Licht zu bringen.'[15]

The case of the genus *Landolphia* illustrates this tendency towards species inflation and the difficulty to make commensurable a crowd of heterogeneous statements from various actors. The *Landolphia* are Apocynaceaes growing in Africa and Madagascar. The first specimen that reached Europe was collected by 1787 in the kingdom of Oware (South West Nigeria) by Palisot de Beauvois, a gentleman-naturalist who accompanied a French military expedition headed by Captain Landolphe. After examination in the herbarium of the Muséum National d'Histoire Naturelle, Palisot de Beauvois described it in 1806 as *Landolphiaowariensis*.[16] In 1844, only three more species were described, but under two different genera (*Landolphia* and *Vahea*) that were only later united (under *Landolphia*) in Bentham and Hooker's *Genera plantarum*.[17] In East Africa, Dr John Kirk had observed some *Landolphia* and their elastic gum during the Zambezi expedition with Livingstone (1858–64). Appointed British consul at Zanzibar a few years later, he studied these plants further and provided Kew with specimens from the coast near Dar-es-Salaam. As a good agent of the new British moral imperialism in Africa, he worked hard to make trade in rubber replace that in slaves on the East coast.[18] The scramble for Africa, and more specifically the search for African rubber-yielding plants, suddenly focused attention on *Landolphia*. Several species produced good rubber and provided a great deal of the African rubber, which represented about one-third of the world trade in the last two decades of the century.[19] *Landolphia* therefore became an object of intense interest on the part of travellers, traders, civil servants in African outposts, etc. During the African rubber boom (1870–1914), when natural rubber resources were wildly depleted, several expeditions were organised in West and Central Africa to find sources of rubber.[20] This brought hundreds of new specimens in European herbaria, and saw many diverse actors – explorers, military men, military physicians, missionaries, traders, colonial officers, etc. – claim to have discovered a new species. *Landolphia* also became a hot topic for closet naturalists at home: parallel to the economic boom of rubber in Africa, one can observe a boom in the botanical literature.[21] But when studying African rubber plants, even those happy few professional botanists who had access to large libraries and herbaria in Kew, Paris and Berlin (the metropolis of ruling colonial powers in Africa) faced great difficulties. They had to 'navigate' in a poorly ordered field of statements – sometimes contradicting each other – about landolphias (concerning their appearance, distribution, specific identity, economic value, etc.). These statements were published in many different places and genre such as travel narrative, administrative reports or economic correspondence, local publications, etc., rather than in the major and readily accessible botanical periodicals.

How to circumscribe species?

As the number of claimed species dramatically increased, botanists had conflicting views on the criteria on which to decide whether two different forms belong to the same species or whether their difference should be awarded a specific value. Some would look for evidence in field observation on the living, some would stress experiment in the garden, and some would insist on the comparison of numerous dried specimens in the herbarium.

The herbarium, the garden or the field?

Jussieu and Ventenat had hoped that the subordination of characters would provide general rules and principles to weight characters and establish affinities through careful morphological observation (made by the eye, sometimes aided by a lens). But it was clear in the early nineteenth century that no one hierarchy of characters could be applied to all plant groups. Subordination of characters had to be done 'locally' because characters had a different taxonomical value depending on the group being considered.[22] The selection of characters providing evidence of relationships, and their weighting, obeyed no general method. John S. Henslow (1796–1861), Professor of Botany in Cambridge University, noted in 1837 that

> there is in short, no law whatever hitherto established, by which the limits of variation to a given species can be satisfactorily assigned, and until such a law be discovered, we cannot expect precision in the details of systematic botany. In this respect, the science is pretty much in the same position which mineralogy occupied before the discovery of the laws of crystallography.

He urged experimental researches, which seemed to be the only scientific approach that could lead to such a law 'for the discrimination of species' which would improve 'the mere empirical rules at present practised'.[23] In the footsteps of Linnaeus and Koelreuter, some botanists (William Herbert (1778–1847), Alexis Jordan (1814–97),[24] Charles C. Babington (1808–95), Joseph Decaisne (1807–82), Charles Naudin (1815–99), etc.) developed, in the 1840s and 1850s, a research programme towards an 'experimental taxonomy' (a phrase that came into use only in the early twentieth century).[25] 'L'histoire naturelle en général, après n'avoir été longtemps qu'une science d'observation, doit tendre à se faire science d'expérimentation', claimed Decaisne, Professor of Culture at the Paris Museum of Natural History.[26] Growing large numbers of varieties and species of the same genus, these experimental taxonomists claimed that only experiment (not herbarium or fieldwork) could decide the species question in a truly scientific way. In short, they claimed they could provide systematic botany with the absolute proof it was lacking. For some of these experimentalists like Alexis Jordan and Charles C. Babington, the persistence of differences – even if they were tenuous – between two forms when cultivated through several generations in a garden could evince the existence of two different species.[27] But many wondered how many years one should grow these plants so as to document the persistence of the differences. Furthermore, Decaisne mentioned the cases of varieties which can breed true (like the whites and the blacks who belong to the same human species). To Herbert and Decaisne, a more strict criterion was to unite in one species the two forms, whenever their crossing generated healthy and fertile offspring. In this way, Decaisne's student Charles Naudin could collate twenty-eight alleged species of honeydew melon into one single species.[28] But even this criterion remained inconclusive because there existed fertile interspecific hybrids and because of an indecisive continuum between 'fertility' and 'sterility'. 'The criterion provided by the crossing to discriminate species is not as absolute as I previously believed', admitted Naudin in 1863.[29]

Other botanists, like Carl W. von Nägeli (1817–91), C.C. Babington or Richard Spruce, advocated the observation *in situ* of the living plant as the best way to detect the minute differences among species. This method was seen as more natural than experimental studies in a garden, and more accurate than the herbarium observation of dried specimens. Herbarium specimens, indeed, showed only a few of the characters that could be used to discriminate plants. Babington claimed, in his *Manual of British Botany*, that 'species exist, and that they may often be easily distinguished amongst living plants, even when separated with difficulty from their allies when dried specimens only are examined'.[30] Robert Wight and G.A. Walker Arnott claimed, in a study on the Indian flora, that 'we have had advantage over the European botanists who have described Indian plants, they having only seen one or two isolated specimens'; Wight and Arnott had observed many exemplars of the same species in 'different localities' and 'in their natural situation'. Contrary to Babington, who invoked the field to advocate the existence of more species than currently acknowledged, Wight and Arnott claimed that field observations allowed botanists to 'cut down species' because they allowed botanists to relativise variable characters like the shape of the leaves and the quantity of pubescence.[31] Promoters of field observation, as well as promoters of experiment (Herbert and Decaisne being lumpers, whereas Jordan was a harsh splitter), hence disagreed on the breadth of the species delimitation.

Finally, neither the comparison of series of dried forms, nor field observations, nor experiment in gardens seemed to provide the ultimate test that would allow botanists to reach agreement on the discrimination of species. Asa Gray sadly observed that there were 'no fixed and philosophical principles for the subordination of characters and the study of affinities in plants'.[32] As H.C. Watson wrote to Darwin,

> The short truth is, that we have no real proof or test of a species in botany. We may occasionally disprove an alleged species by seeing its descendants become another such species, – or we may unite two by finding a full series of intermediate links. Many botanists assume all describable forms, if not (or until) so disproved or united, to be distinct species, ab initio ad finem.[33]

An epistemic crisis

This lack of accepted and conclusive criteria threatened the status of systematics as a science. Systematics seemed unable to cope with observer-dependent knowledge and to achieve communicability and accumulation. Suppressing individual idiosyncrasy was a prominent feature of nineteenth-century science's moral economy and proof-making practices. Aperspectival objectivity (as Daston put it) was asserted through many ways, such as statistical treatment of data, averaging of observers' individual features, self-discipline, and the use of self-recording instruments.[34] In systematics, no such material or literary technology was in sight to produce conclusive and accepted proofs in species delimitation controversies.

More generally, as compared to the rising Laplacian sciences, systematic botanical knowledge seemed unable to discover or demonstrate regular uniform laws in nature. Some botanists complained about the lack of law in botany and saw the natural system as being unable to draw a clear distinction between taxa, using terms such as 'often', 'sometimes . . . at other times', 'rarely', 'almost always', etc.[35] Systematics seemed foreign to causal thinking; it 'explained nothing'. Those passionately concerned to discover the general laws of the vegetable world had shifted to plant geography, or to plant anatomy and physiology. Classifying plants had lost academic interest. In Germany, the 'wissenschaftliche Botanik' emerged in the 1830s and 1840s, and physiological, geographical, microscopical and experimental approaches took the lead in botany.[36] Between 1840 and 1880, systematic botany as a domain owed its survival largely to a single project, i.e. the undertaking of the *Flora brasiliensis*.[37] In France, the Paris Museum of Natural History itself, which had been prominent in the institutionalisation of natural history around 1800, experienced a shift towards experimental approaches and disciplines in the decades after 1838.[38] The Jussieus' chair of systematic botany was suppressed in 1853 and French naturalists complained about natural history being abandoned in the school curricula.[39] In Great Britain, Richard Drayton has shown how, after the break-up of Banks's patronage network, career opportunities in the Empire were vital for the survival of systematics as a science in England.

As no experimental equipment and no conclusive species criteria existed that would mark the boundary between skilled systematists and amateurs, it seemed that systematic botany, a noble and pleasurable science in the eighteenth century, had become a chaos of incommensurable statements because the cost of entrance for a beginner was very low and anybody could claim to have discovered a species. While other sciences experienced an increased professionalisation, botany appeared progressively as a poorly specialised and amateur science connected with horticulture or women's leisure.[40] To many commentators, systematics was a battlefield, where different – and ever opposing – views could never produce any collective and steady knowledge. This was nowhere more dramatically exemplified than in the mid-nineteenth-century debates opposing 'splitters'[41] and 'lumpers'[42] on the circumscription of species.

'Splitters' and 'lumpers'

The splitters (named in French 'l'école analytique') claimed that within Linnaean species, it was possible to find several stable and distinct types which they claimed to be species. Path-breaking works here were those of Carl Wilhelm von Nägeli (1817–91) in 1841[43] and Alexis Jordan (1814–97) in 1847.[44] Though united in their search for smaller and more stable units in nature, the splitters strongly diverged on their ontological views. Some were creationists, like Jordan, others were transformists, like Nägeli. They also differed over the criteria they advocated to circumscribe species: some (like Babington) stressed local field experience, others (like Jordan) experimental gardening, yet others (like Nägeli) microscopical studies. However heterogeneous this movement may have been, this splitting trend in systematics, which flourished between 1840 and 1870, was very powerful and was reinforced by the tendency of local botanists in the peripheries of European empires to create new species. By 1850, a 'splitty' approach informed the writing of many European and extra-European floras and strongly challenged the Linnaean species concept.

The 'lumpers' were not ready to abandon the broad species concept and invoked the authority of Linnaeus. They considered that species were not made of individuals strictly moulded after one single fixed type but were rather a spectra of slightly variable forms clustered around the type. Some, like Joseph Decaisne (1807–82), used experimental cultivation as the criterion to delimit species broadly. Others, like Hooker, Bentham and A. de Candolle, valued the comparison of series of dried specimens as the major criterion for species delimitation.[45] Whether supporters of the morphological criterion or of the experimental criterion, all 'lumpers' worried about a 'wholesale manufacture of species' threatening the whole order of classification.[46]

The controversy on the right delimitation of species developed into numerous skirmishes relating to particular groups and places. The brambles (*Rubus*,

Rosaceaes) controversy was one of the hottest of them. It involved protagonists like A. Jordan, D. Godron, P. Müeller, O. Kuntze, C.C. Babington, G.A. Walker Arnott and G. Bentham. In 1829 Lindley had described twenty-four species of *Rubus*. In 1847, the second edition of Babington's *Manual of British Botany* discriminated thirty-six species. Professor of Botany in Cambridge, Babington created the most extensive British herbarium collection of brambles and grew forty forms that bred true in the botanical garden. On the Continent, Dominique Godron described forty-nine *Rubus* species, whereas the German P.J. Müller (1859) proposed 236 species in France and Germany. This last work seemed much too splitty to Babington, who described forty-five British species in 1869.[47] On the other side, lumpers like G.A. Walker Arnott and Bentham acknowledged no more than half a dozen species, disqualifying other forms as varieties of hybrids.[48]

Three British floras published nearly simultaneously around 1860 made the debate on the delimitation of species very visible in the UK. In their works, C.C. Babington described 1,708 species, W.J. Hooker and G.A. Walker Arnott no more than 1,571, and G. Bentham only 1,285. How could botanists working on the same country and having in principle access to the same material disagree so much?[49] Between five and 236 *Rubus* species, between 1,285 and 1,708 plant species in Britain: where was the truth, and what about the scientificity and certainty of systematic botany?

J.D. Hooker's crusade for the broadly circumscribed species

The impossibility of regulating species claims and reading consensus within the systematic community challenged botany as a science: objectivity (in its new aperspectival sense of observer-independent knowledge), order and cumulativity were at stake. How did leading professional taxonomists working on the biggest herbaria in the world react to these challenges?

Imperial expansion and the development of public herbaria

The nineteenth century saw the rise of the big public natural history collection in general, and of herbaria in particular, as they progressively absorbed and largely superseded private cabinets, and directly acquired holdings as European expansion advanced.[50] The Muséum National d'Histoire Naturelle's herbarium, though instituted as a national collection in 1793, superseded private French collections only in 1857 with one million specimens after Adrien de Jussieu left his herbarium to the Muséum. After absorbing some more private collections and benefiting from French colonial expansion, it contained three million specimens in 1907.[51] The Kew herbarium was established as a national collection in 1854, thanks to Bentham's gift of his cabinet (200,000 specimens), and extended in 1867 with W.J. Hooker's herbarium (already housed at Kew since 1853). By 1860, with 1.2 million specimens perfectly ordered and arranged in a convenient way, Kew already surpassed all other public and private herbaria in the world.[52]

Geneva's herbarium was smaller, though its value was increased by the presence of the specimens used for the writing of the de Candolles' *Prodromus*.[53] By 1898 Kew contained 3.3 million specimens and maintained its first position in the world, although Berlin's herbarium had dramatically increased in size after this city became the capital of the Reich in 1871 (it had four million specimens by 1913).[54]

As these institutions were establishing themselves as centres of a 'world botany', the few men who were in charge of them, or who had a daily access to their resources, such as Joseph Dalton Hooker and George Bentham at Kew, Alphonse de Candolle in Geneva, Asa Gray in Boston or Adolf Engler in Berlin, enjoyed increasing power and influence in the field of systematic botany. Though not agreeing on every matter, they had enough in common to impose their idea that species were variable entities and should be broadly circumscribed on the rest of the botanical community. Among these men, Joseph Dalton Hooker certainly played the leading role.[55] After two decades dominated by splitting trends, in the 1850s he took the lead of a counter-offensive that promoted and imposed the broad species concept.

Hooker's crusade against 'species-mongers'[56]

The son of Sir William J. Hooker, the powerful director of Kew Royal Botanic Gardens and Herbarium, Joseph D. Hooker had from his early years access to the huge resources provided by his father's private herbarium and to his extensive social relations. The latter allowed him to be assistant surgeon and naturalist on the Antarctic expedition from 1839 to 1843 and to make intensive explorations in India and the Himalayas from 1848 to 1851. Aged only 34 when he came back from India, he was one of the most 'travelled' botanists in the world and one of those who had experienced the greatest diversity of plants of the globe, both in the field and in the herbarium.[57] His conception of species, as being more variable and more broadly delimited than most of his contemporaries thought, had already emerged in the 1840s and solidified around 1851–2 while he was working on his floras of New Zealand and of India. In 1853 he estimated that the total number of known species of plant was no more than 50,000, instead of the 70–100,000 currently alleged by others.[58] In his view, broad species were necessary to make classification more sound and stable. As a brilliant plant geographer, a broad delimitation of species was also necessary to draw meaningful comparisons between floras of different regions and explain the distribution of plants (how to study biogeographic relationships between Arctic regions or alpine areas from tropical or temperate regions if similar forms in these different regions are given different names?).

The promising botanist soon felt himself a crusader against 'species-mongers'.[59] 'What with De Vries, Klotzsch, and Steudel we shall have Phanerogamic Botany messed like *Algae*, except we show a bold front', he wrote to Harvey in 1852.[60] Alluding to the *Flora indica*, he wrote in 1853 to Bentham:

> It has . . . been impossible to avoid doing battle with all our predecessors' species, whose utter disregard of one another and of any other part of the world has produced inextricable confusion in many cases . . . I admire your great caution and desire to curb my rabid radicalism: but the tide will turn one day and the reducing of species will go on apace . . . I am a *rara avis*, a man who makes his bread with specific botany . . . What is all very pretty play to amateur Botanists is death to me.[61]

In what J.D. Hooker himself viewed as a fight to the death, the *Florae novae-zelandicae* (1853) and the *Flora indica* (1855) were his first weapons directed against splitters. Their aim was to 'establish the genera and species . . . on a sound and philosophical basis and to unravel their synonymy'.[62]

Within a few years (1853–60), Hooker had convinced most of the leading botanists in the world that it was time to 'show a bold front' against splitters.[63] In France, Joseph Decaisne led the crusade against Jordan's followers. After the *Flora indica* had appeared, Decaisne had first reacted against Hooker's 'lumping' views.[64] A few years later he revised his opinion:

> As many of my colleagues, I had more or less shared this narrow vision of species, but time and experience have modified my opinion, and if I had to write again now the monograph of Plantaginaceae [a family he had treated for the Candolles' *Prodromus*] I would not hesitate to cut down the number of species more than I did.[65]

Hooker may well have influenced in a similar way Alphonse de Candolle. In the early 1850s, although opposed to Jordanian species, de Candolle had not hesitated to publish in his *Prodromus* rather 'splitty' monographs (for instance Dunal's one on Solanaceae in 1852). But he had partly changed his mind by 1860. His 1862 monograph on the *Quercus* genus re-established the specific unity of the European oak tree (*Quercus robur*), established by Linnaeus but split into several species by recent authors.[66]

Installed at Kew with his herbarium in 1854, Bentham also seems to have changed his views after contact with Joseph D. Hooker. Following the publication of his *Handbook of the British Flora* in 1858, which accepted 25 per cent fewer 'good species' than Babington's *Manual of British Botany*, Bentham attacked Babington and strongly objected 'to the elevation to specific rank of forms which traced over a sufficiently wide area are found to be but local or transitional modifications of a species'.[67] J.D. Hooker commented in 1858:

> Bentham's late researches into the British Flora have so greatly modified his views of the limits of species . . . He has completed the MS. of his British Flora having studied every species from all parts of the world, and most of them alive in Britain, France and other parts of Europe. Well – he has turned out as great a lumper as I am! and *worse*.[68]

These crusaders of the broad species concept charged the splitters with killing botany as a science. First they were charged with burdening it quantitatively with new names (Jordan saw in *Draba verna* no less than 200 species!). As Decaisne wrote:

> it would be a great acquisition if the describers of plants would condense their species by reducing them to really stable and natural types instead of dividing and multiplying them *ad infinitum*, as has been the custom for the last 30 years. This opinion is not exclusively my own; it is also that of my excellent friend Dr. J.D. Hooker (Flora Indica, Introductory Essay, etc.), and even that of most serious monographers, who feel instinctively that this way . . . will sooner or later lead to chaos, which would be the death of science.[69]

The narrower delimitation of species was not only criticised as a source of disorder. It was dismissed as 'of no use to the botanist, general or special' and of no philosophical meaning.[70] It was just not science. Splitters were therefore seen as responsible for lowering the status of systematic botany. Their 'puerilities', as Hooker and Thomson put it, repelled the ablest and brightest students, and 'being abandoned by many of those who are best qualified to do it justice, [systematics] fall into the hand of a class of naturalists, whose ideas seldom rise above species, and who, by what has well been called *hairsplitting*, tend to bring the study of these into disrepute'.[71]

The metrological solution: Kew, the Empire, the colonial floras, the *Genera plantarum*, and the *Index kewensis*

How did Hooker and his allies impose their view on the field of systematics? Bentham and J.D. Hooker started in 1858 to work together on a *Genera plantarum*. Bentham had always been reluctant to any joint authorship. Furthermore, he and J.D. Hooker had then different views on species origins (Hooker being converted to evolution by his friend Darwin, whereas Bentham was still fixist). It is only because they shared the same vision on the practical delimitation of species, and on the necessity to reorder systematics, that Bentham agreed to get involved in such a co-authorship.[72] Bentham furthermore agreed to write two floras (Hong Kong and Australia) in Hooker's colonial floras project. This alliance with Bentham allowed Kew to produce decisive enterprises in the crusade against 'species-mongers': the colonial floras, the *Genera plantarum* and, later, the *Index kewensis*. To stabilise taxonomic knowledge and impose their new order on systematics, Hooker and Bentham lacked most of the tools of a perspectivity that transformed other branches of science in the century. But they had the Empire and they instituted Kew as an imperial metrological centre of a world botany.

William J. Hooker took over the direction of Kew Gardens in 1841. Within a few decades he and his son transformed Kew from a princely garden and aristocratic park into Britain's leading botanical institution and a key 'tool of Empire'.[73] Sir William J. Hooker thought of instituting Kew as the botanical advisory centre for Victorian imperial expansion. He worked to reactivate the Banksian colonial network and created in 1849 a Museum of Economic Botany at Kew. He also spearheaded plant-hunting overseas explorations and plant transfer projects. The successful transfer of Cinchona from South America to India in the period 1859–61 was one of these global enterprises. These successful efforts to integrate Kew in the imperial machinery established it as the botanical clearing house for overseas economic enterprises. Kew was at the centre of an active correspondence between London and overseas colonial officers, travellers, planters and missionaries. It served as the head of the network of colonial botanical gardens and a training centre for plant hunters and gardeners. This also led to the centralisation at Kew of botanical collections resulting from the new wave of exploration and surveying that accompanied mid-Victorian imperial and commercial expansion.[74] Initiated with the installation of Bentham's and Hooker's herbaria, Kew's herbarium contained by 1860 more than one million well-arranged specimens supplemented by an extensive library, and was the premier of such institutions in the world.

This allowed the Hookers to launch a wide project to publish colonial floras. 'The want of them', claimed Sir William, 'is a great obstacle to the development of the productive resources of the colonies.'[75] A first step in this direction came in 1851 when W.J. Hooker obtained from the Prime Minister a grant for his son to classify his Indian collections and write a flora. But Hooker and Thomson's *Flora indica* stopped after the first volume, partly for lack of support. In 1857, Sir William obtained support from the Colonial Office for a *Flora of the West Indies*. One year later the green light was given for a flora of Australia. After he became assistant-director to his father at Kew in 1855, J.D. Hooker worked hard on a general plan to publish a flora for each British possession. Through his extensive relations in metropolitan and imperial political circles, W.J. Hooker obtained funding for this general project in 1860. These floras were published very actively in the following decades: *Flora of the British West Indian Islands* (3 vols., [1859]–64), *Florae hongkongensis* (1861), *Flora capensis* (3 vols., 1859–65), *Florae australiensis* (7 vols., 1863–78), *Flora of Tropical Africa* (3 vols., 1868–77), and *Flora of British India* (7 vols., 1872–97).[76]

As well as the Cinchona project, the colonial floras project might be thought of as 'big science'. More than 12,000 pages were published in less than three decades. To secure quick results and wide diffusion, the Hookers broke with the tradition of prestigious, expansive, richly illustrated – and seldom completed – volumes in quarto size. They wanted 'good, but inexpensive, scientific works on the Vegetable productions of the British Colonies' that could be finished in a reasonable time. They had to be 'thoroughly trustworthy in a scientific point of view, and yet not so exclusively scientific in method and language as to be useful to the professed man of science only'.[77] The floras were concise, standard in their presentation, and published in octavo, without illustration. This choice distinguished the colonial floras from the two other contemporaries' big enterprises in systematic botany: the Candolles' *Prodromus* (1824–73) – designed to replace Linnaeus's *Species plantarum* – and the *Flora brasiliensis* (1840–1906) edited successively by Martius, Eichler and Urban.[78] Both took half a century to complete, were written in Latin and were too expensive to reach a wide audience outside professional botanists.

'To secure uniformity of plan', Bentham's *Florae hongkongensis* served as the model for all the following floras.[79] They were all preceded by the same 'outlines of botany' written by G. Bentham to educate local botanists. These outlines were a revision of those given in 1858 by Bentham in his *Handbook of British Flora.* Stating that 'species vary within limits which is [*sic*] often very difficult to express in word', it documented different aspects and causes of variation (such as the influence of dryness on pubescence, and the tendency of the ratio size of flower/size of leaves to increase under light, bright and open conditions, etc.) and presented different cases of accidental aberrations 'which the botanist must always be on his guard against mistaking for specific distinctions'.[80] Such a 'lumpist' manifesto was to be found at the beginning of all colonial floras and helped to shape the views of generations of local amateur and professional botanists.

The *Genera plantarum* was a second mammoth enterprise to reorder plant taxonomy, which Bentham and Hooker undertook from 1858 to 1882. The general format was decided after consulting 'botanical friends in whose judgement we had great confidence' (such as Gray, A. de Candolle, etc.).[81] Although its focus was the groupings at the genus level and above, it had an effect on discussions at the specific level, since the number of species given for each genus tended to be low, in accordance with the broad species concept.

The *Index kewensis* was the third tool that helped both to establish Kew's position as the world centre for systematic botany and to standardise taxonomic practices at the generic and specific level. Started because of a bequest by Charles Darwin, it was undertaken under Sir Joseph's direction in 1885 to provide an index of all ('good') species names. This huge work was completed ten years later, and required the collaboration of no less than twelve librarians and botanists. It has been regularly revised since and remains an essential working tool for botanists.[82] More than a mere index, it was, in its initial volumes (unlike Steudel's earlier index), a guide to correct names, stating clearly which names were to be accepted and which names ('synonyms' or 'bad species' being only a variety of the considered species) were to be discarded.[83] As we shall see later, the *Index kewensis* supported nomenclatural practices which reinforced the 'broad species' standard.

Initiated in the wake of J.D. Hooker's crusade against narrowly delimited species, the colonial floras, the *Genera plantarum* and the *Index kewensis*, establishing species and generic limits on a world-wide basis, turned out to be very successful and powerful weapons. These huge publications provided compelling standards that imposed the broad species concept in practice, and disciplined beginners', local botanists' and travellers' taxonomic practices. In the last third of the nineteenth century they helped to repel narrow species delimitation outside the boundary of dominant professional botany and to reinvent the Linnaean species that remain at the basis of systematics today.[84] 'I have no doubt of the full and entire correctness of the principles you work on and the Kew Floras and the Gen.

Plantarum will more than anything else determine the public botanical opinion and mode of working for the next generations', rightly wrote Asa Gray to Bentham.[85] The Candolles' *Prodromus* also helped to introduce some standard practices. But J.D. Hooker's estimate of the number of species was even lower than that of the Candolles. Furthermore, the *Prodromus* lacked the unity of purpose that informed J.D. Hooker's enterprises: families were treated by a diversity of contributors, some of them having a rather splitty approach. Finally, the *Prodromus*'s audience outside the select circle of full-time botanists was much smaller. On the other hand, the colonial floras were more readily accessible to a greater number of amateur practitioners and contributed to discipline their practices. Many colonial actors (colonial officers, missionaries, military men, etc.) learned the basics of botany while herborising in their outposts and using a Kew Colonial Flora, hence assimilating through 'hands on' experience, rather than through explicit statements, the practice of delimiting species broadly.

Regulating the birth and names of species

Monographs (like those comprised in the *Prodromus*), great floras, and other major publications (*Genera plantarum*, *Index kewensis*, for example) not only promoted the principle of delimiting species (and genera) broadly but also contributed in a very concrete way to the elimination from botanical literature of 'bad species' (seen as mere varieties of broadly circumscribed species) and of 'bad names' (that is, synonyms: names given to a plant which represented a species already described by an earlier author).

The Candolles' *Prodromus* (1824–73) described 58,975 species in 5,134 genera of dicotyledons. From 1862 to 1883, Bentham and Hooker's *Genera plantarum* estimated 95,600 species (monocotyledons and dicotyledons) in around 7,600 genera.[86] That means that – if one takes 1848 and 1872 as the median years of these publications – within a quarter of the century in which many new specimens reached European herbaria in the wake of expansion, botanists could consider with satisfaction that the number of genera was more or less stabilised (which helped the stability of the classification system), but had to accept that the number of known species was still rapidly increasing. 'Birth control' was called for by leading botanists. The leading professional botanists, who had access to the large libraries and herbaria of imperial metropolises, shared similar views on species delimitation and a similar taste for order. In their monographs and their contributions to colonial floras, or the *Prodromus* or *Die natürlichen Pflanzenfamilien*, they severely 'cut down' species proposed by their predecessors, by travellers or by local botanists. In the *Index kewensis*, which compiled all species names, decisions about which species were 'good' and which not were largely based on these monographs (hence legitimating their taxonomic stance), rather than on the minor publications in which travellers and local botanists published. This 'demographic control' exerted by professional botanists in mainstream publications is illustrated in Table 8.1 for some genera (*Erophila*, *Rubus* and *Cirsium*) that had been bones of contention between lumpers and splitters, and for genera of

Table 8.1 Regulating species proliferation in Kew publications: some examples

	*Erophila** (Cruciferae)	*Rubus* (Rosaceae)	*Cirsium* (Composeae)	*Cinchona* (Rubiaceae)	*Landolphia*† (Apocyneae)
Date considered	1885	1885	1885	1885	1902
Proposed species names to this date	≥200	≥1540	≥519	≥205	≥42
Main actors responsible for the numerous species claims in the botanical and travel literature	Jordan	Babington	Nägeli	Pharmacists and several travellers to South America	Several travellers to Africa
Validated species names (as validated by an authoritative Kew publication)	5 (2.5%) (*Index kewensis*)	±670 (43.5%) (*Index kewensis*)	146 (28%) (*Index kewensis*)	67 (33%) (*Index kewensis*)	26 (62%) (*Fl. of Trop. Africa*)

* *Erophila* encompassed the 200 species from *Draba verna Linn.* proposed by Jordan, since *Erophila vulgaris* DC had been collated with *Draba verna Linn.*
† Only species from tropical continental Africa have been counted.

overseas plants (*Cinchona*, *Landolphia*) where monographers repressed travellers' tendency to multiply species.

The controversy between lumpers and splitters on the right circumscription of the species did not reach a closure through any decisive observation or experiment. It was decided in practice by the domination of huge and authoritative publications that influenced most monographers and flora writers in the late nineteenth century. In practice, 'good species' were therefore what 'competent' botanists said they were, and mostly they could create or validate a new species. Closure in general was underdetermined by empirical properties and was secured by the authority of persons, texts and institutions whose power was built on the opportunities provided by travel and empire.

Leading systematists also came to impose names and nomenclatural practices which reinforced – and were reinforced by – the standard of broadly delimited species. The clearest and most famous example is the nomenclatural practice that came to be referred to as the 'Kew rule'. It consisted in considering as valid a binomial name currently in use, even if it had been formed in contradiction with the traditional principle of priority – for instance, when a monographer changed the specific epithet when moving a species from one genus to another.[87] Most late nineteenth-century herbarium botanists, especially in Berlin, Kew and Harvard, followed the Kew rule.[88] In the late nineteenth century the *Index kewensis* was an extremely powerful tool to impose the Kew rule *in practice*. 'Our practice is to take the name under which any given plant is placed in its true genus as the name to be kept up, even though the author of it may have ignored the proper rule of retaining the specific name [i.e. epithet], when transferring it from the old genus to the new', declared B. Daydon Jackson, the compiler of the *Index kewensis*.[89] The Kew rule remained in use in the *Index kewensis* until the third supplement (1901–5) published in 1908, and represented a way by which the *Index kewensis* policed the field of statements in systematics.

The Kew rule, applied in the colonial floras and the *Index kewensis*, tended to favour monographers' work over that of the first describers (often local or colonial botanists, or 'splitters'). For species that were removed to another genus, the original epithet could be changed and the name of the original author would cease to be associated with it as authority. For instance an Apocyneae described by the French botanist Louis-Pierre as '*Ancyclobotryspyriformis* Pierre' in 1899 became '*Landolphia pyriformis* Stapf' when Stapf revised the family for the *Flora of Tropical Africa*.[90] The Kew rule was hence to conform to the kind of taxonomic order Hooker and Bentham wanted to institute. Meanwhile, it had the advantage of reducing bibliographical work to find the correct name of a plant.[91] 'It is only second rate botanists who pride themselves of the number of names, good or bad, to which their initials can be attached', said Bentham.[92] A leading German monographer, Hans Hallier, arguing for the Kew rule, made these points even more explicitly:

> I consider the principle of priority only valid so far as it gives rise to a stable and uniform nomenclature. Therefore I follow the conservative, factual, not-depending-on-the-person, and conditional principle formulated by

> Celakovsky and the Kew botanists [i.e. the Kew rule], rather than the proliferating, depending-on-the-person, and unconditional principle, which was in germ in De Candolle's nomenclatural rules, and which, in its recent inconsiderate mode of application by O. Kuntze and the Americans [i.e. those advocating an integral and retroactive rule of priority], became the progenitor of an overwhelming proletariate of synonyms.[93]

In the mind of Hallier and many of the leading herbarium botanists, the Kew rule was therefore seen as a 'violence légitime' against old customs of priority, necessary to save the taxonomic order from a 'proletariat' of 'bad species' proposed by lower botanists naming too many supposedly new species, this being prompted by the seeking of glory.[94] The steadiness of systematic knowledge, they thought, depended on discarding such impure desires.

Standards: ordering botany and the Empire

What was at stake in this method of manufacturing 'good' (i.e. broad) species was no less than the imposition of an order in the field of botanical knowledge. More precisely, a kind of order in which 'competent' botanists would be able to work, in which their statements would immediately have more visibility and more authority than other statements from various other actors. Only this, claimed the leading professional botanists, would save botany from chaos. Joseph D. Hooker, well aware that the involvement in botany by very diverse social groups with differentiated practices and motivation was threatening the *ancien régime*'s cognitive order of botany, complained about 'beginners refusing to accept the conclusion arrived at by abler botanists', who 'may pause before venturing to institute a genus, [but] it rarely enters into [their] head to hesitate before proposing a new species'. He continued: 'A knowledge of the relative importance of characters can only be acquired by long study; and without a due appreciation of their value, no natural group can be defined.'[95] Contesting both travellers' and pharmacists' legitimacy to produce 'good' systematic knowledge, he claimed that knowledge, however intimate, of one region or of one group of plants was insufficient 'to enable an observer to pronounce upon what characters are of specific importance in that group'.[96] William T. Thiselton-Dyer, J.D. Hooker's son-in-law and successor as director of Kew Gardens, went even further in this boundary work. Giving up the ideals of the republic of letters for that of the domination of a few herbarium botanists, he considered that 'all that can be hoped is a general agreement amongst the staffs of the principal botanical institutions in different countries where systematic botany is worked at; the free-lances must be left to do as they like'.[97]

Local botanists – i.e. botanists with a 'local' outlook, whether at home or in the colonies – and 'species mongers' could well 'discover' hundreds of new species, while leading closet naturalists undertook to control the validation of these species and deny them any contribution to systematics. That was precisely what the broad species concept did. It required wide comparisons in large collections. It contributed to the invention of the 'local botanist', the 'amateur botanist' and of the boundaries that separated them from 'competent' professional botanists. J.D. Hooker revealed this motivation to Asa Gray in 1854: 'You say that we are not to pronounce species the same because they are united apparently by certain forms of each – I grant this fully, but how are we to act upon it & deny local botanists specific value to their small fish?'[98]

Some professional herbarium botanists dreamed of going further, calling for a greater centralisation of botanical publications and a greater control on descriptions of new species and genera.[99] At the International Congress of Botany in 1900, Henri Hua, a botanist at the herbarium of the Paris Muséum, even proposed the creation of a single journal where all new species should be described so as to be accepted by the community.[100]

Kew Gardens and a few other large institutions relegated the splitters and the local (domestic or colonial) botanists to subaltern positions. They imposed the way that legitimate taxonomic evidence had to be administered, favouring the comparison of dry forms over experimental or field criteria. They turned their herbaria into the centres of the order they imposed on the field of taxonomy. They constituted themselves as the metrological centre of the systematic community and attempted to regulate both nomenclature and the species concept. The order of botany needed standards, and standards were necessary for the continuing prosperity of the British Empire. The colonial floras, claimed Joseph Hooker in 1861, 'are indispensable for supplying that fixed nomenclature for their [the colonies'] plants, without which it is impossible for [Kew botanists] or the colonists to carry on a correspondence on these and kindred subjects'. In Sir Joseph's view, the economic expansion of the British Empire and the orderly progress of systematic botany went hand in hand, and both depended on standard names and standard (broad) species delimitation practices.[101] For such economic plants as cinchona, rubber, gutta-percha or timber, to link in a reliable way a botanical name, a description of the plant, and an estimate of the commercial value of the rubber, was essential for the development and stabilisation of imperial trade. Such knowledge was also crucial in determining which of the wild plants could be better cultivated in plantations. For instance, the question whether Assam and Chinese tea belonged to the same species or not was crucial for the development of the Indian tea industry. In Africa many (mostly German) settlers and capitalists wasted a lot of time and money in trying to grow African rubber plants in their plantations instead of the Amazonian *Hevea brasiliensis*. Similarly, Dutch cinchona plantations in Java in the 1850s and 1860s failed because they had (contrary to the British!) introduced a worthless species from South America (but they later took the lead with *Cinchona ledgeriana*!). These famous examples were called for by Kew botanists as illustrations for the need of taxonomic accuracy and a 'central standard of reference', which only a large botanical metrological institution could assure.[102]

Many scholars have depicted Kew as the botanical centre of power within the British Empire, controlling access to career opportunities (especially important in a context where botany was not highly professionalised) and giving leading botanists like the Hookers, Bentham and Thiselton-Dyer the power to impose upon local botanists in the periphery a strict division of work.[103] As we have seen, much more was at stake than the mere institutional hegemony of Kew in the British Empire. It would be an error to interpret the denial to local botanists of the right to institute new species and genera as a mere domination exerted on the field by gate-keepers (herbarium botanists) who monopolised in this way symbolical capital (as in the case of G. Bentham's authorship for the *Florae australiensis*, which overshadowed the merits of the local botanist Ferdinand von Mueller who did much work on it). For Kew botanists, ordering the 'chaos' of botanical knowledge was, for sure, a sincere concern, if not a mission they assigned to themselves. Their ideal was a particular ideal of order, communicability and cumulativity. The taxonomic order they established was in their view necessary to allow accumulation of data and to increase systematic botany's philosophical value (and a greater use for phytogeographical studies), as well as its role within the Empire.

Interestingly, the standards (broad species, taxonomic proof based mainly on observation of dry specimens, Kew rule) which ruled taxonomic practices between 1860 and 1900 became highly contested in the early twentieth century. First there was, in the wake of genetics, a revival of interest in the 'microspecies' or 'jordanons'. Second, ecologists' and experimental taxonomists' experimentally oriented studies, along with those in the field, came to be seen as of greater general and theoretical interest than herbarium studies.[104] Finally, the principle of priority of the original specific epithet was progressively re-established after 1905 in successive international nomenclatural codes.[105] These early twentieth-century challenges found their basis in the changing institutional structure in systematics: the metrological power exerted by botanists at large herbaria such as those of Kew was challenged by those at other (usually smaller and younger) herbaria or laboratories whose scientific agenda focused on the studies of local patterns of variation and field characters.

Acknowledgements

This chapter owes much to Peter Stevens's work and advice, and I thank him gratefully. I also thank Hariet Ritvo, Simon Schaffer and Kapil Raj for their comments on earlier versions. Funding for this research has been provided by the Max Planck Institute for the History of Science in Berlin.

Notes

1 G. Broberg, 'The broken circle', in T. Frängsmyr, J.L. Heilbron and R.E. Rider (eds) *The Quantifying Spirit of the Eighteenth Century*, Berkeley: University of California Press, 1990, pp. 45–71, p. 57.

2 J.E. Lesch, 'Systematics and the geometrical spirit', in Frängsmyr, Heilbron and Rider, op. cit., pp. 73–111.

3 C. Linné, *Species plantarum . . .* , Stockholm: Slavius, 1754. This enabled leading botanists to know and memorise nearly all the genera.

4 A.-P. and A. de Candolle (eds) *Prodromus systematis naturalis historia* . . . , Paris, 1824–73, 17 vols, vol. 17, 1873, p. 313; G. Bentham and J.D. Hooker, *Genera plantarum*, 3 vols, London, Lovell Reeve, 1862–83; P.F. Stevens, *The Development of Biological Systematics: Antoine-Laurent de Jussieu, Nature and the Natural System*, New York: Columbia University Press, 1994, pp. 208, 477.

5 G. Broberg, 'The broken circle', p. 66.

6 H.C. Watson, 'Letter to the editor on the relation between cerebral development and the tendency to particular pursuits – and on the heads of botanists', *The Phrenological Journal*, vol. VIII, 1833, no. XXXV, 97–108. The article also introduces a hierarchy between mere collectors and descriptors, and more philosophically oriented botanists whose 'reflective organs' are more developed.

7 S.M. Walters, 'The name of the rose: a review of ideas on the European bias in angiosperm classification', *The New Phytologist*, 104, 1986, 527–46, pp. 535–7; P. Raven, B. Berlin and D.E. Breedlove, 'The origin of taxonomy', *Science*, 174, 1971, 1210–13.

8 While refraining Darwin from overinterpreting in a realist way the distribution of varieties into species, J.D. Hooker hinted at this bias and stressed the role of 'psychological factors' in delimitating species and genera: 'botanists . . . do not treat large & small genera equally and similarly, & the sum of inequalities thus produced tends to make the species of small genera look more invariable than that of big' (Hooker to Darwin, 14 March 1858, in F. Burkhard and S. Smith (eds) *The Correspondence of Charles Darwin*, Cambridge: Cambridge University Press, vol. 7, 1991, p. 49). He stated that G. Bentham 'will make rather hastily a new species in a large genus, of which a vast number of good species have recently turned up', rather than in a small genus (Hooker to Darwin, 13–15 July 1858, in ibid., p. 131).

9 E. Spary, *Utopia's Garden: French Natural History from Old Regime to Revolution*, Chicago and London: Chicago University Press, 2000.

10 For instances of this diversity see A. Secord, 'Science in the pub: artisan botanists in early nineteenth century Lancashire', *History of Science*, 32, 1994, 269–315 (on artisan botany); Idem, 'Corresponding interests: artisans and gentlemen in natural history exchange networks', *B.J.H.S.*, 27, 1994, 383–408; A.B. Shteir, *Cultivating Women, Cultivating Science*, Baltimore, Md. Johns Hopkins University Press, 1997 (on women in botany); H. Ritvo, *The Platypus and the Mermaid and Other Figments of the Classifying Imagination*, Cambridge, Mass., Harvard University Press, 1997 (on popular zoologies); see also D.E. Allen's path-breaking work, *The Naturalist in Britain. A Social History*, Harmondsworth: Penguin Books, 1976.

11 H. Ritvo, 'Zoological Nomenclature and the Empire of Victorian Science', in B. Lightman (ed.) *Victorian Science in Context*, Chicago: University of Chicago Press, 1997, pp. 335–53.

12 An example of eighteenth-century optimism is J.E. Smith, 'Introductory discourse . . .', *Linn. Soc. of London. Transactions*, I, 1788, 1–55: 'He who determines with certainty a single species of the minutest moss or meanest insect, adds so far to the general stock of knowledge, which is more than can be said of many a celebrated name'; cf. Broberg, 'The broken circle'.

13 J.D. Hooker and T. Thomson, *Flora indica*, London: W. Pamplin, 1855, p. 11; see also a similar statement by B. de Saint-Vincent in 1832 in *Expédition de Morée. Section des Sciences Physiques*, t. III-2, *Botanique*, par MM. Fauché, Brongniart et Bory de Saint-Vincent, Paris: Levrault, 1832, p. 7.

14 To write a monograph one has to find all the original descriptions for all species. On the 'difficulty of obtaining access to the necessary periodicals [that will soon] render the effectual study of botany impossible', see J.D. Hooker and T. Thomson, *Flora indica*, op. cit., p. vi; see P.F. Stevens, 'J.D. Hooker, G. Bentham, A. Gray and F. Mueller on species limits in theory and practice: a mid-nineteenth-century debate and its repercussions', *Historical Records of Australian Science*, 11, June 1997, 345–70, pp. 354–5.

15 J.D. Hooker and T. Thomson, *Flora indica*, op. cit., p. 12; H. Hallier, 'Über Kautschuklianen und andere Apocyneen', *Jahrbücher der Hamburgischen wissenschaftlichen Anstalten*, XVII, 1900, 3: Beiheft (Arbeiten des botanischen Museums), 19–216, p. 19.

16 A.M.F.J. Palisot de Beauvois, *Flore d'Oware et de Benin*, 2 vols (t. I 1804, t. II 1807), t. I, pp. 54–5, tab. XXXIV (description in reality published in 1806). The type specimen is in Delessert's herbarium, now in Geneva.

17 A. de Candolle, 'Apocynaceae', in *Prodromus systematis naturalis regni vegetabilis*, 1824–73, vol. 8, 1844, p. 320 for *Landolphia*, pp. 327–8 for *Vahea*; Bentham and Hooker, *Genera plantarum*, 'Apocynaceae' in vol. II, 1876, 681–728, see pp. 692–3.

18 Anon., 'Sir John Kirk', *Bull. of Miscellaneous Information Kew*, 1922, 49–63; National Portrait Gallery, *David Livingstone and the Victorian Encounter with Africa*, London: National Portrait Gallery Publications, 1996.

19 C. Barlow, S. Jarasuriya and C. Suan Tan, *The World Rubber Industry*, London: Routledge, 1994, table A.3.

20 To mention only West Africa: Chevalier to Soudan (1898–9), Dewèvre to Congo where he died in 1895, and the German 'Westafrikanische Kautschuk-Expedition' to Kamerun and Togo at the initiative of the *Kolonial-Wirtschaftliches Komitee*, 1899–1900. See E. de Wildeman, *Les plantes tropicales de grande culture*, Bruxelles, 1902, t. IV Café, cacao, cola, vanille, caoutchouc, pp. 7–8; R. Schlechter, *Westafrikanische Kautschuk-Expedition, 1899–1900*, Berlin, 1900.

21 There are many mentions of it in travel accounts from the 1860s: in particular studies (local, or focused on a few species, or of mainly economic orientation), often from colonial surgeons and pharmacists; in taxonomical monographs which peaked between 1895 and 1904, which were followed by an almost complete lack of interest. See C. Bonneuil, *Mettre en ordre et discipliner les tropiques: les sciences du végétal dans l'empire français, 1870–1940*, Ph.D. thesis, University of Paris 1997, pp. 118–20. The rubber boom led to gross mismanagement of African forests and depletion of the rubber plants, as well as the overexploitation and murder of African people (mainly in equatorial Africa).

22 See, for instance, the central importance of the characters related to the pod in the classification of Leguminosae whereas fruit characters are less important in other families.

23 J.S. Henslow, 'On the requisites necessary for the advance of botany', *Mag. of Zool. and Bot.*, I, 1837, 113–25, quotations from pp. 116, 120.

24 A wealthy amateur, cultivating more than 60,000 (Jordanian) species in his garden in Lyons.

25 See C. Bange, 'La culture et l'hybridation peuvent seules décider la question de l'espèce: les travaux de Jordan, Decaisne et Naudin assignent une nouvelle fonction aux jardins botaniques', in J.L. Fischer (ed.) *Le jardin entre science et représentation*, Paris: CTHS, 1999, pp. 317–29.

26 J. Decaisne, 'Note sur l'organogénie florale du Poirier, précédée de quelques considérations sur la valeur des caractères spécifiques', *Bull. Soc. Bot. France*, 4, 1857, 338–42, p. 339.

27 C.C. Babington, *The British Rubi: An Attempt to Discriminate the Species of Rubus found in the British Islands*, London, J. van Joost, 1869; A. Jordan, 'Notice sur diverses espèces négligées du genre *Asphodelus*, comprises dans le type de l'*Asphodelus ramosus* de Linné', *Bull. Soc. Bot. France*, 7, 1860, 723.

28 W. Herbert, 'On hybridization amongst vegetables', *J. Hort. Soc. London*, 2, 1847, 7; C. Naudin, 'Essai d'une monographie des espèces et variétés du genre *Cucumis*', *Ann. Sci. Nat., Bot.*, 4e série, 9, 1856, 5–87; J. Decaisne, 'Lettre de M. Decaisne à M. l'abbé Chaboisseau (avril 1860)', *Bull. Soc. Bot. France*, 7, 1860, 261–5, p. 262. See Decaisne's works on the 'polymorphism' of the pear tree (*Pyrus*) from 1852 to 1863, synthesised in J. Decaisne, *Le Poirier*, Paris, 1871.

29 Quoted by C. Bange, 'La culture', op. cit., p. 328.

30 C.C. Babington, *Manual of British Botany*, 5th edn, London: J. van Doorst, 1862, p. iv.
31 R. Wight and G.A. Walker Arnott, *Prodromus florae peninsulae indiae orientalis*, vol. I, London, 1834, p. xxi.
32 Asa Gray (1858) quoted by P.F. Stevens, *The Development of Biological Systematics*, op. cit., p. 123.
33 H.C. Watson to C. Darwin, 19 November 1856, in Burkhard and Smith, *The Correspondence of Charles Darwin*, vol. 6, p. 278.
34 L. Daston, 'Objectivity and the escape from perspective', *Social Studies of Science*, 22, 1992, 597–618; eadem, 'The moral economy of science', *Osiris*, 10, 3–4, 1995, 3–24, pp. 19–20; S. Schaffer, '"Astronomers mark time": discipline and the personal equation', *Science in Context*, 2, 1, 1988, 115–45.
35 Lefébure (1835), quoted by P.F. Stevens, *The Development of Biological Systematics*, op. cit., p. 212.
36 E. Cittadino, *Nature as the Laboratory. Darwinian Plant Ecology in the German Empire. 1880–1900*, Cambridge: Cambridge University Press, 1990, pp. 9–25.
37 C.F.P. von Martius *et al.* (eds) *Flora brasiliensis*, München, Wien, Leipzig, 1840–1906, 15 vols. The fifty-five contributors to this huge flora (20,733 pages) were paid.
38 C. Limoges, 'The development of the Muséum d'Histoire Naturelle of Paris, *c.* 1800–1914', in R. Fox and G. Weisz (eds) *The Organization of Science and Technology in France, 1808–1914*, Cambridge and Paris: Cambridge University Press and Maison des Sciences de l'Homme, 1980, pp. 211–40; C. Schnitter, 'Le développement du Muséum national d'histoire naturelle de Paris au cours de la deuxième moitié du XIXe siècle; "se transformer ou périr"', *Revue d'Histoire des Sciences*, 49, 1, janv.–mars 1996, 53–97.
39 Bonneuil, 'Mettre en ordre', op. cit., pp. 30–1. J. Decaisne, 'Etat de l'enseignement des sciences naturelles en France', 2 February 1879, Archives Nationales F17 3881.
40 See R.H. Drayton, 'Imperial science and a scientific empire: Kew Gardens and the uses of nature, 1772–1903', Ph.D., Yale University, 1993, pp. 190–211, for a detailed study of the British situation (published as *Nature's Government: Science, Imperial Britain, and the 'Improvement' of the World*, Yale University Press, 2000); see also P.F. Stevens, *The Development of Biological Systematics*, op. cit., ch. 7; A.B. Shteir, *Cultivating Women*, op. cit., 1997.
41 Like C.C. Babbington (1808–95, Professor of Botany in Cambridge in 1861) and many local botanists in England; Alexis Jordan (1814–97) and Alexandre Boreau (1803–75), and many local botanists in France; Carl Wilhelm von Nägeli (1817–91) in Switzerland and Munich; Johann F. Klotzsch (1805–60), keeper of Berlin's herbarium (he had studied Himalayan plants before Hooker); Ernst G. von Steudel (1783–1856), physician in Württemberg.
42 Such as J.D. Hooker and G. Bentham in England and J. Decaisne (1807–82) in France.
43 C.W. von Nägeli's first attack on the Linnaean broadly circumscribed species is his Ph.D. thesis dissertation from 1841 ('Die Cirsien der Schweiz', *Neue Denkschriften der Allgemeinen Schweizerischen Gesellschaft für die gesammten Naturwissenschaften*, 5, 1841, viii + 170 pp.), and was followed by studies of algae.
44 A. Jordan, *Observation sur plusieurs plantes nouvelles ou critiques de la France. Premier fragment*, Paris, J.-B. Baillière, 1847. See also idem, 'Remarques sur le fait de l'existence en société, à l'état sauvage, des espèces végétales affines, et sur d'autres faits relatifs à la question de l'espèce', in *Compte-rendu du Congrès de l'Association Française pour l'Avancement des Sciences*, 2e session, Lyon, 1873, 488–505. On Jordan and l'école analytique, see C. Bange, 'La culture', op. cit., and idem, 'Alexis Jordan et les partisans de l'École analytique devant les hybrides végétaux spécifiques', unpublished paper kindly communicated by the author.
45 On J.D. Hooker's views on variation of species and the effects of hybridisation, see *Flora indica*, op. cit., pp. v and 19–36. See also anon. [G. Bentham], '*The British Flora,*

8th ed. 1860 by W.J. Hooker and G.A. Walker Arnott, *Manual of British Botany*, 5th ed. 1862 by C.C. Babington, *Handbook of the British Flora*, 1858 by G. Bentham', *The Natural History Review*, 10, 1863, 34–40, p. 38.

46 Anon [G. Bentham], '*The British Flora*, 8th ed. 1860 by W.J. Hooker and G.A. Walker Arnott, *Manual of British Botany*, 5th ed. 1862 by C.C. Babington, *Handbook of the British Flora*, 1858 by G. Bentham', *The Natural History Review*, 10, 1863, 34–40, p. 38.

47 C.C. Babington, *The British Rubi*; D. Godron, 'Le genre Rubus considéré du point de vue de l'espèce', *Mémoires de la Soc. Royale de Nancy*, 1849, pp.

48 W.J. Hooker and G.A. Walker Arnott, *The British Flora*, 8th edn 1860; G. Bentham, *Handbook of the British Flora*, 1858, pp. 188–91.

49 Anon. [G. Bentham], '*The British Flora*, 8th ed. 1860 by W.J. Hooker and G.A. Walker Arnott, *Manual of British Botany*, 5th ed. 1862 by C.C. Babington, *Handbook of the British Flora*, 1858 by G. Bentham', *The Natural History Review*, 10, 1863, 34–40.

50 This development was made possible by state support to botanical gardens and museums which also had to answer to a growing demand for education from the bourgeoisie and the middle classes. See P.L. Farber, 'The transformation of natural history in the nineteenth century', *J. Hist. Biol.*, 15, 1982, 145–52; S.G. Kolhstedt, 'Essay review: Museums: revisiting sites in the history of the natural sciences', *J. Hist. Biol.*, 28, 1995, 151–66; S. Sheet-Pyenson, *Cathedrals of Science. The Development of Colonial Natural History Museums During the Late Nineteenth Century*, Kingston and Montreal: McGill-Queen's University Press, 1988.

51 Statistics on the Paris herbarium can be found in C. Bonneuil, 'Mettre en ordre', op. cit., p. 42 *bis*.

52 Archives of the RBG Kew Pr 96-007 R. Desmond research material 37 and Archives of the RBG Kew PRO 21. The Paris herbarium could compare in richness, but not in care, organisation and facility of work.

53 Probably a little more than 500,000 specimens in 1850 (A.-P. de Candolle and Delessert's herbaria taken together).

54 Asa Gray's collection at Harvard was small by comparison, although of considerable importance.

55 Alphonse de Candolle deserves more attention than I could pay him in this chapter.

56 This section is much indebted to P.F. Stevens, 'J.D. Hooker, George Bentham'.

57 On Hooker, see W.B. Turill, *Joseph Dalton Hooker, Botanist, Explorer and Administrator*, The Hague, 1963; M. Allan, *The Hookers of Kew, 1785–1911*, London, 1975; anon., 'Sir Joseph Dalton Hooker. 1817–1911', *Bull. of Miscellaneous Information Kew* (1912), 1–14; Drayton, 'Imperial science', 267–368. On J.D. Hooker in the field, see J. Camerini, 'Remains of the day: early Victorians on the field', in B. Lightman (ed.) *Victorian Science in Context*, Chicago: University of Chicago Press, 1997, pp. 354–77.

58 L. Huxley, *Life and letters of Sir Joseph Dalton Hooker*, vol. 1, London: Murray, 1918: 473.

59 J.D. Hooker to A. Gray, 26 January 1854, in ibid., p. 474.

60 Letter to Harvey, 28 March 1852, in ibid., p. 458. Hooker expressed here both his opposition to splitting and to taking microscopical characters too much into account in classification.

61 Hooker to Bentham (s.d. 1853), in ibid., pp. 471, 473.

62 J.D. Hooker, *The Botany of the Antarctic Voyage . . . II Florae novae-zelandicae*, London: Lovell Reeve, 1853–5; Hooker and Thomson, *Flora indica*, p. v.

63 Of course these botanists also arrived at positions similar to Hooker's partly independently, in the context of specific debates (in the families they studied, or in issues debated in their national contexts) they were engaged in. Decaisne was much concerned to fight against Jordan's narrow delimitation of species. Gray adopted a similar position to conform to his position as the director of the Harvard University herbarium, having to discipline local botanists from the West, etc.

64 See Hooker's comments on this in a letter to Darwin (6 December 1857), in Burkhard and Smith, *The Correspondence of Charles Darwin*, op. cit., vol. 6, p. 499.

65 J. Decaisne, 'Note sur l'organogénie florale du Poirier', p. 339.
66 A.-L. de Candolle, 'Étude sur l'espèce à l'occasion d'une révision de la famille de Cupulifères', *Bibliothèque Universelle Archives de Sciences Physiques et Naturelles*, 15, 1862, 211–37.
67 Anon. [G. Bentham], '*The British Flora*, 8th ed. 1860 by W.J. Hooker and G.A. Walker Arnott, *Manual of British Botany*, 5th ed. 1862 by C.C. Babington, *Handbook of the British Flora*, 1858 by G. Bentham', *The Natural History Review*, 10, 1863, 34–40, p. 38.
68 J.D. Hooker to Darwin, 6 December 1857, in Burkhard and Smith, *The Correspondence of Charles Darwin*, op. cit., vol. 6, p. 499 (Hooker's emphasis); see G. Bentham, *Handbook of the British Flora*, op. cit.
69 J. Decaisne, 'Note sur l'organogénie florale du Poirier', p. 339 (translated as 'On the development of the floral organ in the pear', in the *Gardeners' Chronicle and Agricultural Gazette* of 14 November 1857, p. 773). Hooker refers to this remark in a letter to Darwin, 6 December 1857, in Burkhard and Smith, *The Correspondence of Charles Darwin*, op. cit., vol. 6, p. 500. On Decaisne's criticisms of the splitters, see also J. Decaisne, 'Lettre de M. Decaisne à M. l'abbé Chaboisseau (avril 1860)', *Bull. Soc. Bot. Fr.*, 7, 1860, 261–5. On the French debates on species delimitation and variability (Gérard, Naudin, Decaisne, Jordan, Godron, Boreau, etc.), see C. Bange, 'La culture', op. cit., and idem, 'Alexis Jordan'.
70 G. Bentham, 'Address to the anniversary meeting, May 24, 1866', *Proceedings of the Linnean Society of London*, 1865–6, x–li, p. xxxi.
71 Hooker and Thomson, *Flora indica*, op. cit., pp. 12–13.
72 G. Bentham, 'On the joint and separate work of Bentham and Hooker's *Genera plantarum*', *J. of the Linnean Soc.*, 20, 1884, 304–8. Bentham writes (pp. 305–6) that he accepted the collaboration 'notwithstanding my normal aversion for partnership botany' because 'I had always found that I could perfectly coincide with Hooker in his views in scientific botany'.
73 D.R. Headrick, *The Tools of Empire. Technology and European Imperialism in the Nineteenth Century*, Oxford University Press, 1981; Drayton, 'Imperial science', op. cit., pp. 322–46; R. Desmond, *Kew, the History of the Royal Botanic Gardens*, London: Harvill Press, 1995, pp. 205–17; L.H. Brockway, *Science and Colonial Expansion. The Role of the British Royal Botanic Garden*, London: Academic Press, 1979.
74 Drayton, 'Imperial science', op. cit., pp. 324–5.
75 Quoted in ibid., p. 334. See pp. 332–5 on the colonial flora project.
76 G. Bentham, *Florae hongkongensis*, London: Lovell Reeve, 1861; A.H.R. Grisebach, *Flora of the British West Indian Islands*, 3 vols, [1859]–64; G. Bentham, *Florae australiensis*, 7 vols., London: Lovell Reeve, 1863–78; D. Oliver, *Flora of Tropical Africa*, London, L. Reeve and Co., 1868–77, 3 vols (vols 4–10 were published later, from 1896 to 1937, by William Turner Thiselton-Dyer, David Prain and Arthur William Hill); J.D. Hooker, *Flora of British India*, London, 1872–97, 7 vols; W.H. Harvey and O.W. Sonder, *Flora capensis: Being a Systematic Description of the Plants of the Cape Colony*, Caffraria and Port Natal, 1859–1933, 7 vols (3 vols. were published quickly, with vols. 4–7 being published at the turn of the twentieth century under the direction of Thiselton-Dyer).
77 [J.D. Hooker], 'Colonial floras', *Natural History Review*, n.s. 1, 1861, 255–66, pp. 255, 256.
78 C.F.P. von Martius *et al.* (1840–1906). This flora had more than 20,000 quarto pages, and nearly 3,800 plates. A.-P. and A. de Candolle's *Prodromus* (1824–73) had more than 13,000 octavo pages and was not illustrated.
79 [Hooker], 'Colonial floras', op. cit., p. 264; G. Bentham, *Florae hongkongensis*, op. cit., idem, *Handbook of British Flora*, op. cit.
80 Bentham, *Florae hongkongensis*, quotes, resp., p. xxxvi and p. xxxviii.
81 Bentham and Hooker, *Genera plantarum*; Bentham, 'On the joint and separate work', p. 306, for the quotation. On this enterprise, see P.F. Stevens, 'How to interpret

botanical classification – suggestions from history', *Bio Science*, 47, 4, 1997, 243–50, pp. 246–8, and Stevens, *The Development of Biological Systematics*.

82 *Index kewensis*, Oxford: Clarendon Press, 1895 to date; see B.D. Jackson, 'A new "index of plants' names"', *Journal of Botany*, 25, 1887, 66–71 and 150–1, and idem, 'The history of the compilation of the *Index kewensis*', *J. of the Roy. Hortic. Soc.*, 49, 1924, 224–9; and Stafleu (1966).

83 See for instance Otto Kuntze's protestations against Kew's nomenclatural hegemonism in 'Notes on the Index Kewensis', *J. of Bot.*, 34, 1896, 298–307.

84 P.F. Stevens, 'J.D. Hooker, George Bentham', op. cit., p. 360.

85 Gray to Bentham, 21 January 1867, RBG Kew Archives, Asa Gray Corr., fol. 430. I am indebted to P. Stevens for bringing it to my knowledge.

86 Linné, *Species plantarum*; A.-P., A. and C. de Candolle, *Prodromus*, op. cit., Bentham and Hooker, *Genera plantarum*, op. cit.

87 Which was usually only allowed in very particular cases such as the existence in the same genera of the same specific epithet. See de Candolle (1867), pp. 29–30. On the Kew rule see Stevens (1991).

88 Many others opposed it and advocated a consistent application of the principle of priority. Some went as far as asking retroactivity for all names, even at the price of eliminating very widely used names. Successive international congresses in the twentieth century laboriously arrived at the conclusion that the 'Kew rule' should be discarded and the rule of priority ought to be followed (though not retrospectively). On these nomenclatural debates see A. La Vergata, 'Au nom de l'espèce. Classification et nomenclature au XIXe siècle', in S. Atran *et al.* (eds) *Histoire du concept d'espèce dans les sciences de la vie*, Paris, Fondation Singer-Polignac, 1987, pp. 193–225, and D.H. Nicholson, 'A history of botanical nomenclature', *Annals of the Missouri Botanical Garden*, 78, 1991, 32–56. See also G. McOuat, 'Species, rules and meaning: the politics of language and the ends of definitions in 19th century natural history', *Stud. Hist. Phil. Sci.*, 27, 1996, 473–519.

89 Jackson, 'A new "index of plants' names"', p. 69.

90 Otto Stapf, 'Apocynaceae', in W.T. Thiselton-Dyer, *Flora of Tropical Africa*, London, L. Reeve and Co., vol. IV, 1902, section 1, pp. 24–231.

91 Stevens, 'J.D. Hooker, George Bentham', op. cit., p. 355.

92 G. Bentham, 'Notes on Euphorbiaceae', *Journal of the Linnean Society, Botany*, 17, 1878, 185–267, p. 190.

93 Hallier, 'Über Kautschuklianen', p. 39. Note how scientific values of objectivity ('sachlich', 'unpersönlich') are linked to social judgement, being opposed to vulgarity ('Synonymenproletariat'). See also idem, *Das prolifierende persönliche und das sachliche, konservative Prioritätsprincip in der botanischen Nomenklatur*, Hamburg: Lütke & Wulff, 1900.

94 Naming a new species provides a double social advantage. First, your name is quoted at the end of the binomial name (as the authority who described the species first) and can therefore earn a kind of immortality. Second, you can choose a specific epithet which can be a tribute to somebody else you want to please (for instance, your master, or the colonial govenor, etc.). For the disciplining of naming practices in zoology and its limits, see Ritvo, *The Platypus*.

95 Hooker and Thomson, *Flora indica*, resp. pp. 21, 11, 10. See also their emphasis on the necessary knowledge of 'the amount of variation to which organized beings are subject, which alone will render him a sound botanist' (p. 21).

96 J.D. Hooker, 'On some species of *Amomum*, collected in Western Tropical Africa by Dr. Daniell, Staff Surgeon', *Hooker's Journal of Botany*, VI, 1854, 289–99, p. 292.

97 W.T. Thiselton-Dyer, presidential address to the Bot. section of the BAAS quoted by O. Kuntze, 'Notes on the *Index kewensis*', *Journal of Botany*, 34, 1896, 298–307, p. 307.

98 Hooker to Gray, 26 January 1854, in Huxley, *Life and letters*, vol. 1, p. 476. The argument is also made by Stevens, 'J.D. Hooker, George Bentham', p. 353, to which I am indebted.

99 See for instance Hooker and Thomson, *Flora indica*, p. vi; Stevens, 'J.D. Hooker, George Bentham', op. cit., pp. 354–5.

100 H. Hua, 'Établissement d'un organe périodique international destiné à la publication des noms nouveaux pour la science botanique', in *Actes du congrès international de botanique (Paris, 1900)*, Lons-le-Saunier: Declume, 1900, 475–83. But the success of such an enterprise required authority to guarantee that species published elsewhere would not be taken into consideration by anyone, an authority that no single institution possessed.

101 [Hooker], 'Colonial floras', p. 257.

102 W.T. Thiselton-Dyer, *The Botanical Enterprise of the Empire*, London: Eyre & Spottiswoode, 1880, p. 8. On Victorian culture's and the imperial economy's need for standards, see S. Schaffer, 'Empires of physics', in R. Staley (ed.) *The Physics of Empire*, Cambridge: Whipple Museum of the History of Science, 1994, pp. 97–111; idem, 'Metrology, metrification and Victorian values', in B. Lightman (ed.) *Victorian Science in Context*, Chicago: University of Chicago Press, 1997, pp. 439–74.

103 L.H. Brockway, *Science and Colonial Expansion. The Role of the British Royal Botanic Garden*, London, Academic Press, 1979; S. Sangwan, *Science Technology and Colonisation: An Indian Experience 1757–1857*, New Dehli: Ananika Pratashan, 1991; Deepak Kumar (ed.) *Science and Empire: Essays in Indian Context, 1700–1947*, Delhi, 1991; Drayton, 'Imperial science'. See also M.-N. Bourguet and C. Bonneuil, 'De l'inventaire du globe à la "mise en valeur" du monde: botanique colonisation (fin XVIIIe siècle–début XXe siècle). Présentation', *Revue Française d'Histoire d'Outre-Mer*, no. 322–3 (1er semestre 1999), 9–38.

104 Stevens, *The Development of Biological Systematics*, pp. 476–7, n. 61; J.B. Hagen, 'The development of experimental methods in plant taxonomy, 1920–1950', *Taxon*, 32, 1983, 406–16; idem, 'Experimentalists and naturalists in 20th-century botany: experimental taxonomy, 1920–1950', *Journal of the History of Biology*, 17, 1984, 249–70.

105 Stevens, 1991; La Vergata, 'Au nom de l'espèce'.

9 Exploring the margins of precision

H. Otto Sibum

This chapter will take you on a journey through Europe during the years 1875 and 1876. Our traveller is the American civil engineer and instructor of physics at Rensselaer Polytechnic, Henry A. Rowland. He was sent (and accompanied for part of the journey) by the President of Johns Hopkins University, Daniel Coit Gilman, with the aim of studying the material culture of science in Europe. In particular he was to survey workmanship in laboratory physics, the quality of instrument making and the most important research topics pursued by European physicists. Equipped with substantial funds and looking forward to becoming the founding director of the new physics laboratory at Johns Hopkins University in Baltimore, the 27-year-old physicist expected to do his most interesting and challenging fieldwork. Historians of science have paid little attention to this European trip, let alone to integrate Rowland's experiences as a traveller with an understanding of his early laboratory physics at Johns Hopkins.[1]

Instead historians of physics customarily begin their accounts of Rowland's career as a physicist with his famous optical investigations, the production and distribution of diffraction gratings, which represent the height of precision spectroscopy.[2] When it is noticed at all, Rowland's first major research 'On the Mechanical Equivalent of Heat' (a 124-page text which appeared in the *Proceedings of the American Academy of Arts and Sciences* in 1880) is seen simply as an attempt to replicate a thirty-year-old experiment to achieve an exact value for the mechanical equivalent of heat in absolute measures.[3] By reconstituting his European experiences in great detail we will provide a hitherto neglected perspective on this experiment, and simultanously give new insights about an important moment of change in the development of what has been called the culture of precision in the nineteenth century.

During the second half of the century precision measurement became an icon for the exact sciences, but it was clear to many that their striving for precision using evermore sensitive devices pushed the work of physics to its limits. In private correspondence with Rowland the German doyen of precision measurement, Friedrich Kohlrausch, brought out an important problematic with his discussion of the standards of accuracy that an envisaged 'international laboratory' would force on physicists' experimental work. He conceded that

> if an international laboratory wants to bestow unity upon science and technology this laboratory has to work much more precisely than physicists have currently done . . . To bring a bunch of experienced physicists for this purpose permanently together seems impossible. Every one of us certainly knows that he would work best at home.[4]

The tension between the local and global expressed in this remark points to a pressing issue of the time: the embodied character of experimental knowledge. Experimentalists' insistence on working best at home or maintaining physical contact with objects and instruments in producing new effects and experimental knowledge ran counter to the modern view of science, which implied that scientfic knowledge could only be achieved if it was independent of the competencies of individual scientists or the peculiarities of a specific place. This chapter will take a close look at a very delimited period of time in which precision measurement became an ultimate technique to discriminate between what counted as local and what as scientific knowledge.[5]

When he met British scientists, Rowland experienced at first hand the work required to establish a 'coherent system of units' with its absolute measures, as it was being promoted by the British Association for the Advancement of Science.[6] But he soon recognised that the aims of this programme were far from being fully realised. Rowland found that the precision instruments and the measurement techniques used in different places – together with the theories of electrodynamics and thermodynamics that were entangled with these measurements – differed from country to country. Moreover, he saw that the different values obtained by experimentalists in their measurement of fundamental units shook the emerging British energy physics to its foundations.[7] Rowland regarded the new regime based on the concept of energy and precision measurement to be the most fruitful project for the future, but thought it would require a new kind of physics integrating the achievements of industrial production and engineering. For Rowland, the screw – and the drive to perfect that simplest of machines – became an emblem of the accuracy and precision for which science should strive. But precision measurement remained ambiguous. On the one hand it provided the means to standardise and guarantee a coherent system of units as an important working basis for an international expert culture. On the other hand it also made possible the creation of new effects that could only be detected at specific sites and often violated accepted standards. During his trip through the scientific landscape of Europe, Rowland realised the strategic importance of a reinvestigation of the mechanical equivalent of heat. In the light of the new energy physics, and the connected visionary system of absolute units, he regarded the determination of the mechanical equivalent of heat to be an investigation of 'one of the most important constants of nature'. Only an exact value in absolute measures would set the science of energy on solid grounds and only absolute measures (guaranteed through the use of precision instruments) would, according to him, lead to a true representation of nature and therefore provide the sinew of the rising international community of modern scientists.

Exploring British measures of unity

On 15 July 1875 the steamship *Bothnia* reached Queenstown in Ireland, one of the central departure points for European emigrants to the United States. Just as hundreds of those emigrating believed the US to be the promised land, Henry A. Rowland arriving on European shores believed that everything he would see would be better. Passing through Dublin he went directly to Edinburgh in order to visit one of the few people with whom he had already established contact, the physicist James Clerk Maxwell. During the semester breaks Maxwell preferred to return from his position in Cambridge to his family home in Scotland. Rowland was eager to hear at first hand how the Cavendish Professor of Experimental Physics and author of *A Treatise on Electricity and Magnetism* approached the formation of physical theory in his new laboratory. And he was excited to meet the man who had helped him publish his first paper.[8] Rowland used the occasion to discuss at length several urgent topics concerning electricity and magnetism, particularly the creation and measurement of magnetic effects.[9]

From the *Treatise* Rowland knew that according to Maxwell the whole universe was filled with an immaterial ether; and that all electromagnetic phenomena could be described on the basis of a continuum theory. Rowland was fascinated by the engineering character of this most pure and consistent electrodynamical theory. A subtle immaterial and apparently unmeasurable ether was the basis for all known electrical phenomena, but it behaved like an invisible machine. Maxwell's model of a mechanical ether even allowed him to predict the propagation of electromagnetic waves in space, whose velocity was dependent on the elasticity and density of the system. For the case of empty space this velocity could be calculated from the ratio of two units of charge, the electrostatic and the electromagnetic. The need for this distinction had both a theoretical and practical significance. Maxwell made this clear in his work for the British submarine cable project: assuming the capacity of a conductor such as the Atlantic cable to be very large, then this would have a demonstrable effect on the conducting process which it was possible to measure exactly. In order to be able to transfer from the electrostatic to the electrodynamic system of units Maxwell required this ratio, which had the dimension of a velocity, and which he designated with v. The close connection between Maxwell's electrodynamics and technical applications could be guaranteed practically through the development of exact working galvanometers. And it is not surprising that Maxwell worked on a very costly project for the exact experimental determination of the unit of electrical resistance, the ohm, which formed part of a comprehensive programme for realising an absolute system of units.[10]

The British Association for the Advancement of Science had begun to develop an absolute system of units in the 1860s, in order to bring an end to the chaos of different units of measure that prevailed within their own and other countries. In order to be able to establish such an absolute system of units, all quantities had to be traced back to the dimensions of length, mass and time. The foot-pound was introduced as the basic quantity for mechanical force, the measure of mechanical work. Electric, magnetic, thermal and chemical phenomena now had to be brought into correspondence within this 'coherent system'. Three fundamental laws, those of Ohm, Faraday and Joule, set out the central relations between the units of these systems. Indeed, only by making use of the mechanical equivalent of heat, which expressed a constant relation between mechanical work and heat, was it possible to establish 'a truly absolute system'. This decisive number represented the necessary connection between mechanical work and the other forces working in nature.

The energy physics proposed by William Thomson, later Lord Kelvin, assumed its material form with this system of units, and Maxwell's theory of electrodynamics was also closely bound to it.[11] With his experimental procedure for the exact determination of electrical resistance on the basis of the absolute system of units, Maxwell delivered the first working standard of resistance for science and industry – and could at the same time hope to provide the true value for the ratio of the electrostatic and electromagnetic units. If the number he found corresponded with the value for the velocity of light, this would be a great success, for the identity of light and electromagnetic waves would thereby be established, bringing at least an indirect proof for the existence of the ether. Rowland was impressed by this representative of an emerging experimental culture, in which rational methods, forms of representation and working techniques were an integral part of his moral system. Indeed, for Maxwell, precision measurement embodied a very high moral value: this kind of research not only promised the unification of theories, it also yielded a trustworthy basis for scientific communication. Absolute measures were true representations of nature. Accuracy thereby became the basic virtue of a new, evermore fully established exact science, which also accorded well with the values of Victorian England.[12]

However, Rowland also recognised that Maxwell's science of electricity was still only local knowledge, and that despite its publication in a book it could only be understood and extended gradually and through immense effort on the part of its followers.[13] Rowland's own journey could therefore play an important role in establishing scientific exchange between experimental cultures that were still very distinct. He undertook this task gladly, yet saw already that to pursue his own form of physics he would have to take a slightly different tack on the matter. He wrote to his President:

> After seeing Maxwell I felt somewhat discouraged, for here I met a mind whose superiority was almost oppressive but I have since recovered and believe I see more clearly the path in which I hope to excel in the future. It lies midway between the purely mathematical physicist and the purely experimental, and in a place where few are working.[14]

The great importance that leading Victorian physicists invested in precision measurement also meant that differences in measured values could lead to fundamental conflicts, and Rowland was soon to experience a controversy of this kind. From the publications of the British Association for the Advancement of

Science the reader could infer that James Joule from Manchester had unleashed just such a conflict. In order to provide an indirect measure of the mechanical equivalent of heat, Joule heated water with a battery and a copper conductor. Thereby, he determined the resistance of the wire by means of the standard of resistance established by Maxwell's team. The result of his experiment differed from that obtained from his earlier, direct measures of the equivalent, however. Joule maintained that the new experiment was carried out with more precision than the earlier experiment; either the ohm or the mechanical equivalent of heat had been falsely measured. Maxwell saw here a fundamental challenge, which questioned his credibility as an experimentalist. He had to reconsider his own measurement of resistance, but at the same time he asked colleagues to carry out direct determinations of the mechanical equivalent. Finally he planned to carry out himself such an experiment in Cambridge but wasn't able to do so. Joule's replication of his own experiment had led to a threat to the energy concept underlying the theories of thermodynamics and electrodynamics. Correspondence between Joule and William Thomson shows that Joule planned to repeat his old direct determination of the mechanical equivalent of heat in order to better safeguard the dynamic theory of heat. Joule thus wrote in 1869: 'I am for one thing going to overhaul the old experiments. This is surely the time for new proofs now the hermetical conclave are preparing to put down the dynamical theory as heretical.'[15]

The observation of minor deviations in measured values had led to a deep questioning of the credibility of the experimentalists, their practices and theories. Through further improvement of standards of precision measurement the collectively created link – the absolute system of units – could also bring discord between laboratories. Above all this would provide the opponents of the new energy physics with new arguments. The bone of contention was the mechanical equivalent, Royal Society fellow George Gore argued: one had to clearly distinguish between the 'principle of equivalency of forces and the principle of energy conservation'. Now only the former (i.e. a constant relation between cause and effect) had been demonstrated in a few cases through Joule's equivalent. But whether it was valid for all kinds of correlation of natural forces Gore doubted. For him the principle of energy conservation was valid only in mechanics.

Experimental cultures are certainly very strongly dependent on the standards of technical development, and this was the reason Rowland examined instrument collections in the different places he visited, looking for suitable instruments. On his way to the Continent he next visited laboratories in London and Oxford. But to his dismay he discovered that England was not the place he had thought it to be: 'I came over here with the notion that I should find everything better than in America but one month has wrought an entire change.' His visits led Rowland, for example, to regard English investments in instruments as quite problematic compared with his own standards:

> I was considerably disappointed with the apparatus [held at King's College, London] as the whole collection had more the appearance of a museum of

antiquities than of a working cabinet of instruments, but still there were some instruments of value . . . as I wandered around [the Oxford physical laboratory] the same old feeling of disappointment came over me. For as usual the architect had got the best of the physicist. There are some fine pieces of apparatus but as a whole they might far better have increased their stock than have spent as they did 10,000 pounds for useless ornamentation.[16]

Although in his few weeks of stay he had still not seen much of the glorious Empire, he wrote confidently that 'the time is coming when England will find herself behind, though in many things we cannot say she is at present.'[17] Within a month of beginning his tour Rowland was already expressing a wish to return to America in order to be able to start 'work of value'. And finding that 'None of the dealers keep a stock on hand except of the more common instruments', he was very concerned that he had not yet ordered any apparatus. But he finally decided to continue his trip, attending the BAAS meeting in Bristol where he met many of the elite of British science, being introduced at a formal dinner by William Thomson as an American who had even measured occult magnetism. Occult phenomena, such as animal magnetism or ghost appearances, were a major concern among leading Victorian scientists, and William Crookes – who had been working on the quantitative determination of occult forces and their relation to magnetism – was among the participants as well.[18] The determination of equivalences between magnetism, electricity and heat was also pursued by George Gore with whom Rowland had conversations that may well have reinforced his interest in redetermining the mechanical equivalent of heat. In the early 1870s the Reverend Henry Highton had drawn Gore's experimental work on the magnetism of electrodynamical spirals into controversy in a critical paper On 'the Relations Between Chemical Change, Heat, and Force, with a Special View to the Economy of Electrodynamic Engines'.[19] According to Highton, the determination of the mechanical equivalent of heat did not apply to electrodynamic engines because Gore's experiments, for example, had shown 'that there is no connection between heat and other forms of energy in [a current-carrying] wire'. Therefore 'conservation of energy is [only] true in mechanics', and according to Gore one had to distinguish carefully between the principles of 'conservation of matter and energy' and the 'equivalency of forces'.[20] We do not have any literary evidence that Rowland was informed about these concerns about 'Joule's equivalent' by Gore himself, but notebook entries he made a few days later while en route to Germany via Paris show that Rowland started to sketch plans for improved versions to redetermine the mechanical equivalent of heat by converting electrical or magnetical forces. Figure 9.1 shows an indirect method involving a rotating magnet, over which a copper disc is hung whose mechanical moment could be measured. In a drawing dated 30 September 1875 we see an adapted version of the paddle-wheel experiment Joule had developed thirty years earlier (Figure 9.2). In the middle, a copper pot filled with water in which a paddle wheel stands can be seen. In contrast to the Joule version, the pot itself revolved in order to allow an exact determination of the moment of force and

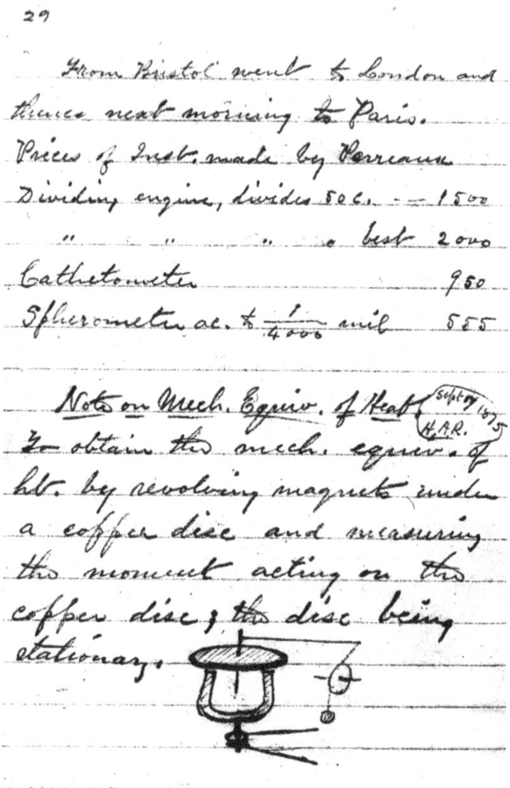
29

From Bristol went to London and thence next morning to Paris.

Prices of Inst. made by Perreaux

Dividing engine, divides 50 c.	1500
" " " best	2000
Cathetometer	950
Spherometer ac. to $\frac{1}{4000}$ mil	555

Notes on Mech. Equiv. of Heat (Sept 7 1875 H.A.R.)

To obtain the mech. equiv. of ht. by revolving magnets under a copper disc and measuring the moment acting on the copper disc, the disc being stationary.

Figure 9.1 Note of mechanical equivalent of heat by Henry A. Rowland, 7 September 1875. In 'European Tour Notebook.'
Source: Henry A. Rowland Papers MS 6, Special Collections, Milton S. Eisenhower Library, The Johns Hopkins University.

inertia that arose. Above in the picture is an electromagnetic device designed to ensure the uniform motion of the axle. In Figure 9.3 the paddle wheel in the middle of the pot may be identified. Rowland drew this as he knew it from the publication, with solid arms attached to the inner walls and arms attached to the axle which stirred the water. But Rowland clearly held this design to be unsatisfactory – writing 'poor' beside it.[21]

Berliner Spielräume

These plans for work were interrupted by the further course of his journey. On 29 October 1875 Rowland reached Berlin, a rising centre of science, industry and political life. Germany was a state in rapid transition. Letters sent by the engineer Franz Reuleaux from the World Exhibition in Philadelphia and pub-

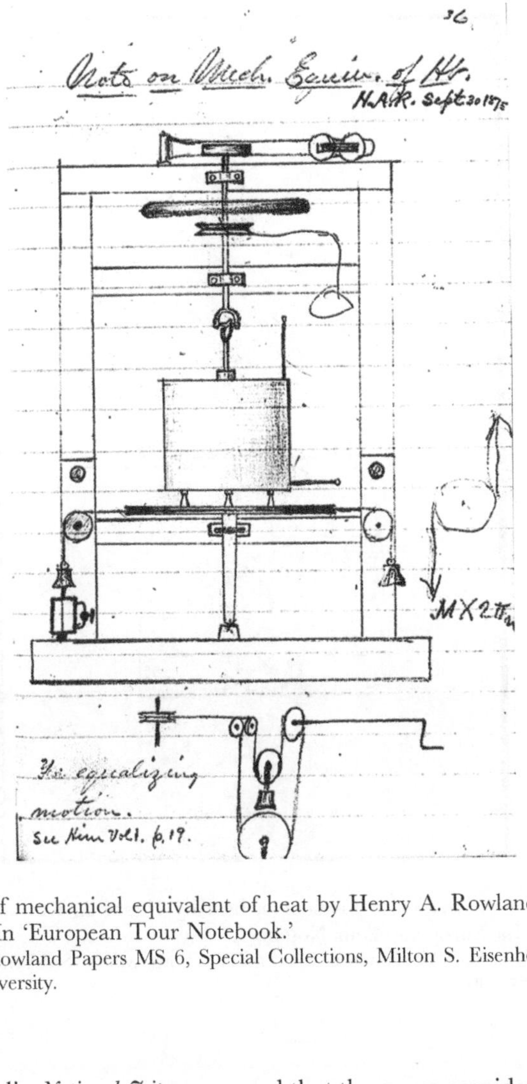

Figure 9.2 Note of mechanical equivalent of heat by Henry A. Rowland, 30 September 1875. In 'European Tour Notebook.'

Source: Henry A. Rowland Papers MS 6, Special Collections, Milton S. Eisenhower Library, The Johns Hopkins University.

lished in the Berlin *National Zeitung* ensured that there was considerable sensation: Reuleaux reported that the German technology on display there was cheap and nasty, and above all that its exhibition presentations were regarded as militaristic. Such an assessment of the state of the nation from the respected head of the Gewerbeakademie exacerbated the mood of depression and criticism in the country.[22] With the foundation of the Empire, and the development of new technological systems such as the railway and telegraphy, Germany was indeed on the way to becoming a leading industrial nation, but this transition involved massive changes in everyday culture. Rowland was led to recognise such problems above all in his restless search for good instruments. He wrote impatiently about the poor network of instrument makers: 'The German professors, strange to say,

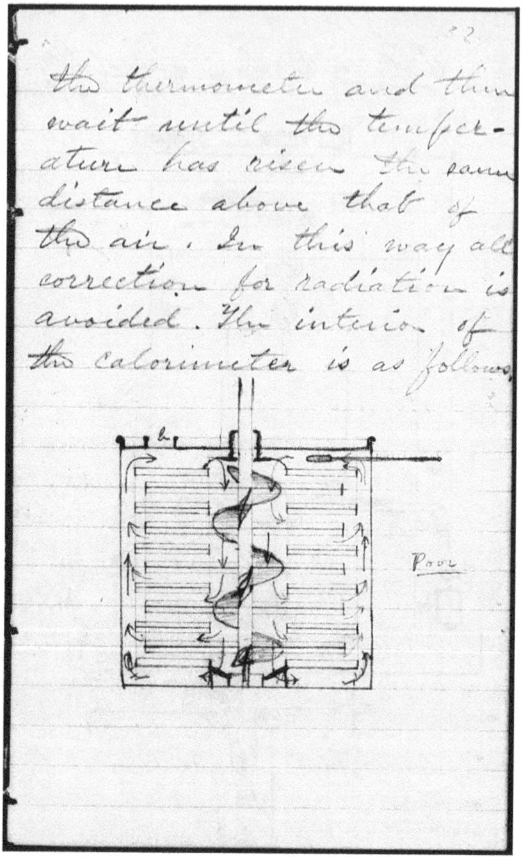

Figure 9.3 Note of mechanical equivalent of heat by Henry A. Rowland, 30 September 1875. In 'European Tour Notebook.'

Source: Henry A. Rowland Papers MS 6, Special Collections, Milton S. Eisenhower Library, The Johns Hopkins University.

know little about any but local makers and so I find the buying of German instruments about as difficult as learning German.'

However, as a trained engineer Rowland had great opportunities to see significant traces of improvement. In the Gewerbeakademie Reuleaux had set up a vast collection of mechanical models in order to teach his students theoretical kinematics.[23] Furthermore, his pioneering publication in advanced engineering had just left the printers and was about to be published in English in 1876.[24] The book marked an important change in the history of engineering, claiming to be able to present the kinematical essence of the entire previous development of steam engine technology. In the preface to the German edition, Reuleaux distinguished between descriptive analysis ('beschreibende Analysirung') and kinematic synthesis ('kinematische Synthese'), taking them to be the engineers' techniques to understand old and invent new machines.[25] This new field of research belonged without doubt to the exact sciences and was an attempt 'to make the science of machinery deductive'.[26] Reuleaux described all previous human endeavours in inventing machines as being to a certain extent unconscious acts of production, which he called tinkering or 'Pröbeln'. His *Theoretische Kinematik* had for him the character of a science because it allowed engineers to understand and create elements and mechanisms of a machine 'without the aid of invention, present or past'.[27] Reuleaux thought that for the first time his theoretical analysis provided a scientific tool for generating new machines through kinematical synthesis. But he was keen to demonstrate that his 'science of invention' was not meant to denigrate more unconscious invention:

> If I attempt to show that [and how] new mechanisms and machines could be developed through a scientific method, I am not thinking of degrading the value of invention in this machine area. I seek to provide the ingenious head with new and effective means, I would like to accelerate his mental activity, but I do not degrade these activities. Everywhere in my book I have done justice to the genius who does big steps without knowing the rule.[28]

In fact Reuleaux argued for a labour-saving method of induction which would change the troublesome 'empirical procedure of groping in the dark for a solution, which often depends on great luck'. *Theoretische Kinematik* would transform the engineer's capability of induction into a higher mental effort.[29] This civilised engineering – Reuleaux's science of machinery – should become the logical foundation for a machine philosophy that was urgently needed in modern society.

In his chapter 'Sketch of the History of Machine Development' he even equated the progress of civilised societies with the degree of perfection of machines. In ancient times, he argued, workmanship was determined largely by the skill of the worker and depended less on the quality of the machine, and he concluded that the history of civilisation was to a large extent the history of human efforts in perfecting their machines – a developmental history of mechanics. From the engineering point of view that Reuleaux represented, perfection meant improving the 'pair of elements'; that is, diminishing the free play, the 'Spielraum', between the teeth of two engaged gears. The term 'Spielraum' needs further explanation: Reuleaux remarked that, significantly, the English had named it 'freedom' instead of 'Spielraum', commenting that 'therefore they recognised it as closer to the freedom inherent in the universe'.[30] In the German language the word 'Spielraum' was used in various contexts. Literally it meant a 'room to play in' – and in all its uses it referred to a circumscribed freedom to act. A carpenter for example may have described a good working drawer as needing 'Spiel' (i.e. free play in its frame). In political or economical negotiations the actors wished to have a certain degree of free play in their actions.[31] The engineer Reuleaux was concerned with the 'Spielraum' of the gear's teeth; that is, with their clearance: a perfect machine required a set of working gears without any play:

> Although [toothed wheels] have been known for thousands of years, their improvement to-day is still essentially in the direction of excluding force-closure, that especially which has remained with the 'clearance' or 'freedom' allowed between the surfaces of the teeth, and which has often enough made itself disagreeably felt. In the Chinese winding mill (gin) and in the similar machine used by the Egyptians, and worked by water (the Sakkiah), there is a large amount of play left between the teeth, . . . But we see that during the Middle Ages, and in the last few centuries, the freedom has been more and more reduced, as greater care has been taken to find the kinematic condition to be fulfilled by the form of the teeth-profiles, until we have now succeeded in reducing it to a very small fraction of the pitch . . . I believe that in a few decades it will be the rule to employ spur-wheels working without any clearance between the teeth.[32]

In his historical analysis Reuleaux distinguished between the 'history of a development' and 'history of the ordinary kind'. The latter 'gives us in chronological order a series of individual phenomena, which may be retrogressive as well as progressive. The developmental history seeks only to find the steps by which some known position has been reached, – it repeats itself anew with each nation's growing civilisation, it even reflects itself in the development of each single individual.' Reuleaux exemplified developmental history by means of sketching the development of the simplest machine – the screw – from the caveman's practice of making fire down to his own days. Reuleaux's history of the development of the machine was the model for the development of civilisation, measured by the degree of free play. We read that

> The whole inner nature of the machine is . . . the result of a systematic restriction; its completeness indicates the increasingly skilful constrainment of motion until all indefiniteness is entirely removed. . . . If we look for a parallel to it elsewhere we may find it in the great problem of human civilization. In this the development of machinery forms indeed but one factor, but its outline is sufficiently distinct to stand out separately before us. Just as the poet contrasts the gentle and lovable Odyssean wanderers with the untamable [*sic*] Cyclops, the 'lawless-thoughted monsters', so appears to us the unrestrained power of natural forces, acting and reacting in limitless freedom, bringing forth from the struggle of all against all their inevitable but unknown results, compared with the action of forces in the machine, carefully constrained and guided so as to produce the single result aimed at. Wise restriction creates the State, by it alone can its capacities receive their full development; by restriction in the machine we have gradually become masters of the most tremenduous forces, and brought them completely under our control.[33]

When Rowland arrived in Berlin, the headquarters of this new kind of engineering, Reuleaux was visiting the World Exhibition in Philadelphia. The American engineer and physicist asked for permission to work in the laboratory of the physicist Hermann von Helmholtz, a request that was initially declined on the basis of lack of space. What could move the great physicist to engage him, Rowland asked himself? A few years previously he had begun research on the development of electrodynamic theory. He had to find something there that would show his capabilities as a manually dexterous engineer and young Maxwellian, and that promised Helmholtz something for the theoretical clarification of electrodynamics. He proposed an experiment he had already devised in the United States, but had not yet found an appropriate place to perform because of the very sensitive measurements it required. He proposed to prove whether a rotating, electrically charged metal disc gave rise to a magnetic effect and whether this could be measured. Helmholtz was enthusiastic. He immediately granted Rowland a working place in the cellar of the laboratory at the Humboldt University, and admitted him to the team of Berlin experimental physicists.[34]

Helmholtz's sudden change of mind can only be understood if we consider more closely the situation in the development of electrodynamics in Germany. Very few knew Maxwell's theory of electrodynamics, and most German physicists worked with Weber's electrodynamics, a completely different approach. Where the Scot saw an ether machine, adherents to Weber's theory saw empty space. Or, better expressed, Weberian physicists were not interested in how the electric or magnetic forces crossed the space between two charges localised in different places. In order to describe an example of induction phenomena, Maxwellians used mechanical models which satisfactorily represented the mechanisms of translation of the ether between two conductors. Wilhelm Weber described the interaction between two infinitesimally small parts of the conductors only through their change of spatial position in relation to each other. The main equation of this theory based on action at a distance was an extension of Coulomb's law with a velocity and acceleration component.[35]

From the standpoint of the history of science the contemporaneous existence of two very different theories is extremely interesting. Historical research carried out to date justifies the opinion that these theories stemmed from two different experimental cultures, whose specific features have still not been satisfactorily demonstrated. Nevertheless Rowland performed an important intermediary role in the process of reciprocal clarification, all the more since Helmholtz had just begun to formulate a third, modified theory of electrodynamics. He wanted to retain the concept of action at a distance characteristic of Weber's theory, but wished to bring the theory, among others, into agreement with the principle of the conservation of energy. Rowland's experiment gave Helmholtz the prospect of producing a decision between the three theories. Yet although Rowland could demonstrate such a magnetic effect, and could even 'measure it with tolerable accuracy', it turned out that the result could just as well be accommodated by all three theories.[36] For our interest in discerning the meaning of Rowland's travelling experience for his planning and execution of a redetermination of the mechanical equivalent, it is only important to mention that in this phase of his trip Rowland recognised the immense importance of exact standard units. Since he had measured the surface charge of the metal disc in electrostatic units, but the magnetic force was expressed in electromagnetic units, he had to apply an exact conversion factor between the two systems of units, which was only given by Maxwell's ratio. He could only say with certainty that the magnetic effect existed if he knew exactly the margins of error of the units that appeared in the calculations. But these very standards had been sent tottering by the advanced dexterity in play. Although Weber and others had also determined exact values, discrepancies still existed and Rowland trusted the value provided by Maxwell, because this physicist had most impressed him. Even in this most exact of sciences with its precision instruments, trust in particular people was often decisive for the choice of experimental methods and results. But by this time at the latest Rowland had decided to clear these perpetual discrepancies out of the way. He set out a future plan of research which was intent upon first measuring the mechanical equivalent of heat through the direct method (that is, the friction of water) in order to determine afterwards the value of the ohm. Only an absolutely reliable unit could guarantee the possibility of proving the existence of this subtle ether through exact measurement. He turned once more to improving Joule's method in drawings in his diary. Only thus could he prove the most important term in the coherent system of absolute units. He had already initiated contact with his colleague Edward Pickering in Harvard University in Boston in order to carry out with him this experiment on his return to America. Already in January 1876, while carrying out his electrical experiments in Helmholtz's laboratory, Rowland had thanked Gilman for sending the first $200 in assistance of determining the mechanical equivalent of heat. The results of the previous month had left him very impatient, he wrote, because he would like to carry out the experiment directly in Berlin, but the following grounds spoke against it: 'I find that it could not be advantageously carried out without the use of a steam engine and a room specially prepared for it. The people are so slow here also that I do not feel like attempting any more experiments.'[37]

These experiences guided his aims for the rest of his journey, but also shaped his view of the kind of physics he would like to practise in the future. While still in Germany he was able to obtain from a foundation another $800 that, meanwhile, he saw necessary for performing his heat experiments in Baltimore. He was extremely baffled, however, by the fact that just those physicists in the United States who were still not ready to see the importance of practising physics with absolute measures had been elected to give permission for his purchase of instruments:

> My dear Prof. Gilman, Your brief letter of Feb. 28th containing Prof. Newcomb's letter reached me a few days ago in Vienna. The reason of my laying so much stress upon the instruments whose purchase Prof. Newcomb criticises is easily stated. I am surprised that Prof. Newcomb remarks that in Physics only relative measures are required. To be sure such *was* the case once; but one of the most important changes now taking place is the introduction of the absolute system into *all* departments of Physics, and particularily in magnetism and electricity. There is a wider field open in this direction than in any other and it can only be entered by one having the

most perfect instruments. I am perfectly satisfied that with the class of instruments which I have so far ordered results can be obtained which will bring the Physical Institute of the J.H.U. before the world.[38]

Visiting London and Cambridge

On his way home he visited the world exhibition of scientific instruments in London. He found there not only Reuleaux's working models, but also, in one of the showcases, James Joule's paddle-wheel experiment, but in a wholly different form from the one he recognised from the published representation. Figure 9.4 shows Rowland's notebook entry of the paddle wheel as displayed at the exhibition. As was discovered in a project to rebuild and perform Joule's experiment, one can only reproduce it with knowledge of this design. The very different illustration in Joule's publication from 1850 was not meant to give a faithful representation of the actual apparatus.[39] Rowland made precise drawings because he believed that the friction shown in Joule's published article was too small to be confirmed, as Figure 9.4 (bottom left) shows.

One of his last visits was to the Cavendish Laboratory in the University of Cambridge. This was an educational institution in which mathematics was an essential part of a liberal education, which served the formation of the spirit. Students had to reproduce laws, proofs and theorems from memory and solve problems with speed and accuracy – the maxim of the Mathematical Tripos. Until the late 1860s thermodynamics and electrodynamics were regarded as progressive sciences, whose principles were still not generally recognised, and therefore could not be taught. Only the permanent sciences, i.e. knowledge that was established, could be referred to in the syllabus. Most lecturers refused to give experimental lectures, for they felt even the use of logarithms to determine a degree of longitude was too practical a turn in the Cambridge academic form of life:

> I think the examiners are taking too *practical* a turn; it is a waste of time to calculate *actually* a longitude by the help of logarithmic tables and lunar observations. It would be a fault not to know *how*, but a greater to be handy at it.[40]

Cambridge was committed to an oral tradition of knowledge, and its students were trained in fast and accurate mental performances. Rowland quickly perceived some of the constraints against which Maxwell was forced to work: the integration of experimental physics in Cambridge indeed required many clever manoeuvres, of which one will be discussed here – Maxwell's efforts to surmount this apparent cleft between mental and bodily work through a special teaching programme. Only a few students came to Maxwell's experimental classes, and it was extremely difficult to keep them there. Engagement with experimental work deeply offended the values and norms of academic culture that were imparted through college life. Only a handful of students taking up experimental work as an avocation were prepared to experience what Maxwell meant with the expression

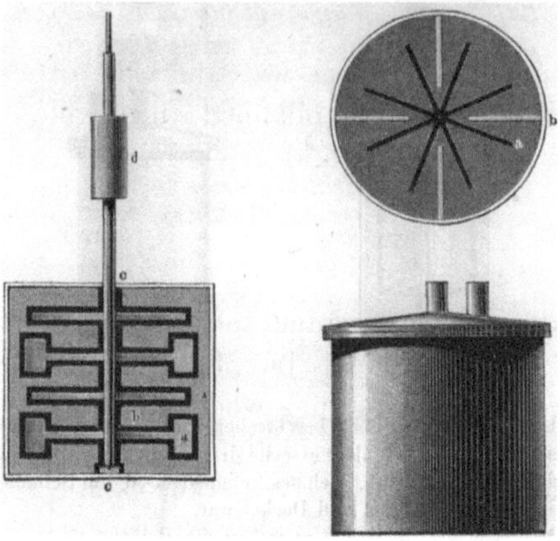

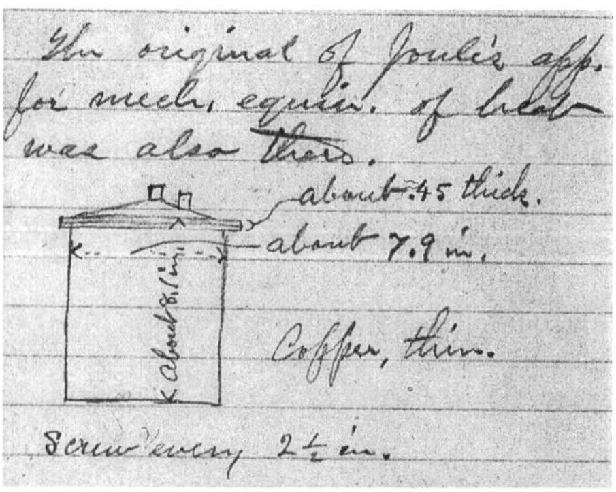
The original of Joule's app.
for mech. equiv. of heat
was also there.
about .45 thick.
about 7.9 in.
about 8.1 in.
Copper, thin.
Screw every 2½ in.

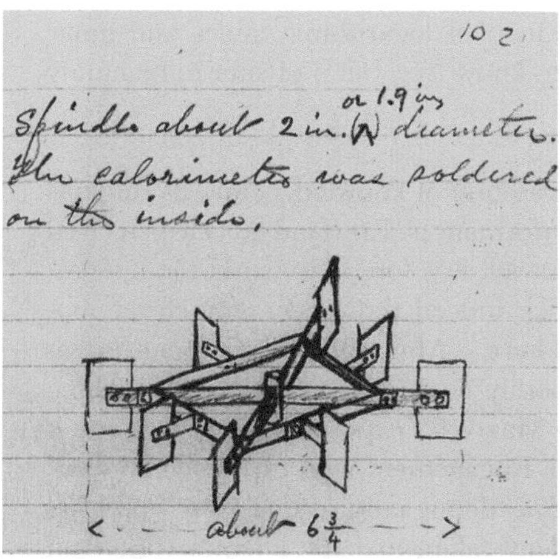
10 2
Spindle about 2 in. (or 1.9 ins) diameter.
The calorimeter was soldered
on the inside.
about 6 3/4

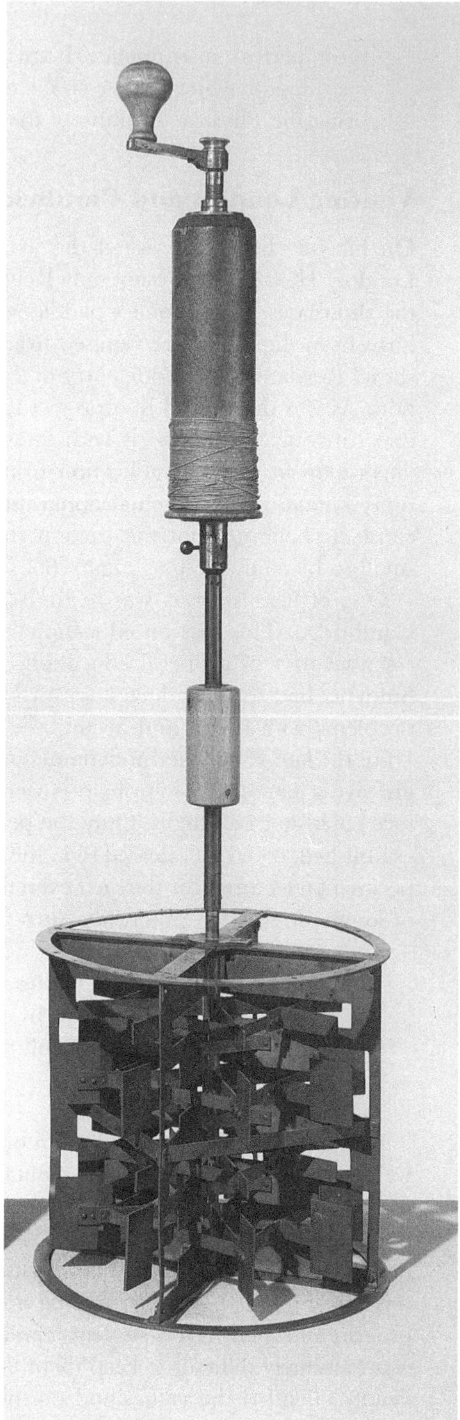

'The facts are things which must be felt, they cannot be learnt from any description of them.' He entrusted students with the replication of historical experiments or he obtained instruments from the museum in order to bring them to understand that one could only become familiar with this experimental knowledge through the active use of one's senses. Maxwell sought to make it clear to these students that there was no principal difference in acquiring practical knowledge or theoretical knowledge, since the distinction itself was not appropriate. In both cases the senses were involved and in both forms of knowledge one could attain speed and accuracy of performance. Maxwell actually translated the often particular language of experimentalists into the mathematical language with which he was familiar.[41] Rowland was impressed with Maxwell's epistemologically motivated introduction to experimental physics, but could not surmount his dismay at the social friction evoked by Maxwell's work to establish an experimental culture in Cambridge. He recognised that this was still not the kind of place where he could practise his own experimental culture with its armoury of perfect instruments.

It should not come as a surprise that in the period in which experimental physics laboratories were first being established in European universities, different cultures realised this integration in ways appropriate to their traditions. Rowland's European trip is very telling in this regard. He noted that 'In France and England they had experiments of illustration and research, but in Germany they only did experiments of investigation.' Rowland designated the form of experimental practice Maxwell had described as experiments of research, which indeed belonged to the progressive sciences, but were limited exclusively to the exact measurement of standards of units and constants. Experiments of investigation, on the other hand, involved a procedure which, to date, he had only come across in Helmholtz's laboratory, and which he himself very much intended to realise in America. There theory and practice, measurement, and qualitative research went hand in hand. Although the most abstract problems came under discussion, they were given a practical turn on the laboratory bench, and the form and manner of the inquiry lay completely within the hands of the researcher, be it through exact measurement or through independent constructions of experimental apparatus. Inversely, effects were often produced against

Figure 9.4 (*opposite*) Different perspectives of James Joule's paddle-wheel experiment from 1850. *Top left*: Visual representation of the paddle wheel of the authorised reprint from 1872 (James Prescott Joule, *Das mechanische Wärmeäquivalent: Gesammelte Abhandlungen*. German translation by J.W. Spengel, Braunschweig: Vieweg, 1872). *Middle and bottom left*: Notebook entries taken by Henry A. Rowland while at the Loan Exhibition where Joule's paddle-wheel experiment was displayed. *Right*: Visual representation of the stirrer of James Joule's original paddle-wheel device held at the London Science Museum. (Reprinted with permission of the Board/Science and Society Picture Library.)

Sources: (top left) as indicated; (middle and bottom left) Henry A. Rowland Papers MS 6, Special Collections, Milton S. Eisenhower Library, The Johns Hopkins University; (right) London Science Museum.

which the existing theories had to prove their strength. For Rowland it was this 'art of induction', the probing of free play in action, which was the true source for innovations in science and art. Helmholtz had created a place for physical research in Berlin which was drawn from the community of learned craftsmen, or, as the British physicist and visitor Arthur Schuster put it, 'we profit not only through the individually imparted advice of the master, but also through that which the other colleagues have done and we harvest thereby the advantage of the detailed study of a special problem without losing the all-embracing culture, developing interest in a series of different questions'. In Cambridge, Maxwell also aimed at practising this art of induction, but experimental physics there did not meet Rowland's standards of precision. In this respect the American traveller was too much of an engineer, for whom British experimentation with 'extemporised apparatus' was no longer sufficient. Rowland found that the lack of modern engineering he diagnosed was not only characteristic of practical work in experiment in the university, but, even more significantly, the merging of the art of the engineer with the canon of physics was still commented upon in the scholarly world. For instance, Robert Moon, a fellow of Queens' College, Cambridge, wrote:

> It may be an object of curiosity to some of my readers to know how the above measure of work came to be adopted . . . If my memory serves me rightly, Helmholtz, in his paper in Taylor's 'Scientific memoirs' for 1852, treats it a thing 'well known'; and in Thomson and Tait's 'Natural Philosophy' it appears to be regarded in the same light. Professor Maxwell informs us that 'If a body whose mass is one pound is lifted one foot high in opposition to the force of gravitation, a certain amount of work is done, and this amount of work is known among engineers as a foot-pound.' If this passage may be taken to afford the true clue, we can hardly fail to be struck with the originality of the suggestion, that a great philosophical principle may be established by 'engineering evidence'.[42]

Rowland, who knew Moon's publications, saw at once the consequences and wrote in his diary 'were his [Moon's] views to be received there would have to be a general transfer of many of those books which we have hitherto accounted as standard from our libraries to the waste basket'.[43] When he reached Glasgow and learnt there from William Thomson that Joule had just begun to redetermine the mechanical equivalent of heat in order to place the dynamic theory of heat on a better foundation, he could not begin quickly enough his own work of engineering evidence for this most important 'constant of nature'.

Translating European values

Rowland's year of travel had led him through an extremely heterogeneous scientific landscape. He soon sensed that energy physics seemed to be built on very swampy grounds. And it made matters worse that it was the very masters of precision measurement who had so impressed him who attested to the poor state of the foundations of this exact science. But he had also experienced the explorative potential that research resting on these measuring techniques possessed. At least in Helmholtz's laboratory he had been successful in measuring direct effects of the imponderable ether. So it should also be possible for him to show that beyond the many vague determinations of equivalencies between various natural forces he could establish the absolute measure of the mechanical equivalent of heat – 'one of the most important constants of nature', as he once put it. Only then would the science of energy rest on solid natural grounds. In the attempts of his European colleagues, he saw their material culture as well as their working attitude as the key issue to reform. As a new professor at Johns Hopkins University he explained to the Board of Trustees that in modern times 'it is useless to expect anything from extemporized apparatus'.[44] Modern physics needed the highest standards from the workshop, its tools and machinery, and in particular excellent craftsmanship from the instrument maker, to provide the researcher with precision instruments of outstanding quality. Furthermore a course in physics, well balanced between mathematical and experimental methods, was required to train 'physicists of precision'. Equally important would be space for experiments of investigation, which could be guaranteed through the provision of research fellowships. Only such a regime would be able to make 'the modern order of things' work, he argued, on the occasion of the opening of the new physics laboratory. According to him the modern order of things began with Galileo who was the first researcher and who only trusted his own reasoning because he had tested it experimentally.[45] In Rowland's laboratory the new generation of students would have to learn

> to test their knowledge constantly and thus see for themselves the sad results of vague speculation; they must learn by direct experiment that there is such a thing in the world as truth and that their own mind is most liable to error. They must try experiment after experiment and work problem after problem until they become men of action and not of theory. This, then, is the use of the laboratory in general education, to train the mind in right modes of thought by constantly bringing it in contact with absolute truth . . .[46]

Rowland regarded the practices of precision measurement which accompanied the different branches of the new science of energy as the most vulnerable link in the work of European scientists of the 1870s. He was clearly convinced that applying his qualifications as an engineer and physicist to this vexed subject of precision measurement, so relevant for thermodynamics, electrodynamics and optics, would bring Johns Hopkins to a leading role in science. Rowland thus translated his European experiences into an advanced regime of modern physics whose work required the establishment of an absolute system of units with reliable values. Determining the 'absolute value' of the mechanical equivalent of heat had the highest priority because every other constant, and therefore every important theory, hinged on this accurate value. He even told his students that 'conservation of energy is equality between different kinds of energy and could not be discovered before energy was measured'.[47] As shown above during his fieldwork he met important visionaries of the science of energy but had learned that the material culture and the actual practices of European science were not as reliable as he had thought. Moreover, some physicists and engineers even appreciated American engineering and industry. But at the same time many scientific techniques developed in Europe, such as the graphical method and Lagrangian analysis, appeared to Rowland to be highly useful for his project of transforming European values of scientific work. Furthermore he hired a German-born instrument maker, Theodore W. Schneider, to help build the most important instruments and machines and provide assistance in performing the work.[48]

The significance of the European experience for his laboratory physics at Johns Hopkins is indicated by his importation of scientific instruments from Europe; this experience also paved the way for his project on absolute measures in modern physics and the science of energy in particular. In this regard he had identified the direct method of determining the mechanical equivalent of heat as the most urgent to pursue. In what follows, however, we do not describe in full detail the various practices of this investigation. Instead we concentrate on the appropriation and further development of one single technique created and employed in his experiment to determine the mechanical equivalent of heat, and which can be identifed as characteristic of all of Rowland's most important works performed in the founding period of the laboratory: *the engineering technique of diminishing free play*. In Rowland's laboratory this technique shaped his research in significant ways, and it can most bluntly be described as exploring the means of perfecting the screw – this most simple but key machine in modern physics.

From the engineer's perspective, this 'pair of elements' had a long developmental history and its perfection was a measure of civilised societies and guaranteed to remove any indefiniteness. For the engineer-physicist Rowland the perfection of the screw (i.e. reducing free play) would provide certainty and advance the culture of physics. In order to improve Joule's experiment on the mechanical equivalent of heat, Rowland took great care to increase stability by perfecting the working mechanisms. He employed a chronograph in order to control the accuracy of steady motion of the shaft on which the calorimeter was mounted. This graphical method produced inscriptions which display fine lines of extremely narrow but equal distances (Figures 9.5, 9.6). By this means the experimenter could even visually check how accurately the cogwheels and screws used in the experimental set-up were manufactured and how uniform the rotation of the calorimeter was. Second, this inscription served as a fine time-grid on which the duration of the experimental trial was marked. This grid was used to mark significant transits of the mercury over the divisions of the thermometer. In doing so, Rowland succeeded in increasing the sensitivity of his measurements without decreasing their accuracy. Rowland's work with this method of preventing errors up to 1/100,000th of an inch finally culminated in his manufacture of diffraction gratings to improve optical researches. The ruling engine Rowland constructed could produce gratings on glass plates with 14,000 lines to an inch. With

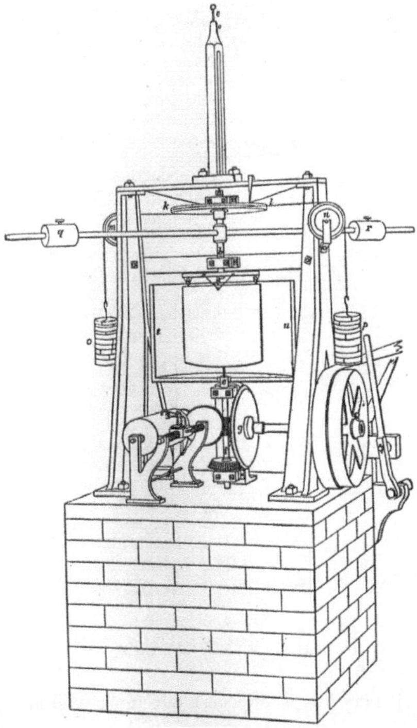

Figure 9.5 Rowland's set-up of the experiment to determine the mechanical equivalent of heat as published in the *Proceedings of the American Academy of Arts and Sciences* of 1879/80, p. 159.

a view towards diminishing the free play in any machine he also published an article on the 'screw' in the *Encyclopaedia Britannica* in which his error-controlling method is laid out in detail.

In making gratings for optical purposes the periodic error had to be very perfectly eliminated, since the periodic displacements of the lines only one-millionth of an inch from their mean position produced 'ghosts' in the spectrum. This was indeed the most sensitive method of detecting the existence of this error, and it was practically impossible to mount the most perfect screw without introducing it. A very practical method of determining this error was to rule a short grating with very long lines on a piece of common, thin plate glass cut in two with a diamond and superimpose the two halves with the rulings together and displaced sideways over each other one-half the pitch of the screw. On looking at the plates in a proper light so as to have the spectral colours show through it, dark lines appeared, which were wavy if there was a periodic error and straight if there was none.[49]

By providing an experimental technique which employed ghosts in order to prevent periodic errors created either through 'drunken screws or eccentric heads', Rowland delivered another masterpiece in precision measurement. In retrospect

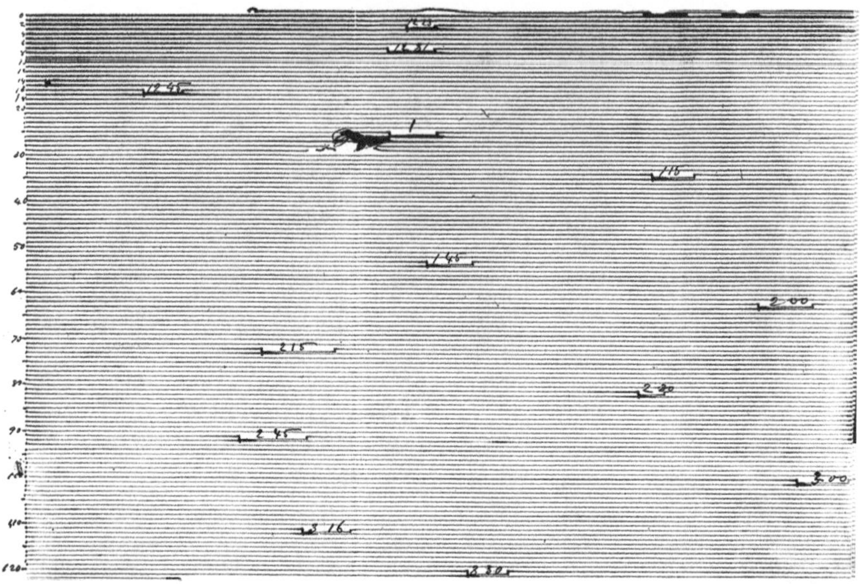

Figure 9.6 Inscription of a radiation experiment taken with a chronograph as depicted as detail in Figure 9.5.
Source: Henry A. Rowland Papers MS 6, Special Collections, Milton S. Eisenhower Library, The Johns Hopkins University.

his early experimental research in creating and measuring thermodynamic and electrodynamic effects appears as the beginning of this programme of engineering certainty in physics. In an elaborate theoretical and practical study on the construction of the dividing engine for the manufacture of gratings he developed the necessary 'Theory of Ghosts', with the view to discriminate quantitatively between the intensities of spectral lines.[50]

Rowland's famous remarks on ghosts, created through imperfect human practices in standardising material culture but finally controlled through human intellect, was a brilliant summary of his European fieldwork. It was no longer necessary for him to worry about 'ghosts' and other 'lawless-thoughted monsters' – they simply resulted from a lack of precision in the practice of scientific reasoning. He had further shown that modern physics was able to improve on the high standards of industry. In his essay on education he had indicated that 'human reason is too weak to go far without a guide and so we must appeal to experiment'. But he concluded that the modern physicist had to develop a specific attitude:

> All the facts which we have considered, the liability to error in whatever direction we go, the infirmity of our minds in their reasoning power, the fallibility of witnesses and experimenters, lead the scientist to be specially sceptical with reference to any statement made to him or any so-called knowledge which may be brought to his attention.[51]

It was precisely this attitude which was expressed in his scientific publications. In his paper on the determination of the mechanical equivalent of heat, for example, he gave a detailed account of all previous attempts in history to measure the mechanical equivalent of heat. It was an unprecedented attempt to write the developmental history of this 'golden number of that century', as Joule's constant at the turn of the century was called. He judged all previous attempts according to his new standard of workmanship in modern physics. In this survey only James Joule passed the test as a high-rank experimentalist and 'physicist of precision', even becoming the role model for the physicists of the new millennium. And yet Rowland's push for more precision and control simultaneously led to new and quite unexpected effects: Joule's mechanical equivalent of heat was valid only at the temperature range of 15.5 to 16.1 degrees Celsius and by using a specific thermometer. In a letter to Maxwell, Rowland wrote:

> My dear Professor Maxwell, I have now almost finished my research on the mechanical equivalent of heat with the curious result that the specific heat of water decreases from 0 to about 35 °C after which it increases. This is according to the air thermometer, but according to the mercurial thermometer it is almost constant . . .[52]

Rowland drew attention to the important role precision instrumentation, here thermometry, played in judging the validity of this scientific fact. With his new regime of measuring devices allowing absolute measures to be made, he hoped to personally settle this debate over the value of the mechanical equivalent of heat. And by setting up a standard heat laboratory at Johns Hopkins, announced in the University Circular and sent throughout the country, Rowland indicated that his laboratory was now prepared to make the most reliable comparisons of standards in the world:

> In order to secure uniformity throughout the country in certain physical standards, and to facilitate the use of the absolute system of heat measurement, it has been thought advisable to organize in the physical department of this University a subdepartment, where comparisons of standards can be made.[53]

But it should be mentioned that despite its importance Rowland's measures did not settle the debate over the correct value of 'the golden number of that century'. The process of how this was realised has, however, to be left to another historical investigation.[54]

Conclusion

This chapter has shown that the traveller Rowland experienced a very hetereogeneous field of local knowledge traditions rather than a unified culture of the exact sciences. His visit to major European laboratories made him an important witness of and mediator in a historical epoch in which the sciences not only varied distinctively in their rational procedures, experimental practices and moral outlook, but were also on the point of a significant change. Participating temporarily in these changing worlds of physics enabled him to identify a core issue modern science had yet to resolve: the provision of sufficient evidence for one of nature's most important constants, the mechanical equivalent of heat, which had already become the building block of the science of energy and the absolute system of units. Rowland's fieldwork in Europe and his dual qualifications as an engineer and a physicist convinced him to bring this matter to an end through precision measurement. The modern standards of Rowland's physics broke with traditional forms of physical investigation and created a close connection between industry, engineering and science. His impressive publication on the determination of the mechanical equivalent of heat in absolute measures strengthened modern scientists' belief in the universality of this constant, and was in fact the beginning of an era of engineering the final value of this 'golden number of that century'. His technique of reduction of free play, visible in its various forms in Rowland's researches on heat, electricity and optics as well as in his improvement of precision screws, became a leading motif for a significant portion of subsequent physical research. But it also represents the complex nature of precision measurement, its potential to advance control and liberation. In all his researches in which precision measurement was employed as the technique of control, the practices and results show that the advancement in precision measurement had always unexpected pay-offs which opened new fields of research.[55]

Acknowledgements

I would like to thank the participants of the workshop, the members of the independent research group 'Experimental History of Science' at the Max Planck Institute, and Christoph Bonneuil, Bob Brain, Simon Schaffer and Richard Staley in particular for their helpful comments. Richard Staley also translated substantial parts of this chapter. I quoted from manuscripts and correspondence held at the Eisenhower Library archive at Johns Hopkins University, Baltimore, at Cambridge University Library (UK), and at Berlin-Brandenburg Academy of Sciences archive, and am grateful for permission to use this material.

Notes

1 For this chapter I made particular use of H.A. Rowland's 'European Tour Notebook', held at the Johns Hopkins University in the Milton S. Eisenhower Library, Special Collections, Rowland papers, MS 6, Box 22 (archive material hereafter cited as JHU, Rowland papers, MS . . .) and the relevant private and official correspondence by him held in JHU. The few accounts which refer to Rowland's travel describe it either as a masterpiece of self-education (S. Rezneck, 'The education of an American scientist: H.A. Rowland, 1848–1901', *American Journal of Physics*, 28, 1960, 155–62) or convincingly show relations between his travel and his later electro-magnetic research (see J.D. Miller, 'Rowland and the nature of electric currents', *Isis*, 63, 1972, 5–27.

2 See for example G.K. Sweetnam, *The Command of Light: Rowland's School of Physics and the Spectrum*, Philadelphia: American Philosophical Society, 2000. Idem, 'Precision implemented: Henry Rowland, the concave diffraction grating, and the analysis of light', in M.N. Wise (ed.) *The Values of Precision*, Princeton, N.J.: Princeton University Press, 1995, pp. 283–310; Klaus Hentschel, 'The discovery of the redshift of solar Fraunhofer lines by Rowland and Jewell in Baltimore around 1890', *Historical Studies in the Physical Sciences* (*HSPS*): 23, 1993, 219–77.

3 H.A. Rowland, 'On the mechanical equivalent of heat, with subsidiary researches on the variation of the mercurial from the air thermometer, and on the variation of the specific heat of water', *Proceedings of the American Academy of Arts and Sciences*, vol. XV, 1880, 75–200, reprinted in *The Physical Papers of Henry Augustus Rowland*, Baltimore: Johns Hopkins University Press, 1902, pp. 343–468; idem, 'Appendix to paper on the mechanical equivalent of heat, containing the comparison with Dr. Joule's thermometer', *Proceedings of the American Academy of Arts and Sciences*, vol. XVI, 1881, 38–45, reprinted in *The Physical Papers of Henry Augustus Rowland*, pp. 469–80.

4 'Ich meine daß die Verbürgung auf höchstens 1/1000 Fehler vorausgesetzt werden muß, wenn ein internationales Laboratorium die Einheit der Wissenschaft und der Technik octroyren will. Zu diesem Zwecke aber müßte dieses Laboratorium wesentlich genauer arbeiten, als die Physiker bis jetzt gethan haben . . . Eine Anzahl erfahrener Physiker aber zu diesem Zwecke dauernd zu vereinigen, wird unmöglich sein. Es weiß ja auch jeder von uns, daß er zu Hause am besten arbeitet', F. Kohlrausch to H.A. Rowland 28 April 1882, JHU, Rowland papers, MS 6.

5 Historians of science have investigated the nineteenth-century culture of precision in a number of ways. See for example T. Porter, *Trust in Numbers: the Pursuit of Objectivity in Science and Public Life*, Princeton, N.J.: Princeton University Press, 1995; M.N. Wise, *The Values of Precision*, op. cit.; K.M. Olesko, *Physics as a Calling: Discipline and Practice in the Königsberger Seminar for Physics*, Ithaca, N.Y.: Cornell University Press, 1991; L. Daston and P. Galison, 'The image of objectivity', *Representations*, 40, 1992, 81–128; L. Daston, 'The moral economy of science', *Osiris*, 10, 1995, 3–24; S. Schaffer, 'Accurate measurement is an English science', in M.N. Wise, *The Values of Precision*, op. cit., pp. 135–72. This chapter investigates the meaning of experiment within this context.

6 See BAAS, *Report of the Committee appointed by the British Association on Standards of Electrical Resistance* (Newcastle upon Tyne, 1863), London, 1864, pp. 111–76. Cf. S. Schaffer, 'Late Victorian metrology and its instrumentation: a manufactory of ohms', in R. Bud and Susan E. Cozzens (eds) *Invisible Connections. Instruments, Institutions, and Science*, Bellingham: SPIE-Press, 1992, pp. 23–65.

7 In the 1860s indirect determinations of the value of the mechanical equivalent of heat by Joule led to deviating results which had a strong impact on energy physics in general. See H.O. Sibum, 'An old hand in a new system', in J.-P. Gaudillière and I. Löwy (eds) *The Invisible Industrialist – Manufactures and the Production of Scientific Knowledge* (Houndmills: Macmillan, 1998), pp. 23–57, and C. Smith, *The Science of Energy: A Cultural History of Energy Physics in Victorian Britain*, Chicago: University of Chicago Press, 1998.

8 J.C. Maxwell, *A Treatise on Electricity and Magnetism*, 2 vols., 3rd edn, Oxford: Clarendon Press, 1891, republication by Dover Publications (New York, 1954), the hereafter quoted version; H.A. Rowland, 'On magnetic permeability and the maximum of magnetism of iron, steel and nickel', *Philosophical Magazine*, 46, 1873, 140–59. Rowland had sent this publication first to the *American Journal of Science* but it was not accepted for publication. See letters J.D. Dana to H.A. Rowland 20 June 1873, 2 July 1873, and Maxwell's letter to Rowland 9 July 1873 in which he recommends immediate publication in the *Philosophical Magazine* because the *Transactions* of the Royal Society would be too late; see also Dana's excuse, Dana to Rowland, 3 August 1873, in N. Reingold, *Science in Nineteenth Century America*, New York: Hill & Wang, 1964, pp. 264–6.

9 According to John David Miller, Rowland had discussed with Maxwell his early plans for an experiment to produce magnetic effects by means of a rotating non-conducting disc. If one could make the surface density and the velocity of this electrified surface so great that the magnetic force is a measurable quantity, Maxwell stated in his *Treatise*, 'we may at least verify our supposition that a moving electrified body is equivalent to an electric current', in J.C. Maxwell, *A Treatise on Electricity and Magnetism*, vol. II, op. cit., p. 370. See J.D. Miller, 'Rowland and the nature of electric currents', *Isis*, vol. 63, no. 216, 1972.

10 Schaffer, 'Late Victorian metrology', op. cit.; B. Hunt, 'The ohm is where the art is: British telegraphic engineers and the development of electrical standards', *Osiris*, 9, 1994, 48–63; J. Bennett *et al.*, *Empires of Physics. A Guide to the Exhibition*, Cambridge: Whipple Museum of the History of Science, 1993.

11 C. Smith and M.N. Wise, *Energy & Empire. A Biographical Study of Lord Kelvin*, Cambridge: Cambridge University Press, 1989; C. Smith, *The Science of Energy*, op. cit.; see in particular BAAS, *Report* (Newcastle upon Tyne), op. cit., pp. 111–76.

12 S. Schaffer, 'Accurate measurement is an English science', op. cit.

13 For a detailed study on the Cambridge world of physics see A. Warwick, *Masters of Theory: A Pedagogical History of Mathematical Physics at Cambridge University, 1760–1930*, Chicago University Press, forthcoming.

14 Rowland to D. Gilman, 14 August 1875, JHU, Gilman collection, MS.

15 Joule to Thomson, 29 December 1869, Cambridge University Library (hereafter CUL), MSS Add 7342, J 287.

16 Rowland to Gilman, 14 August 1875, JHU, Gilman collection, MS.

17 Ibid.

18 R. Noakes, 'Imponderables in the balance: rewriting the history of Victorian physics and psychical research', unpublished paper held at HSS Vancouver 2000.

19 G. Gore, 'On the magnetism of electrodynamic spirals', *Philosophical Magazine and Journal of Science* 40, 1870, 264–8. H. Highton, 'On the relations between chemical change, heat, and force, with a special view to the economy of electrodynamic engines', *The Quarterly Journal of Science*, vol. I, 1871, 77–94. This latter journal was edited by William Crookes.

20 G. Gore, *The Art of Scientific Discovery or the General Conditions and Methods of Research in Physics and Chemistry*, London: Longmans, Green and Co., 1878, pp. 164–5. According to him Joule's experiment has given evidence for the 'equivalency of forces (or of cause and effect)'. He is quite hestitant to claim the universality of that principle when discussing Joule's conversion experiment: 'Transformations of other forces have been effected, and their equivalents determined.'

21 Figure 9.1, H.A. Rowland, *European Tour Notebook*, 'Notes on mechanical equivalent of heat H.A.R., Sep. 17, 1875. To obtain the mechanical equivalent of heat by revolving magnet under a copper disc and measuring the moment acting on the copper disc the disc being stationary', p. 29. Figure 9.2, p. 36, Figure 9.3, pp. 37–8.

22 Reuleaux's letters created an image of the American culture which was much advanced in automation and standardising machinery, which would guarantee high quality of goods. Compare F. Reuleaux, *Briefe aus Philadelphia*, Braunschweig: Vieweg, 1877. The Gewerbeakademie became in 1879 the Technical University in which Reuleaux became head of the department of Maschineningenieurwesen and in 1890/1 the Rektor of the Technical University.

23 This collection of mechanical models was known as the 'Reuleaux Sammlung'. It was destroyed during the Second World War, but must have amounted to several hundred models.

24 F. Reuleaux, *Theoretische Kinematik. Grundzüge einer Theorie des Maschinenwesens*, Braunschweig: Vieweg und Sohn, 1875. English translation, by A.B.W. Kennedy, *The Kinematics of Machinery. Outlines of a Theory of Machines*, London, 1876. Reprint, with a new introduction by E.S. Ferguson, New York: Dover Publications, 1963.

25 F. Reuleaux, *Theoretische Kinematik*, 1875, pp. XIII–XIV.
26 F. Reuleaux, *The Kinematics of Machinery*, reprint 1963, p. 22. In the German edition the sentence runs 'Es handelt sich darum, die Maschinenwissenschaft der Deduktion zu gewinnen' (1875, p. 26).
27 Ibid., p. 20; German edition, 1875, p. 23: 'So lange sie nicht auch ohne Erfindungsgeschichte zu den Elementen und Mechanismen der Maschinen zu gelangen vermag, darf sie den Karakter einer Wissenschaft nicht für sich in Anspruch nehmen.'
28 F. Reuleaux, *Theoretische Kinematik*, 1875, p. XIV; English translation by H.O. Sibum; in the English edition the preface by Reuleaux is replaced by a different one written by the translator A.B.W. Kennedy.
29 A complementary step is the sign language he developed in order to overcome problems in engineering practices by means of visual representations based on geometry. He compared its importance with the sign language of chemistry.
30 'Von den Engländern für uns bedeutungsvoll "Freiheit" [freedom] genannt, also als "der kosmischen Freiheit näher stehend," erkannt' (F. Reuleaux, *Theoretische Kinematik*, 1875, asterisk on p. 233).
31 The German expression is 'Handlungsspielraum'.
32 Reuleaux, *The Kinematics of Machinery*, 1876, pp. 237–8.
33 Ibid., pp. 241–2.
34 Letters from Rowland to Helmholtz, 13 November 1875, Berlin Brandenburg Academy of Sciences archive, Rowland to Gilman, Berlin, 19 November 1875, JHU, Gilman collection, MS. For a discussion see Miller, 'Rowland and the nature of electric currents', op. cit.
35 For a detailed comparision of Weber's and Maxwell's electrodynamics see J. Meya and H.O. Sibum, *Das Fünfte Element. Wirkungen und Deutungen der Elektrizität*, Reinbek bei Hamburg, 1987, pp. 195–232. For a recent discussion see J.Z. Buchwald, 'Electrodynamics in context: object states, laboratory practice, and anti-romanticism', in D. Cahan (ed.) *Hermann von Helmholtz and the Foundations of Nineteenth-Century Science*, Berkeley: University of California Press, 1993, pp. 334–73; W. Kaiser, 'Helmholtz's instrumental role in the formation of classical electrodynamics', in D. Cahan, *Hermann von Helmholtz*, pp. 374–402; O. Darrigol, 'Baconian bees in the electromagnetic fields: experimenter-theorists in nineteenth-century electrodynamics', *Stud. Hist. Phil. Mod. Phys.*, 30: 3, 1999, 307–45.
36 A report and interpretation of Rowland's experiment was published by H. von Helmholtz, 'Bericht betreffend Versuche über die elektromagnetische Wirkung elektrischer Convection, ausgeführt von Hrn. Henry A. Rowland', *Annalen der Physik und Chemie*, 158, 1876, 487–93. H.A. Rowland, 'On the magnetic effect of electric convection', *American Journal of Science and Arts*, 15, 1878, 30–8. For a detailed account of Rowland's experiment see J.D. Miller, 'Rowland and the nature of electric currents', op. cit. For a further discussion see J.Z. Buchwald, *From Maxwell to Microphysics. Aspects of Electromagnetic Theory in the Last Quarter of the Nineteenth Century*, Chicago: University of Chicago Press, 1985, pp. 73ff.; W. Kaiser, 'Helmholtz's instrumental role', pp. 374–402.
37 Rowland to Gilman, Berlin, 6 January 1876, JHU, Gilman collection, MS.
38 Rowland to Gilman, Graz, 26 March 1876, JHU, Gilman collection, MS.
39 H.O. Sibum, 'Reworking the mechanical value of heat: instruments of precision and gestures of accuracy in early Victorian England', *Studies in History and Philosophy of Science*, 26: 1, 1995, 73–106.
40 Sir F. Pollock to De Morgan, 1869, in W.W. Rouse Ball, *Cambridge Papers*, London: Macmillan, 1918, p. 288.
41 See H.O. Sibum, 'Working experiments: a history of gestural knowledge', *Cambridge Review*, 116, 1995, 25–37; J.C. Maxwell to F. Jenkin, 18 November 1874, CUL, MSS Add 7655/242.

42 R. Moon, 'On the measure of work in the theory of energy', *Philosophical Magazine*, April 1874, 291–4, p. 293.

43 H.A. Rowland, 'Measure of work in the theory of energy. A reply to articles by R. Moon, M.A.', in the *Phil. Mag.*, September 1873 and April 1874. Unpublished manuscript, Rowland Archive, JHU, Rowland papers, MS 6. Ser. 5, Box 39.

44 H.A. Rowland, Report to the Board of Trustees, undated, JHU, Rowland papers, MS 6, pp. 17–18. Although he was a great admirer of Faraday and other British scientists, Rowland wanted to make it clear that the 'science of the future' had to go beyond experimental practice, based as that was on improvising performances.

45 H.A. Rowland, lecture notes prepared for the opening of the physics laboratory, JHU, Rowland papers, MS 6, Ser. 5.

46 H.A. Rowland, 'The physical laboratory in modern education', in *The Physical Papers*, op. cit., pp. 614–618, p. 617.

47 H.A. Rowland, 'Progress of science from qualitative to quantitative', in *Notebook on Thermodynamics*, 1877, JHU, Rowland papers, MS 6.

48 According to my enquiries in the lists of passengers arriving at the United States it is very likely that Schneider arrived with his family as a young boy in 1857. I am very grateful to Wolfgang Grams who introduced me to the archival material of the 'Forschungsstelle deutscher Auswanderer' at the University of Oldenburg. For T.W. Schneider's importance in the work of making diffraction gratings see K. Hentschel, 'The discovery of the redshift', op. cit., pp. 232f.

49 H.A. Rowland, 'Screw', *Encyclopaedia Britannica*, 9th edition, vol. XXI, and in Rowland, *The Physical Papers*, op. cit., pp. 506–11, p. 510.

50 'The effect of small errors of ruling is to produce diffused light around the spectral lines. This diffused light is subtracted from the light of the primary line, . . . Thus the effect of the periodic error is to diminish the intensity of the ordinary spectral lines (primary lines) from the intensity 1 to $J_0^2(b\mu a_1)$, and surround it with a symmetrical system of lines called ghosts', H.A. Rowland, 'Gratings in theory and practice', in *The Physical Papers*, op. cit., pp. 525–44, p. 538.

51 H.A. Rowland, 'The highest aim of the physicists', in *The Physical Papers*, op. cit., pp. 668–78.

52 Rowland to Maxwell, 1877, CUL MSS Add 7655/II/7184; for a detailed account see H.O. Sibum, 'An old hand in a new system', op. cit.

53 *Physical Laboratory: Comparisons of Standards*, Johns Hopkins University Circulars, 3, 1880, 31. H.A. Rowland, *The Physical Papers*, op. cit., pp. 477–80, p. 477.

54 In the last two decades of the nineteenth century the publications concerning the correct value of the mechanical equivalent of heat increase in number and size. See for example R.T. Glazebrook, 'The value of the mechanical equivalent of heat, deduced from some experiments performed with the view of establishing the relation between the electrical and mechanical units; together with an investigation into the capacity for heat of water at different temperatures', *Philosophical Transactions of the Royal Society of London* (A), 184, 1894, 361–504; O. Reynolds and W.H. Moorby, 'On the mechanical equivalent of heat', *Philosophical Transactions of the Royal Society of London* (A), 190, 1898, 301–422.

55 See in particular R. Staley's chapter in this volume. For the relationship between modern science and metrology see S. Schaffer, 'Modernity and metrology', in L. Guzzetti (ed.) *Science and Power: The Historical Foundations of Research Policies in Europe*, Florence, 8–19 December 1994, 71–91.

10 Travelling light

Richard Staley

There are a few key episodes from Albert Michelson's early career that are often described, including especially the story of his months in Berlin and Potsdam in 1881, trying to get an ether-drift experiment to work while on leave from the US Navy, followed by his successful collaboration with Edward Morley in Cleveland in 1887, again failing to detect an ether wind with a more precise version of the same experiment. Through the 1880s and 1890s Michelson was a scientist on the move, building a career around the performance of a few brilliantly designed experiments, and the further development of the instrument that lay at the heart of his ether-drift apparatus. It has often been noted that both Michelson and the Nobel Prize committee that honoured him in 1907 placed more emphasis on the interferometer and its applications than on the ether-drift experiment, a fact that has seemed puzzling in the light of the experiment's importance in the development of electrodynamics and special relativity. This chapter will explore Michelson's engagement with instruments in order to deliver new insight into his experimental work. Our histories of this traveller have focused too narrowly on single experiments to discern the many links tying both the origins of his ether-drift apparatus, and the uses he later gave it, to instruments and concerns that stem from his earliest research on the velocity of light. They have also been too celebratory to recognise some of the deep challenges that underlay Michelson's search to give a distinctive stamp to his research and reputation. In the second part of the chapter I focus on Michelson's research on standards of length, outlining the productive relations between his scientific work and precision industry in the US and showing how fully Michelson, like Rowland, exemplifies the role of an 'academic mechanician', that Evans and Warner have identified as a typically American phenomenon, bridging science and engineering. This is also the research that most fully explains why Michelson received the Nobel Prize when he did, a question unanswered in earlier historical accounts.[1]

The origins of an instrument

In September 1880 Michelson set sail for a year's study leave in Europe, thankfully released from the requirement to serve at sea after four years instructing physics and chemistry at the Naval Academy in Annapolis. The previous two years had seen him develop an identity and reputation as one of America's most promising young research scientists, and so far his search for the professional position to match it had been unsuccessful.[2] But he would have been glad to take the opportunity that paid leave offered – perhaps the chance to learn at the feet of some of Europe's well-known physicists would be even more valuable than continuing teaching. Michelson carried letters of introduction provided by Edward Pickering, head of the Harvard Observatory, and began his tour with a stay in Helmholtz's newly constructed laboratory in Berlin.[3] He did not have much to learn from Helmholtz's laboratory classes (and so attended theoretical physics classes and worked on mathematics and mechanics at home), but was soon pouring a great deal of energy into an experiment to crack one of the toughest nuts in optical research.[4] Michelson went to Helmholtz with an idea – conceived somewhere between the east coast of the US and Germany – to address a view James Clerk Maxwell had expressed in a letter to David Todd at the Nautical Almanac Office (later published in *Nature*). Maxwell wrote that the velocity of the earth with respect to the (stationary) ether alters the velocity of light by a quantity that, depending on the square of the ratio of the earth's velocity to that of light, was 'quite too small to be observed'.[5]

Michelson had won his reputation by turning a demonstration experiment for his class into a new determination of the velocity of light by terrestrial measurement, joining the distinguished French experimentalists Fizeau, Foucault and Cornu. Improving on Foucault's revolving mirror method with an arrangement that was elegant in its simplicity, and which significantly increased the accuracy of the determination by increasing the magnitudes to be measured, Michelson sent a beam of light back and forth between a measuring station, revolving mirror and lens, and a mirror placed some distance away (see Figure 10.1). During early trials in 1878, his commanding officer Captain Sampson introduced Michelson to Simon Newcomb, who was in the final stages of planning a more elaborate velocity of light experiment.[6] Then Director of the Nautical

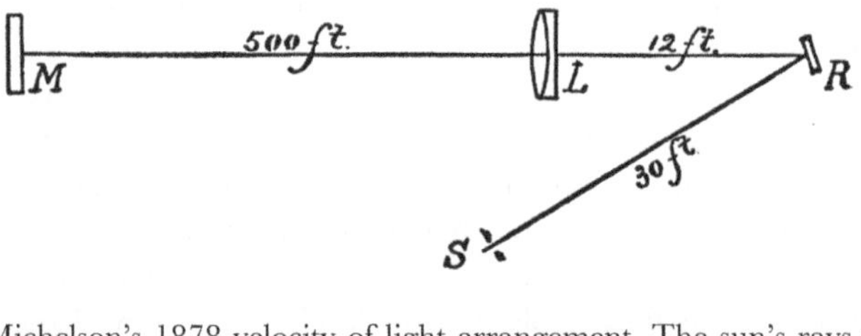

Figure 10.1 Michelson's 1878 velocity of light arrangement. The sun's rays pass through the slit *S* to the revolving mirror *R*, and are reflected to a fixed plane mirror *M*. An image of the slit is formed on the surface of *M* by means of the lens *L*, and the light is reflected to retrace its path. When *R* is at rest the return image coincides with the slit itself; when *R* rotates rapidly the image is displaced through an angle twice the displacement of the mirror during the time required for the light to travel from *R* to *M* and back again.

Source: A.A. Michelson, 'Experimental determination of the velocity of light', *Proceedings of the American Association for the Advancement of Science* 27 (1878), 71–7, Fig. 1.

Almanac in nearby Washington, the astronomer was one of America's most prominent and influential scientists. The success of Michelson's trials helped Newcomb gain funding, and after Michelson completed his experiment (with improved instruments) in 1879, he was transferred to the Nautical Almanac Office to assist Newcomb's work.[7]

Together with Newcomb's patronage, Michelson's need to ensure he had the best possible instruments brought him into contact with an ever-widening circle of scientists, and he soon complemented his focus on the speed of light with wider studies in optics. Significantly, a letter from Pickering shows that he introduced Michelson to experiments in interferential refractometry in August 1879, after Maxwell's letter was received in the Almanac (where Michelson might first

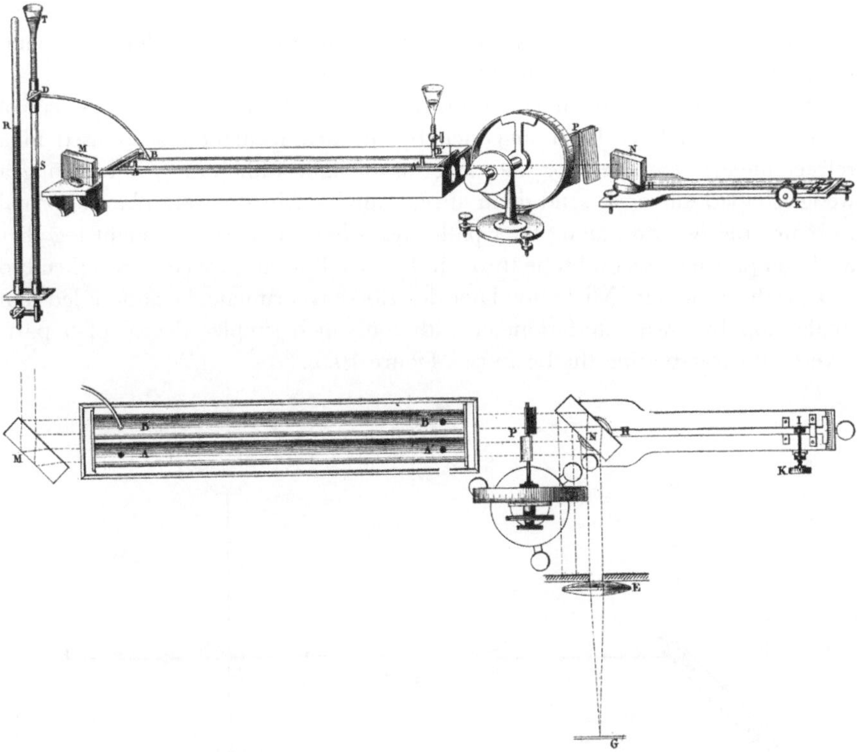

Figure 10.2 The Jamin refractometer, using two equally thick plane parallel glass plates M and N silvered on their rear surfaces. Light from an extended source incident on M at about 45° gives rise to two beams, one reflected from the front surface of M and rear surface of N, the other reflected from the rear surface of M and the front surface of N; the two are recombined to give interference patterns in the focal plane of the telescope E. Chambers of gases or liquids are inserted in AA' and BB', and P carries compensator plates.

Source: J. Jamin, 'Mémoire sur les variations de l'indice de réfraction de l'eau à diverses pressions', *Annales de chimie et de physique* 52 (1858), 163–71.

have learnt of it), but before it was published. Pickering followed up an earlier conversation by referring Michelson to papers in which E.E. Mascart followed the example of Jules Jamin, and used an arrangement of parallel mirrors to split a beam of light into two components that travelled different pathways – parallel to one another and in close proximity – that could then be recombined, forming interference patterns (see Figure 10.2).[8]

With an arrangement of this kind it was possible to measure relative differences in the velocity of light passing through different media placed in each arm, that were of the order of fractions of the wavelength of light. Thus his work on the velocity of light gave Michelson an immediate interest in the subject of Maxwell's letter; and his awareness of refractometer experiments gave him the confidence that a second order result was 'easily measurable'.[9] Since the difficulty was that all known methods of measuring the velocity involved the necessity of returning the light over its path (leading to a compensation minimising the effect sought), Michelson looked for a means of sending two rays on different round journeys perpendicular to each other.

At first he tried an arrangement using a masked lens to form two pencils of light (this could have been suggested by the arrangement Fizeau used in a refractometer experiment on the dragging of light with water), introducing a wedge-shaped mirror to send them at right angles. Mirrors were placed normal to return the two rays along their pathways, where they were brought together again on passing a second time through the lens. But this proved very difficult to work with, something Michelson later described as fortunate, because it led him to develop his own interferometer with the much simpler device of a part-silvered mirror splitting the beam (see Figure 10.3).[10]

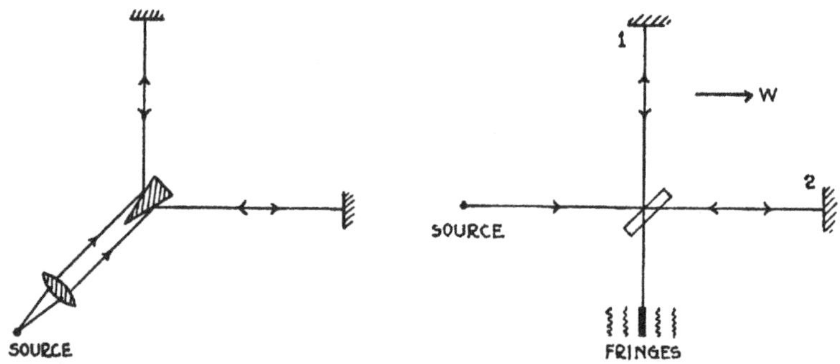

FIG. 2

FIG. 3

Figure 10.3 The two arrangements with which Michelson attempted to produce interference of light beams that travel parallel and transverse to the velocity *w* of the apparatus through the ether. The device Michelson tried unsuccessfully is depicted in his Fig. 2.

Source: A.A. Michelson, 'Report, conference on the Michelson–Morley experiment', *Astrophysical Journal* 68, no. 5 (1928), 342–5, p. 343.

Previous historians have speculated on the provenance of this feature, attributing it to Michelson's awareness of the work of Fizeau, or his familiarity with the naval sextant.[11] But his papers show that he had already used such a mirror himself. In Foucault's velocity of light experiment the returning light beam was deflected just 0.7 mm. To make it possible to measure the position of a beam so close to the original without breaking the light path, Foucault had inserted a piece of glass, observing the faint reflection of the returning ray. Michelson briefly used an improved version, part silvering the glass to strengthen the reflection (see Figure 10.4). But when a new lens allowed the light path to be lengthened from the 500 feet of his first trials to 2,000 feet the returning beam was displaced some 5 cm. This made it possible to measure the position of the beam without the observer's head getting in the way. So Michelson dispensed with the mirror (using instead a combined slit and scale).[12] But it is likely that in 1878 he observed the cross of light the device produced, and decided in 1881 to use both arms in a new, table-top velocity of light measurement. The continuity between the two arrangements is emphasised by the fact that in both cases the partially

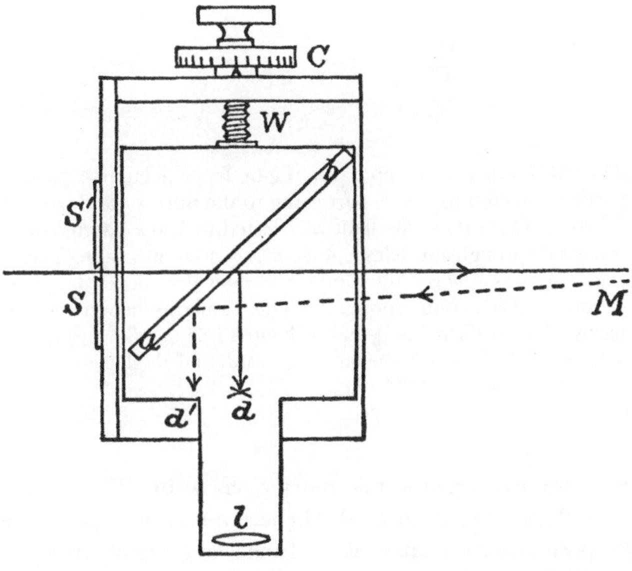

Figure 10.4 The slit and micrometer Michelson used in 1878. *S* is the slit; *ab* a piece of plane glass partly silvered. The light proceeds in the direction *SM* and is returned in the opposite direction, part being reflected from *ab*, forming an image of the slit made to coincide with cross-hairs at *d*. When the mirror revolves the image is displaced to *d′*. *ab*, *d*, and *l*, the lens for viewing the image, were moved by the screw *w* in the direction *S′S* until the displaced image again coincided with the cross-hairs. The distance moved, equal to the displacement, was read on the divided circle *c*.

Source: A.A. Michelson, 'Experimental determination of the velocity of light', *Proceedings of the American Association for the Advancement of Science* 27 (1878), 71–7, Fig. 3.

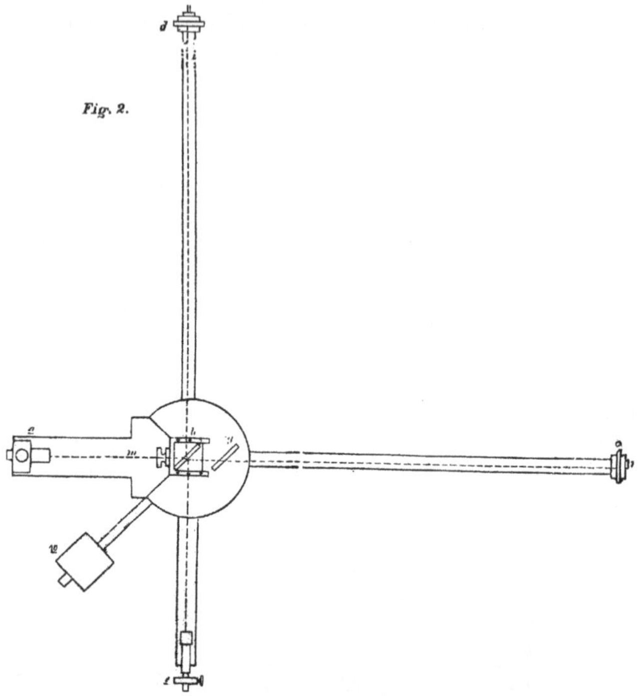

Figure 10.5 The 1881 ether-drift apparatus. Light from a lamp *a* passes through the partially silvered mirror *b*, part going to the mirror *c* and part reflected to the mirror *d*. From these the light is reflected to *b*, the interference fringes being observed through the telescope at *e*. *g* is a compensating glass cut from the same piece as *b*; *m* is a micrometer screw which moves the plate *b* in the direction *bc*; *w* is a counterpoise. Note the similarity between the housing of the partly silvered mirror and that in Figure 10.4.

Source: A.A. Michelson, 'The relative motion of the earth and the luminiferous ether', *American Journal of Science* 22 (1881), 120–9, p. 123.

silvered mirror was carried on a micrometer screw. In 1878 this allowed measurement of the displacement; in 1881 Michelson used it to provide adjustments in the width, position or direction of the fringes (see Figure 10.5).

Thus, paying close attention to Michelson's instruments shows practical connections that have not been visible to historians focused on the ether-drift experiment alone: Michelson's instrumental breakthrough came from pushing to the centre of the apparatus a device that had played a necessary but ancillary role in his earlier experimental work.

When Helmholtz heard of Michelson's plan he expressed doubt as to its feasibility, believing temperature effects might swamp the interference phenomena, but nevertheless made the facilities of the laboratory available.[13] Newcomb helped find funding from Alexander Graham Bell ($2,000), and Michelson asked the respected instrument makers Schmidt & Haensch to construct the apparatus, swinging the cross of light on a revolving turntable. As the apparatus was turned, shifts in the interference fringes would disclose relative differences in the velocity of light in different orientations. A letter to C.A. Bell (who administered the funds his uncle provided) illustrates Michelson's view of his reputation at the time:

> I took the liberty of questioning the prudence of the offer which Prof. Bell was kind enough to make, for as I then said, I was young and therefore liable to err, and my entire reputation as a 'scientist' was based upon a single research, which fortunately was successful. Since, however, he was pleased to differ with me, I gratefully acknowledge the generous gift.
>
> I have not however undertaken this work without consulting several prominent scientific men. The answer they all give is that if successful the success would be grand; but that it is at present impossible to say whether the experiments are practicable or not. All, however, advise me to try. So that, with this assurance and that of Prof. Bell, that in either case he will consider the money well spent, I shall begin the work with greater confidence.[14]

When Michelson set up the apparatus on a stone pier in the laboratory, he found it impossible to observe the interference bands on account of vibration – and so obtained permission to move from busy Berlin to the astronomical observatory then under construction in nearby Potsdam. Ordinary stone piers there were too unstable for his purposes also. Only the cellar under the great dome, with its massive foundations, provided conditions quiet enough to perform the experiment.[15] When Michelson came to publish his work this series of relocations provided vivid, if imprecise, indications of the sensitivity of the apparatus.[16] The apparatus was very difficult to use, and had to be returned to the makers for improvements. But the result of the experiment seemed clear – Michelson decided the small shifts observed were due solely to experimental imperfections. This 'null result' clearly shaped Michelson's perception of the experiment, and the apparatus. He had expected to be engaged in a kind of cosmic meteorology, establishing the movement of the earth in relation to the solar system and stars by testing the extent of the drift in different orientations and times of the year. His failure to detect the ether led Michelson to doubt that it was stationary, but did not prompt him to question its existence.

Michelson left Potsdam and Berlin with a result, which he published in the *American Journal of Science*, and a new instrument. After spending the summer studying with Quincke and Bunsen in Heidelberg, he travelled to Paris, the home of so many of the physicists he had drawn on in developing both his experiments. His velocity of light measurement meant Michelson's reputation preceded him: in Paris Alfred Cornu expected to be meeting a much older man. But attempting to put his second, ether-drift experiment on show, Michelson discovered that it did not travel well. One disappointment was recognising that he had neglected to take into account the small effect of the passage of light transverse to the direction of a possible ether wind. Alfred Potier (and later H.A.

Lorentz) pointed out this error.[17] To establish the absence of a second-order effect satisfactorily, the light would need to travel a significantly longer distance. But it was also difficult simply to get the apparatus working again. On his first attempt Michelson found it impossible to demonstrate the interference fringes, a fact that did not surprise Cornu, who maintained that Fresnel had attempted the same thing fifty years ago without success. Only after four days' work was Michelson able to show his audience the bands of light and darkness the instrument was capable of delivering.[18] Clearly it would take considerable work to add the lustre of a second experiment, or even a new instrument, to the single research on which Michelson's reputation rested to date.

Singular experiments and the culture of repetition

While Michelson travelled Europe, Simon Newcomb had begun obtaining preliminary results. When it became clear that his value for the velocity of light would be as much as 200 km/s lower than Michelson's (and even further away from Cornu's figure), the worried astronomer interrogated both his own procedures and those Michelson had followed in order to track down the reason for the discrepancy. Newcomb engaged in an extensive theoretical deconstruction and, as far as possible, a material reconstruction of Michelson's arrangement. Had it been possible to replicate the experiment he would certainly have done so. Newcomb sent countless urgent queries trying to catch up with what Michelson had done in 1879, in letters that followed the young scientist across Europe. Their correspondence makes visible a number of features of physics in the period: the working network that existed between particular American experimentalists, the instrument makers and standards they trusted, and a kind of economy of precision instruments that brought them together.

Michelson had used a scale from William A. Rogers at the Harvard Observatory, which had been returned to Rogers for rechecking.[19] Michelson advised Newcomb that the Fauth firm had made the mirror, but that he would recommend using Alvan Clark instead.[20] The nagging problem of the whereabouts of the revolving mirror was finally solved when it was found much later in the home of Michelson's father-in-law.[21] The central issue of Newcomb's enquiries was one of distance measurements, both of small increments and large distances. Newcomb, for example, wondered from which point on the revolving mirror Michelson had measured – the centre of the mirror as the publication assumed, or from its top. Michelson conceded that it was the latter and the appropriate correction hadn't been made. His number went down a few kilometres per second.[22] Then there was the question of the tape Michelson used in measuring the baseline. Were the measurements made from the metal ring at the end, or the first division, and by whom? Newcomb got different results than Michelson when making a rough comparison with the standard metre scale and so compared his own tape against Michelson's, independent of any hypothesis concerning the scales of measure.[23] These exchanges show a common but careful recourse to standards, that, however, was not sufficient to answer the questions at issue.

For Newcomb, meaningful measures could only be established by linking specific instruments with specific protocols. This was a lesson he had learnt early in his observational career, trying to co-ordinate the results obtained by different astronomers (even within the Naval Observatory alone).[24] He also made it central to his critique of John Stallo's attack on science in 1882, and it underlay his view that scientific specification of the content of concepts would transform economic theory.[25] But in 1881, even after all his work reconstructing as precisely he could Michelson's methods of measurement, there remained an 'embarrassing' discrepancy between their results.[26]

No longer so clearly a single event in time, Michelson's first research – itself involving trials and performances over nearly two years – was also no longer so clearly successful. But even a revised result was not enough for Newcomb. Seeking further assurance he obtained funding for Michelson to make a second determination on returning to the United States. Taking up a new professorship in physics at the Case School of Applied Science in 1882, Michelson set up his instruments along the tracks of the New York, Chicago and St Louis Railroad Company in Cleveland.[27] This time his number came out 50 km/s lower, and finally Michelson and Newcomb had results that (just) came within their error estimates. Newcomb gave two figures, one using only results supposed to be nearly free from constant errors of 299,860 km/s (with a variation of less than 10 km/s between the separate measures) and one, including all determinations, of 299,810 ± 50. Michelson's figure from his Cleveland trials was 299,853 ± 60 km/s, and he reduced his Annapolis result by 34 km/s to read 299,910.[28] It was crucial for Newcomb that it be made clear that Michelson's repetition was an independent experiment despite the many connections between their work. Newcomb wrote in an introductory note to Michelson's paper 'no instructions and suggestions had been sent to him except such as related to the investigation of possible sources of error in the application of his method'.[29] For Newcomb the stakes of velocity of light measurements were extremely high, for they delivered a value that would rewrite the astronomical tables, thereby influencing even precision surveying.[30]

Pushed by Newcomb, Michelson's extended work on the velocity of light provided a particularly deep initiation in the rigours of experimental practice. His repetitions undoubtedly helped him refine his approach to practical experimental work and the requirements of writing reports.[31] Yet while Michelson surely welcomed some features of this experience, his later work bears the marks of a reaction against it also.

Repeating velocity of light measures in different places enabled a consolidation of results. Travel played a more complex role in the early history of the ether-drift apparatus, first enabling the experiment to be performed at all (with the move from Berlin to Potsdam) and then (from Potsdam to Paris) rendering its shortcomings dramatically visible. Further, while a highly active culture of experimentation existed around velocity of light determinations, the situation was very different for ether drift. Michelson's first research entered a field marked by the work of scientists in France, the United States and soon Britain, employing a variety of arrangements and constantly refining techniques. In contrast we have no record of other scientists repeating the ether-drift experiment, let alone engaging in the detailed exchanges of challenge and reconstruction that took place between Newcomb and Michelson.

In his first years at Case, Michelson was primarily concerned with teaching and velocity of light but was also clearly interested in making more of his Potsdam experiment, canvassing the opinions of Lord Rayleigh and Sir William Thomson in particular.[32] The two giants of British physics told Michelson the best course of action would be to repeat Fizeau's 1851 experiment on the dragging of light in water, followed by his Potsdam experiment.[33] These experiments stood in potential opposition to each other. The Fizeau experiment implied that the ether was not dragged with a medium, while the most natural interpretation of a negative result to the ether-drift experiment was that the ether was not stationary but completely carried by the earth. If both were to be established, they created a conundrum. Michelson enlisted the aid of Edward Morley.[34] A chemist at the nearby Western Reserve University, Morley made the facilities of his laboratory available to the younger scientist, collaborated in planning the Fizeau experiment, and then completed the measurements when Michelson suffered a nervous breakdown that incapacitated him for several months.[35] When their results confirmed Fizeau's findings, Rayleigh once more encouraged a repetition of the Potsdam experiment. Michelson explained that he had been disheartened by the lack of response to his earlier work, writing:

> I have repeatedly tried to interest my scientific friends in this experiment without avail, and the reason for my never publishing the correction [due to Potier and Lorentz] was (I am ashamed to confess it) that I was discouraged by the slight attention the work received, and did not think it worthwhile.
>
> Your letter has however once more fired my enthusiasm and it has decided me to begin the work at once.[36]

In his long letter Michelson outlined the possibility of ether being carried within depressions in the earth's surface, which would render laboratory experiments unlikely to reveal an ether wind and make performance of the experiment in an exposed situation, such as a mountain, important. He also described improvements to lengthen the light path and decrease the problem of vibrations by mounting the apparatus on a concrete slab that floated in a trough of mercury.

Both experiments involved research repetitions, and the terms Michelson and Morley used to justify the endeavour suggest they were carried out with a particular ideology. In the first paper they wrote that the uncertainty of the results and interpretation of Fizeau's experiments 'must be our excuse for its repetition'.[37] Implicit is the idea that ideally an experiment should only need to be executed once. This is illustrated very clearly in the comments Morley wrote to his father to explain the breakdown Michelson suffered. Morley wrote: 'I can only guess at the stresses which brought about his illness. Overwork – and the ruthless discipline with which he drove himself to a task he felt must be done with such perfection that it could never again be called into question.'[38] We can see how deeply Newcomb's insistent inquiries haunted Michelson, with his desire to enter an active field – with a definitive result. The extent to which he sought, or was at least content to perform the ether drift experiment as a singular event, is highlighted by the fact that although Michelson and Morley wrote that it would be carried out at different times of the year (and should also be repeated at different elevations), they did not complete these protocols, publishing instead what had been intended to be merely preliminary results from just a few days of observations.[39] Despite the confidence this expresses in the experiment design and sensitivity, their decision not to eliminate uncertainty by repetitions across time and space is curious given the perceived importance of the experiment, the lengths to which Newcomb and Michelson were prepared to go to consolidate velocity of light measures, and also the kind of exhaustive chemical meteorology Morley had undertaken previously, testing the oxygen content of air samples taken under different conditions.[40] But we shall see that there were many reasons for moving on to work with other – related – problems.

Interferometers and their uses

To date I have discussed Michelson and Morley's work in 1886 and 1887 largely in the terms in which they are most commonly described, as experiments, and describing (however briefly) their theoretical significance. Now we turn to a less familiar view. Alongside the unfinished singularity we have observed in the Potsdam repetition, other facets of Michelson's work attest to the elaboration of a new culture of instrumentation around the technique of interferential refraction, and to a long-term continuity of research concerns. By speaking of a culture of instrumentation I mean that we can see Michelson identifying a continuity of apparatus across a number of experiments and beginning to work consciously with creating variations and modifications for different purposes. In 1881 Michelson had spoken of his apparatus as a new form of refractor. In the 1886 Fizeau replication, Michelson and Morley considered the refractometer arrangement Fizeau had used, and then introduced a modified form after practical trials. Just as he had in 1881, rather than dividing the wave front by using only two portions of the light passing through a lens, Michelson again used a partially silvered mirror to split the beam, with additional mirrors arranged to allow a greater distance between the pathways of light (see Figure 10.6). Repeating the ether-drift experiment Michelson and Morley used a different arrangement of mirrors to achieve the perpendicular form required, multiplied their number to increase the path length and transferred the micrometer movement to a mirror at the end of the arms (see Figure 10.7). In a paper published just after the ether-drift experiment, they described a simpler arrangement more reminiscent of 1881 and outlined how this could be used as a comparator to establish light as a standard of length.[41] The partially silvered mirror allowed a far greater flexibility of arrangement than the lenses and glass plates used in earlier refractometers.

It was standards of length that absorbed most of Michelson's research efforts over the next six years. His work to overcome the significant practical problems

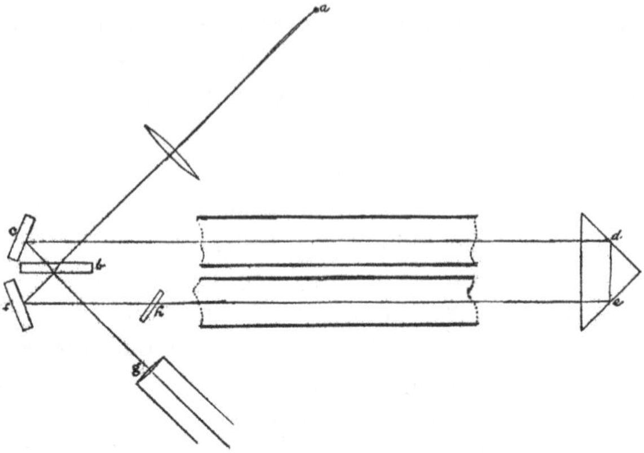

Figure 10.6 Plan of the refractometer used in Michelson and Morley's version of the Fizeau experiment. Light from the source *a* is divided on the half-silvered surface *b*, one part following the path *b c d e f b g* and the other the path *b f e d c b g*. Interference patterns are observed in the telescope *g*. Fizeau had used a lens similar to that shown in Figure 10.3, where Michelson and Morley used *b*, *c*, *f* and the compensating glass *h*.

Source: A.A. Michelson and E.W. Morley, 'Influence of motion of the medium on the velocity of light', *American Journal of Science* 31 (1886), 377–86, p. 381.

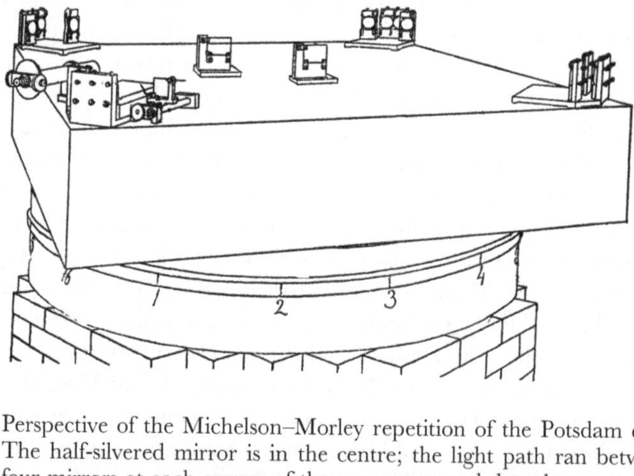

Figure 10.7 Perspective of the Michelson–Morley repetition of the Potsdam experiment. The half-silvered mirror is in the centre; the light path ran between sets of four mirrors at each corner of the apparatus; and the telescope and a mirror on a micrometer screw are shown on the front-left corner. The stone block rested on a wooden float contained in a cast-iron trough.

Source: A.A. Michelson and E.W. Morley, 'On the relative motion of the earth and the luminiferous ether', *American Journal of Science* 34 (1887), 333–45, p. 337.

involved proved extremely fruitful, not least for suggesting many other applications of interferential methods, which Michelson did not hesitate to explore and publicise: if he could not interest others in the ether-drift experiment, he might interest them in an instrument. In a vice-presidential address to the physics section of the American Association for the Advancement of Science in 1888 Michelson issued 'A plea for light waves', urging other scientists to work with interference fringes despite the more seductive appeal of a field like electricity. It would soon be possible, he said, to use a light wave itself as a unit of measuring minute distances, and a refractometer could test the flatness of a plane or the volume of a cubic decimetre with an error of less than one part in a million. His new techniques of length measurement could therefore be used to establish a new standard of mass.[42] Michelson also focused on the implications of the broadening of spectral lines for the conception of radiation in solids and liquids.[43] This was a phenomenon that both he and the German physicist Hans Ebert had investigated interferometrically. In this, his first opportunity to address the physics discipline more broadly, Michelson made no mention of his ether-drift experiment. It was one of the 'many other interesting and important applications of light-waves' that he feared 'would be wearisome even to enumerate'.[44] By 1890 Michelson was ready to systematise his work on variations and modifications of the apparatus he had made his own in a major paper on the topic of measurement with light waves. There he instituted a comparison of the properties of microscopes, telescopes and refractometers – showing that the latter sacrificed resolution and definition to achieve far greater accuracy of measurement – and gave a taxonomy of the various types of refractometers possible (see Figure 10.8).[45]

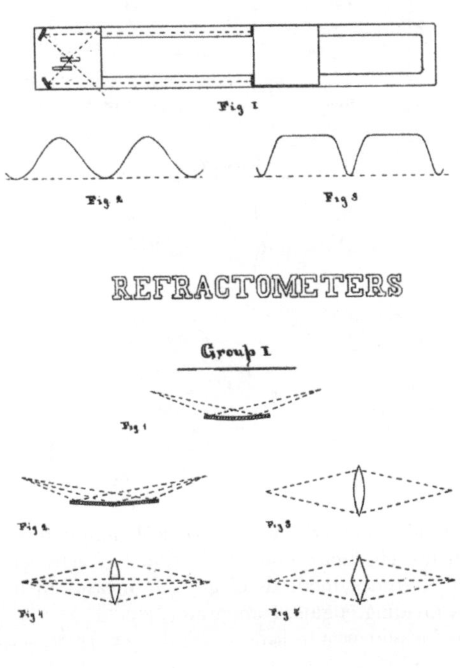

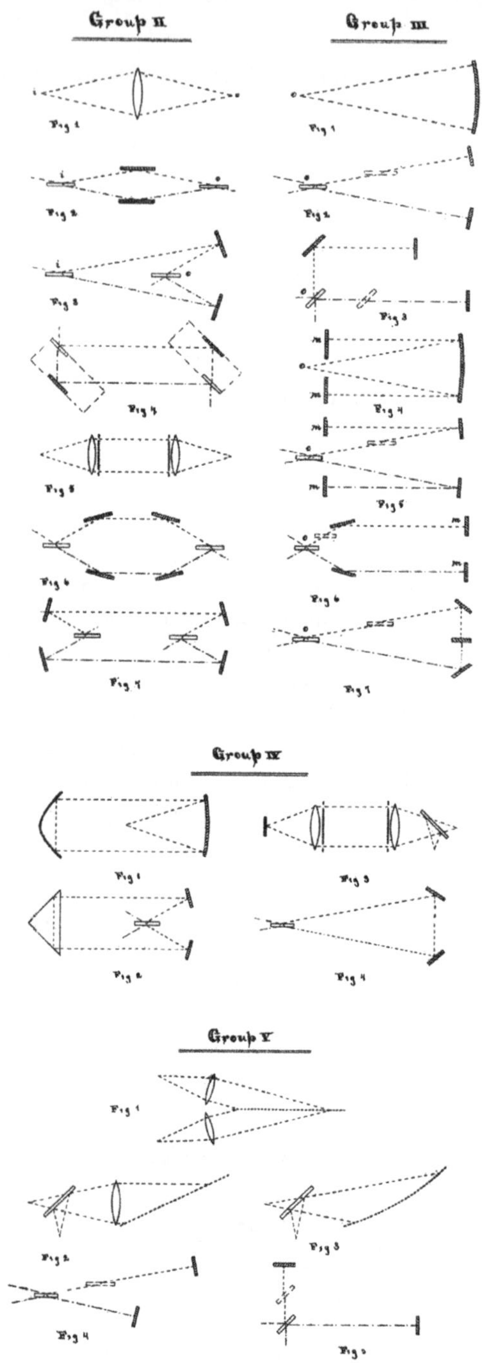

Figure 10.8 Michelson's illustration detailing the optical paths of five groups of refractometer arrangements. His figure 1 (above the title 'Refractometers') shows the application of interference methods of angular measurement to the testing of the ways of a dividing engine, comparator, etc.

Source: A.A. Michelson, 'Measurement by light-waves', *American Journal of Science* 39 (1890), 115–21, plate 3.

Based on lectures given at MIT, his first book, *Light Waves and Their Uses*, provided an even more extensive overview of the uses Michelson had found for interferometers (that term gradually replaced the earlier name given the family of devices).[46]

Standards of length in science and industry

It is significant that Michelson devoted so much attention to rendering light as a standard of length. In doing so he responded creatively to the issues raised in his dialogue with Newcomb, and developed a typically distinctive approach to a field of considerable interest to contemporary science and industry. Newcomb's critical concerns dated from at least 1876, when the Transit of Venus Commission borrowed three steel standard measuring rods from the Coast and Geodetic Survey Office of Standard Weights and Measures and found they did not correspond to the standard rules of Darling, Brown & Sharpe.[47] His velocity of light paper emphasised the hope that such experiments could be conducted with a precision that would enable them to test the invariability of the standard of length.[48] Another astronomer Michelson contacted in the course of his first experiment had recently made standards of length the primary subject of his research. At the time he checked Michelson's scale, William Rogers was in the midst of a series of comparisons of the standard measures of the US, Britain and France.[49] Finally, C.S. Peirce at the Coast Survey, Mace de Lepinay in Marseilles and Louis Bell at Johns Hopkins University all linked light and length in this period, with the American scientists advocating the possibility of using diffraction grating measurements to establish the wavelength of sodium as a standard of length.[50]

Within US industry the precise measurement of length was an equally pressing issue. Darling, Brown & Sharpe was one of the most respected machine-tool firms on the East Coast who themselves supplied the Coast Survey with metal rods for the construction of standard rules and furnished standards to the State of Massachusetts. They were deeply worried by uncertainty over the relations between the metre and the yard in the Coast Survey, and their constant need to return to the Coast Survey for comparisons. As they wrote on one occasion, all their ruling operations depended on the results of the comparison.[51] Such problems were widespread. In 1876 the New York, Lake Erie and Western Railroad Company approached Pratt & Whitney on finding that nuts cut at one of their own in-house shops would not fit bolts made at another. Investigating the products of different companies showed that the manufacturers of taps and dies had been working to different standards. The Master Car Builder's Association selected Pratt & Whitney to furnish standard United States thread screw gauges. An 1882 committee report described the situation in colourful terms:

> Like Diogenes with his lamp, in search of an honest man, this company went to and fro in the land in search of a true inch, a true foot, or a true yard. They procured from different sources what they supposed were the most reliable standards of measurement, and found that none agreed. They

> had the same standards measured by what were considered the most reliable measuring machines and instruments in the country, and found that no two of these would measure the standard alike.[52]

For Pratt & Whitney, running the gamut of gun, sewing machine, and then bicycle and automobile manufacturing in the years from 1860, the provision of accurate gauges was the indispensable key to the growing industrial manufacture of interchangeable parts. Like Michelson, they were put in contact with Rogers in 1879. Assembling a mixture of mechanical expertise and scientific creditability they brought their own Mr Bond together with Rogers to work with the Coast Survey scientists in conducting a series of comparisons. Bond helped Rogers design and build a universal comparator to make a variety of end and line measure comparisons, and two exemplars were made. One Rogers–Bond comparator was installed in the standards room of the Harvard Observatory, while its pair stood in the Gauge Division of Pratt & Whitney and the firm's catalogues advertised it among their products (see Figures 10.9, 10.10).[53] Most importantly Rogers pressed the Coast Survey to allow the critical comparison between the two Pratt & Whitney bars and the US standard to be made both using the Coast survey apparatus and the new comparator, which he would transport to Washington for the purpose. He wanted to take readings himself and establish cross-comparisons with the measurements John Clark made using the Coast Survey apparatus.[54] The Coast Survey complied, permitting an unusual transparency that enabled an assurance of continuity of protocols across the standards institution, the scientist, and the manufacturer.

Pratt & Whitney's effort left Darling, Brown & Sharpe vulnerable should there be any discrepancy between the standards available commercially, a situation they could not tolerate. Darling spent many months considering the problems of length comparisons, took out patents on new reading scales, and like Rogers asked to visit the Coast Survey Office in order to inspect their arrangements.[55] But while Darling sought transparency from the Coast Survey, he jealously guarded his own ruling machine, even within his own firm. The company operated three different dividing engines long into the twentieth century, and during his lifetime Darling protected his engine from the scrutiny of his partners within an off-limits room.[56]

The achievement of reliable standards of length involved a delicate teasing apart of realms of work often regarded as private, and required close associations across industrial and scientific realms. In the same period that the gauges of railway lines run by different companies began to be unified, and railway time began to be standardised in regions across the US, efforts were initiated to make the nuts and bolts of the industry interchangeable also. These changes gradually reconfigured commercial competition with a demand for technical uniformity across different companies, both reflecting and facilitating new business practices in transportation and industrial manufacturing. Michelson's solution to the standards problem was to be characteristically distinctive – even singular. Indeed, despite advocating the method he developed as 'practical', his principal comparison of

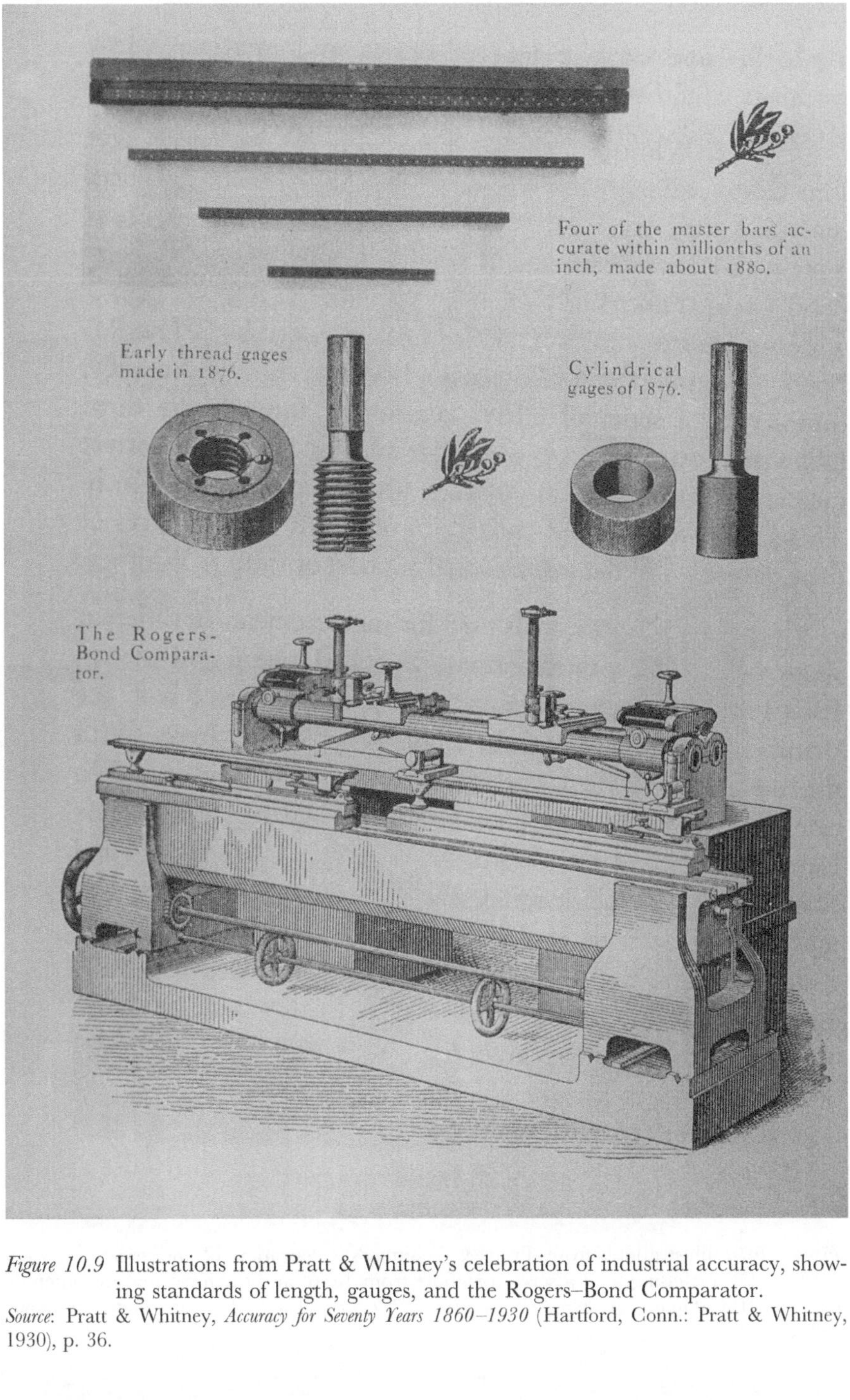

Figure 10.9 Illustrations from Pratt & Whitney's celebration of industrial accuracy, showing standards of length, gauges, and the Rogers–Bond Comparator.

Source: Pratt & Whitney, *Accuracy for Seventy Years 1860–1930* (Hartford, Conn.: Pratt & Whitney, 1930), p. 36.

Figure 10.10 Illustrations from Pratt & Whitney's celebration of industrial accuracy, including here a screw machine from 1876, and the firm's interferometer.

Source: Pratt & Whitney, *Accuracy for Seventy Years 1860–1930* (Hartford, Conn.: Pratt & Whitney, 1930), p. 38.

the international metre with light waves was not to be repeated by another scientist for fifteen years. Nevertheless his research opened up a rich range of scientific problems, and also led to the development of methods of testing and machining that gradually transformed practices in the precision optical and mechanical industries, a potential Michelson foresaw as he began his work.

Light and length

Perhaps – working with sodium light in 1881 – it was noticing the regular shift of interference fringes with minute turns of the screws carrying his mirrors that led Michelson to believe he could institute a new approach to distance measurement. But he would have recognised particular problems very quickly. How far would the homogeneity of different light sources allow a moving mirror to track light waves: millimetres, centimetres or decimetres? In June 1887, after completing the Fizeau experiment but before repeating the ether-drift experiment, Michelson and Morley conducted preliminary experiments with apparatus like that employed in Potsdam.[57] Since the instrument didn't need to rotate it could be kept stable on a brick pier relatively easily, and the screw micrometer movement was transferred to a mirror on one of the arms. Slowly moving this mirror, Michelson and Morley proceeded to determine how many wavelengths they could count. (With one complete turn of a screw of one hundred threads per inch, the mirror traversed nearly 1,000 wavelengths.[58]) Fizeau had observed interference only up to a path-difference of 50,000 wavelengths. With a lower density of sodium vapour Michelson and Morley were able to track light waves more than four times this distance. But interference patterns alternated with uniform illumination as a result of the superposition of the two series of waves in sodium light, so they followed the fringes of light from thallium, lithium and hydrogen as well.

They later described how small distances – of a countable number of wavelengths – could be used to step off an intermediate standard decimetre, with an interferential comparer being employed to step off the whole metre. But before publishing on standards, Michelson and Morley conducted their Potsdam repetition on 8, 9, 11 and 12 July. This was important. Since the standards programme rested upon the assumption that the wavelength and velocity of light did not vary with location or orientation (or that any variation could be established extremely precisely) it was imperative that a more accurate version of the ether-drift experiment be established first. Given the null result achieved, Michelson's conceit (only expressed strongly subsequently) was that the use of light waves would establish an absolute, natural and practical standard. Describing the successful completion of his work comparing the international metre with the wavelength of cadmium in 1893, Michelson wrote:

> We have thus a means of comparing the fundamental base of the metric system with a natural unit with the same degree of approximation as that which obtains in the comparison of two standard metres. This natural unit

> depends only on the properties of the vibrating atoms and of the universal ether; it is thus, in all probability, one of the most constant dimensions in all nature.[59]

Light waves would allow a standard of length to be established independently of temporal or geographical travel – without referring measures to a particular metal rod held in Sèvres.

From 1887 Michelson and Morley continued with the standards programme rather than completing the ether-drift experiment. Extending their ability to get mirrors to travel along machined ways with the precision required to track fringes over ever-greater distances was already proving extremely fruitful. Using harmonic analysis to extrapolate backwards from the varying visibility of the fringes to the structure of the light source, the refractometer disclosed a fine structure not revealed by the spectroscope. In the case of the red hydrogen line, for example, the interference phenomena disappeared at about 15,000 and 45,000 wavelengths, indicating it must be a double line with components about one-sixtieth as distant as the sodium lines.[60] Michelson's instrument was now a tool of discovery, but such complexity rendered many sources unsuitable for use as a standard.[61] The director of the International Bureau of Weights and Measures suggested that Michelson and Morley continue their analysis until they found a homogeneous source, and then establish a comparison with the International Metre held in Sèvres.[62] By the time Michelson was ready to do so in 1892 – now working without Morley, having left Case for a position as founding professor of physics at Clark University – he could report that most of the bright lines of the spectrum were double, triple or still more complex in constitution.[63] Cadmium vapour, however, disclosed three lines (red, green and blue) that were almost ideally simple: a suitable source had been found. Michelson's search had shown that the great resolving power of the refractometer enabled its use in spectroscopy, and established the practical basis for his important two-part paper 'On the application of interference methods to spectroscopic measurements' published in the *Philosophical Magazine* in 1891 and 1892.[64]

There were also significant problems of a mechanical and optical nature to be solved building the apparatus – and the sensitivity of interferential techniques could be used to check the instrument's components. The job of constructing an interferential comparer and intermediate standards went to the instrument makers Warner & Swasey in Cleveland and the optician John A. Brashear in Pittsburgh (see Figure 10.11).[65] While the ether drift apparatus required upwards of 135 hours of attention, Warner & Swasey spent more than 590 hours making the mechanical apparatus for the determination of the metre (and the process destroyed Morley's patience with the firm).[66] A letter Brashear wrote to Edward Holden at the Lick Observatory on 5 April 1889 shows how far providing the glass plates stretched one of America's premier optical workshops:

> [W]e have completed two of the most difficult pieces of work ever undertaken by us. One was a set of 16 plane and parallel plates for Prof. Michelson

Figure 10.11 Perspective of the apparatus used to determine the length of the metre in terms of the wavelength of cadmium light. The microscopes used to read the reference lines on the metre are visible, and the handles which turn the screws to move the reference plane and intermediate standard are on the left.

Source: A.A. Michelson, *Light Waves and Their Uses* (Chicago: University of Chicago Press, 1902), p. 103.

> for his research on the absolute value of a sodium wave length, and its transference to a steel bar in a freezing mixture, (i.e., multiples of the wave length) which is proposed as a new standard of measurement.
>
> The limiting error allowed was .05 sodium wave length for every plate . . . Hastings [after testing it] wrote us an unsolicited letter in which he said . . . that 'it was the most perfect piece of optical work he had ever seen, and that nowhere else in the world could he have obtained it.' . . . After Michelson tested them he used two adjectives – 'Exquisite' – 'Superb' – both underscored. So they must be good . . . It did take a vast amount of time to work and test them finally. You couldn't look at them [the fringes] without disturbing them . . .[67]

In this case the regime produced long-term changes in the workshop. Later Brashear wrote to Lord Rayleigh and noted that 'the prism surfaces are all tested by interference methods and made plane within 1/10 λ. We make so many accurate planes for refractometers that we carry the same accuracy into our prism work.'[68] The accuracy required in making plane parallel ways, however, went beyond the instrument-maker's capabilities, requiring the experimentalist to do the final grinding himself. Michelson and Wadsworth foresaw the possibility that the optical techniques of testing planeness and parallelity, and the grinding techniques they employed, might be of use in industrial manufacture.[69]

This brief sketch will suggest the range and depth of Michelson's work with interferometric techniques from 1886 onwards. The invitation to establish a determination of the international metre in Sèvres undoubtedly constituted the apotheosis of the new instrument, and was perceived as such by others. Benjamin Gould highlighted the significance of the fact that an American professor using an American instrument maker had been chosen for the task. In an echo of Michelson's claims for the unvarying veracity of the technique, when requesting Clark University to release Michelson for his journey to France, he wrote:

> It is my conviction that the assent of Clark University will not only redound to its high honor, and be gratefully recognized throughout the civilized world, but will constitute an unending title to remembrance, and full appreciation in the history of science.[70]

Michelson found the metre to contain 1,553,163.5 wavelengths of the red line of cadmium.[71]

The interferometer and the diffraction grating

Michelson had established a wide range of new applications for interferometric measurements and their extraordinary accuracy largely through his own research. When Pieter Zeeman found that exposing the light source to a magnetic field produced a broadening of the spectral lines emitted, Michelson was quick to examine the phenomenon with the interferometer.[72] In 1899 a dispute erupted that usefully highlights the instrument's vulnerability even after some ten years of development. Thomas Preston, a spectroscopist working on the Zeeman effect in Dublin and long-time observer of Michelson's work, chose the pages of *Nature* to mount a major challenge. At issue was priority in the establishment of the anomalous Zeeman effect.[73] Preston's attack reveals the extent to which the reputation of the interferometer and its inventor were related. He wrote:

> Investigations demanding special attention are those of Prof. A.A. Michelson, both on account of his reputation as an original investigator and by reason of the nature of the apparatus which he employed . . . Prof. Michelson concluded some years ago that the spectral lines themselves instead of being, as ordinarily supposed, narrow bands of approximately uniform illumination from edge to edge, are on the contrary in most cases really complexes, some of them being close triplets, and so on. This structure has never yet been observed by means of any ordinary form of spectroscope, and accordingly it has been suggested that it does not exist in the light radiated from the source, but is imposed on the spectral lines by the apparatus used, namely, the interferometer. Be this as it may, the application of this instrument to the study of radiation phenomena in the magnetic field is highly interesting.[74]

Michelson had concluded that the separation of the triplets produced by the magnetic field was independent of both the spectral line and the substance. Since in all other observations the separation of the components varied considerably, Preston suggested Michelson's instrument suffered some peculiarity that had not yet been taken into account. The challenge was serious, repeated when Michelson discounted Preston's suggestion that diffraction might be the source of the difficulty. Preston wrote that if Michelson maintained the law he advocated, either the cause of the error had to be determined in the instrument, or the interferometer had to be standardised so that it could be employed as a measuring instrument.[75]

At this point Rayleigh intervened with a supportive comment that nevertheless highlights Michelson's exposed situation. Rayleigh wrote:

> The questions raised by Mr Preston can only be fully answered by Prof. Michelson himself; but as one of the few who have used the interferometer in situations involving high interference, I should like to make a remark or two. My opportunity was due to the kindness of Prof. Michelson, who some years ago left in my hands a small instrument of his model.

Rayleigh also argued diffraction was not a factor, and left room for flexibility by allowing that the estimation of the visibility of the bands and the deduction of the structure were 'rather delicate matters'.[76] Preston's next letter showed how the dispute was resolved: Michelson admitted the law had probably been generalised from insufficient data, thereby re-establishing Preston's confidence in the instrument.[77]

Preston's demands were provocative, especially his call to standardise the interferometer. Even if the possibility of rendering light waves a standard of length could be granted, the fact that this measurement relied upon establishing relatively simple observational conditions, such as using a homogeneous light source, left the performance of the interferometer even more vulnerable in the complex situations typical in spectroscopy. There Michelson would have to show his results cohered with those achieved by other means, or risk the charge that they were an artefact. Reputation alone would not establish the value of the instrument. And Preston's question about standardisation was particularly pointed, for Michelson's determination of the metre had given an absolute wavelength for cadmium different from that established spectroscopically by Rowland – and even more worryingly, gave different values for the wavelengths of cadmium relative to one another.[78] Which instrument was to be trusted, the grating familiar to so many or Michelson's interferometer?

Over the next ten years this was a matter of increasingly open concern within the spectroscopic community – with confidence in the accuracy of grating work gradually weakening as discrepancies between spectroscopic measurements of the wavelength of solar and terrestrial elements became apparent from the 1890s. By the early twentieth century spectroscopists recognised a crisis, and, contra Preston, it was how best to standardise Rowland's spectroscopic wavelength tables against Michelson's absolute measurements that was in question. Again it became crucial to supplement Michelson's single research – this time on standards of length – with repetition.[79] The debate was strongly shaped by the invention of a new form of interferometer by the French physicists Charles Fabry and Alfred Pérot in 1896. Their instrument used the multiple reflection of light between two plane parallel half-silvered mirrors to produce interference characterised by extremely sharp maxima and minima.[80] For the next decade Fabry and Pérot used the instrument in spectroscopic and metrological research, standardising both grating wavelength tables and Michelson's interferometric measurements.[81] In 1907 Fabry redetermined the length of the metre in terms of the wavelength of the red line of cadmium, obtaining a value of 1,553,164.13. These debates in the spectroscopic community provide the most important immediate context for the award of the Nobel Prize to Michelson, for research and instruments that had for so long been so singular.

Michelson himself had two characteristically instrumental answers to the challenge Preston articulated. The first was his 1898 invention of a new instrument using parallel plates of glass to form a spectroscope of very high resolving power that nevertheless avoided some of the disadvantages of the interferometer (Preston ordered an echelon spectroscope but was unable to use it before he died at age 40 in 1900). Even more fundamentally, in 1902 Michelson began work to better Rowland's gratings by using the interferometer to construct a more accurate screw and ruling engine. Again Michelson credited the standards programme with suggesting yet another application for interferential refractometry:

> The idea of utilizing the extraordinary accuracy which may be obtained with the interferometer for determining and correcting the errors of the screw, occurred to me during the measurement of the meter in light waves; and it was hoped that by its means it would be possible to construct a ruling engine which should rule gratings having a resolving power of one million – that is, which could separate doublets whose distance was a thousand times smaller than that between the sodium lines.[82]

Achieving this would constitute a most effective demonstration of the superior accuracy of interferometric techniques. Michelson devoted much of the rest of his career to attempting this goal, and although he was only partially successful, long after his death his two ruling engines were modified by G.R. Harrison at MIT and Bausch & Lomb to incorporate interferometric control of the ruling process itself.[83] Michelson's machines, in other people's hands, ruled a new generation of diffraction gratings.

Notes

1 C.J. Evans and D.J. Warner, 'Precision engineering and experimental physics: William A. Rogers, the first academic mechanician in the US', in S. Goldberg and R.H. Stuewer (eds) *The Michelson Era in American Science: 1870–1930*, New York: American Institute of Physics, 1988, pp. 2–12. For biographies see D.M. Livingston, *The Master of Light: A Biography of Albert A. Michelson*, New York: Charles Scribner's Sons, 1973, and B. Jaffe, *Michelson and the Speed of Light*, New York: Anchor, 1961. An important source of information on his early work is Michelson's correspondence with Newcomb, held

in the Simon Newcomb Papers (SNP) and Naval Historical Foundation Collection (NHFC) in the Library of Congress, and the Department of the Navy, Records of the Naval Observatory, Record Group 78 in the National Archives and Records Administration (NARA RG 78). Much of this correspondence is published in N. Reingold (ed.) *Science in Nineteenth-Century America: A Documentary History*, New York: Hill & Wang, 1964, pp. 275–306, and Livingston quotes some additional letters. The Michelson Collection (MC) in the Nimitz Library of the US Naval Academy holds material initially gathered for the Michelson Museum in China Lake, including copies of much correspondence held elsewhere (for an early guide see W.B. Plum, 'The Michelson Museum', *American Journal of Physics* 22, 1954, 177–81). The Nimitz Library also holds Livingston's sources and notes.

2 Abandoning hope of being appointed a Navy professor because of his inability to cope with the examination in mathematics and astronomy required, Michelson then sought a university position. See Michelson to A.M. Mayer, 26.6.1880, in Reingold, *Science*, pp. 286–7, Wolcott Gibbs to Julius Hilgard, 19.7.1880, in Livingston, *Master*, pp. 65–6, George Barker to Newcomb, 3.12.1880, SNP 15, and A.A. Michelson to E.C. Pickering, 7.7.1880, Harvard University Archives (HUA) HUG 1690.15, E.C. Pickering Private Letters 1850–83.

3 Pickering provided letters himself, and wrote to Gibbs asking for his help introducing Michelson to Fizeau, Cornu and Jamin. See E.C. Pickering to Michelson, A.C. Ranyard and A. Auwer all on 8.9.1880, and Pickering to Michelson and J.W. Gibbs on 16.9.1880, in HUA UAV 630.14 Harvard College Observatory/E.C. Pickering, Director/Letterbooks (outgoing)/B2, July 1880–October 1883.

4 Michelson to Newcomb, 22.11.1880, in Reingold, *Science*, pp. 287–8.

5 Maxwell wrote a letter to the Director of the *Nautical Almanac* that was later published. Maxwell to D.P. Todd, 19.3.1879, *Nature* 21, 1880, 314–15. Newcomb had a similar view. See Newcomb to Michelson, 2.5 and 2.6.1881 in Reingold, *Science*, pp. 290–1, 292 and 296–7.

6 An enthusiastic report on the progress of Michelson's plan was sent to Newcomb from the Department of Physics and Chemistry at the Naval Academy (perhaps by Sampson) on 25.3.1878, SNP 45, General Correspondence 23 January 1856–31 December 1879. The earliest published discussion was A.A. Michelson, 'On a method of measuring the velocity of light', *American Journal of Science* 15, 1878, 394–5.

7 Newcomb had first written of the need for an accurate determination of the velocity of light in 1867, and in 1876 began making preparations for a measurement with George Barker of the University of Pennsylvania. See Simon Newcomb, 'Measures of the velocity of light made under direction of the Secretary of the Navy during the years 1880–82', *Astronomical Papers Prepared for the Use of the American Ephemeris and Nautical Almanac* 2, 1882, 107–230, pp. 113–20, on p. 120.

8 Pickering referred to Mascart, 'Sur la réfraction des gaz', *Comptes Rendus* 78, 1874, 617–21; Mascart, 'Sur la dispersion des gaz', *Comptes Rendus* 78, 1874, 679–82, and idem, 'Sur la réfraction de l'eau comprimée', *Comptes Rendus* 78, 1874, 801–5. Pickering to Michelson, n.d., circa 21.8.1879, HUA UAV 630.14 Harvard College Observatory/E.C. Pickering, Director/Letterbooks (outgoing)/B1, February 1877–July 1880.

9 A.A. Michelson, 'The relative motion of the earth and the luminiferous ether', *American Journal of Science* 22, 1881, 120–9, p. 121.

10 See A.A. Michelson, 'Report, conference on the Michelson–Morley experiment', *Astrophysical Journal* 68, no. 5, 1928, 342–5, p. 343. Michelson's comments have not been noticed by previous scholars.

11 On Fizeau as a source see L.S. Swenson, *The Ethereal Ether: A History of the Michelson–Morley Aether-Drift Experiments, 1880–1930*, Austin: University of Texas Press, 1972, pp. 61–2. On the sextant see Darwin H. Stapleton, 'The context of science: the community of industry and higher education in Cleveland in the 1880s', in Goldberg and Stuewer, *The Michelson Era*, pp. 13–22, p. 16. Others have described it as following

from a study of Jamin's apparatus. R.S. Shankland, 'Michelson–Morley experiment', *American Journal of Physics* 32, 1964, 16–35, pp. 18–19, and Livingston, *Master*, p. 77.

12 See A.A. Michelson, 'Experimental determination of the velocity of light', *Proceedings of the American Association for the Advancement of Science* 27, 1878, 71–7, on pp. 75–6, and idem, 'Experimental determination of the velocity of light', *U.S. Nautical Almanac Office Astronomical Papers* 1, *no. Part III*, 1879, 115–45, on p. 116.

13 See Michelson to Newcomb, 22.11.1880, in Reingold, *Science*, pp. 287–8.

14 Michelson to C.A. Bell, 22.12.1880, as quoted in Livingston, *Master*, p. 76.

15 This was not a good sign for the lab.: the unusually large sum of 310,000 marks had just been spent on its foundations. See Paul Forman, John Heilbron and Spencer Weart, 'Physics circa 1900: personnel, funding and productivity of the academic establishments', *Historical Studies in the Physical Sciences* 5, 1973, 1–185, p. 111.

16 Michelson, 'Relative motion', p. 124.

17 Potier's insight was acknowledged in A.A. Michelson, 'Sur le mouvement relatif de la terre et de l'ether,' *Comptes Rendus* 94, 1882, 520–3. H.A. Lorentz, 'De l'influence du mouvement de la terre sur les phénomènes lumineux', *Archives Néerlandaises* 21, 1886, 297ff., in H.A. Lorentz, *Collected Papers,* The Hague: Martinus Nijhoff, 1937, vol. IV, pp. 153–214, on pp. 204–8.

18 Livingston, *Master*, pp. 87–8.

19 On Rogers see Evans and Warner, 'Precision engineering', and Edward W. Morley, 'Biographical memoir of William Augustus Rogers', *Biographical Memoirs – National Academy of the Sciences* 4, 1902, 187–9.

20 Michelson to Newcomb, 15.6.1882, in Reingold, *Science*, pp. 303–4.

21 Ibid.

22 See Newcomb to Michelson, 15.8.1881 and Michelson to Newcomb, 29.8.1881, in Reingold, *Science*, pp. 296–7 and 298–300, on pp. 297 and 298. See below for Michelson's final corrections.

23 Newcomb to Michelson, 2.5.1881, in Reingold, *Science*, pp. 290–1, on p. 291. Michelson used Würdemann's copy of the standard. On Würdemann see S. Turner, 'William Würdemann: first mechanician of the U.S. Coast Survey', *Rittenhouse* 5, no. 20, 1991, 97–110.

24 A.L. Norberg, 'Simon Newcomb's early astronomical career', *Isis* 69, 1978, 209–25, p. 218.

25 See A.E. Moyer, *A Scientist's Voice in American Culture: Simon Newcomb and the Rhetoric of Scientific Method,* Berkeley, Los Angeles and Oxford: University of California Press, 1992, p. 151. Moyer regards Newcomb to have anticipated Percy Bridgman's often-quoted introductory summary of 'the operational character of concepts' in *The Logic of Modern Physics*, 1927.

26 Newcomb wrote 'I have been much embarrassed at the result I am going to get for the velocity of light': Newcomb to Michelson, 2.5.1881, in Reingold, *Science*, pp. 290–1, on p. 290. Reporting to the National Academy he allowed that he hesitated to publish the results of his first three series of results because of their divergence from Michelson's value, but highlighted Michelson's measurement of the distance the light travelled: 'his description of this process is not so explicit that a scientific jury could pronounce the measure absolutely free from the possibility of error': S. Newcomb, 'Report of progress in the experiments on the velocity of light', *Report of the National Academy of Sciences, 1881*, pp. 17–19, on p. 18.

27 Barker (Newcomb's first collaborator in velocity of light plans) was instrumental in obtaining Michelson's position. See Barker to John Stockwell, 22.3.1881, Case Western Reserve University Archives. The trustees gave Michelson $7,500 to purchase instruments for the new laboratory, enabling him to take advantage of the excellence of European, and particularly French, instrument makers.

28 A.A. Michelson, 'Supplementary measures of the velocities of white and colored light in air, water, and carbon disulphide, made with the aid of the Bache Fund of the

National Academy of Sciences', *Astronomical Papers Prepared for the Use of the American Ephemeris and Nautical Almanac* 2, 1882, 231–58. For the corrections to the earlier result see pp. 243–4.

29 Simon Newcomb, 'Introductory note', to Michelson, ibid., p. 235.

30 See Newcomb to Com. R.W. Thompson, Secretary of the Navy, 12.12.1878, in SNP 57.

31 Writing from Europe Michelson admitted it would have been helpful if Newcomb had been able to study his first paper thoroughly before it was published. Michelson to Newcomb, 29.8.1881, in Reingold, *Science*, pp. 298–300 on p. 300.

32 Michelson met the two British physicists at the BAAS meeting in Montreal, and attended Thomson's Baltimore lectures. In Montreal Michelson spoke on velocity of light and was supported by Rayleigh in a dispute with the British physicists Young and Forbes over the velocity of red and blue light. A.A. Michelson, 'On the velocity of light in carbon disulphide and the difference in velocity of red and blue light in the same', *Reports of the British Association for the Advancement of Science*, 1884, 654. For a description of the relationship between Michelson and Rayleigh see R.S. Shankland, 'Rayleigh and Michelson', *Isis* 58, 1967, 86–8.

33 See Michelson's discussion of his plans in a letter to J.W. Gibbs: Michelson to Gibbs, 15.12.1884, in Reingold, *Science*, pp. 307–8. The papers were published as A.A. Michelson and E.W. Morley, 'Influence of motion of the medium on the velocity of light', *American Journal of Science* 31, 1886, 377–86; and 'On the relative motion of the earth and the luminiferous ether', *American Journal of Science* 34, 1887, 333–45. See also Michelson and Morley to Thomson, 27.3.1886, and Thomson to Michelson and Morley, quoted in S.P. Thompson, *The Life of William Thomson, Baron Kelvin of Largs*, London: Macmillan, 1910, vol. II, p. 857.

34 On Morley see H.R. Williams, *Edward Williams Morley: His Influence on Science in America*, Easton, Pa., 1957.

35 It was initially thought this would end his active career. See Livingston, *Master*, pp. 111–15.

36 Michelson to Rayleigh, 6.3.1887, Rayleigh Archives, Research Library, Hanscom Air Force Base, Bedford, Mass.

37 Michelson and Morley, 'Influence of motion', p. 380. In the case of the second experiment, they wrote that they sought a theoretical effect that was much too large to be masked by experimental errors, in the light of the oversights that had undermined the result of the earlier attempt. Michelson and Morley, 'Relative motion', p. 335.

38 E.W. Morley to S.B. Morley, 17.9.1885, Edward W. Morley Papers (EWMP), Library of Congress, Manuscript Division.

39 Michelson's next ether-drift paper was published ten years later, using a light path that extended up the side of the laboratory at the University of Chicago. A.A. Michelson, 'The relative motion of the earth and the ether', *American Journal of Science* 3, 1897, 475–8.

40 See E.G. Spittler, 'Morley, Edward Williams', in C.C. Gillispie (ed.) *Dictionary of Scientific Biography*, vol. 9, New York: Charles Scribner's Sons, 1974, pp. 530–1.

41 A.A. Michelson and E.W. Morley, 'On a method of making the wave-length of sodium light the actual and practical standard of length', *American Journal of Science* 34, 1887, 427–30.

42 A.A. Michelson, 'A plea for light waves', *Proceedings of the American Association for the Advancement of Science* 37, 1889, 1–14. George Hale first met Michelson on this occasion. George E. Hale, 'Some of Michelson's researches', *Publications of the Astronomical Society of the Pacific* 43, 1931, 174–85, p. 176.

43 Michelson, 'Plea', pp. 75–8.

44 Ibid., p. 78. Another application Michelson did not discuss was the measurement of the diameter of astronomical objects, which he began in 1890 and then resumed with

his famous measurement of the diameter of Betelgeuse in 1920. See D. DeVorkin, 'Michelson and the problem of stellar diameters', *Journal for the History of Astronomy* 6, 1975, 1–18.

45 A.A. Michelson, 'Measurement by light-waves', *American Journal of Science* 39, 1890, 115–21.

46 A.A. Michelson, *Light Waves and Their Uses*, Chicago: University of Chicago Press, 1902.

47 Newcomb found that the rods were shorter than the standard rules of Darling, Brown & Sharpe by 0.10 inches in 35 feet. Newcomb to J.E. Hilgard, 17.4.1876, NARA RG 167, Entry 7 Box 3. I have not yet located Hilgard's reply.

48 See Newcomb, 'Measures', pp. 111 and 204.

49 See note 19; also, William A. Rogers, 'On the present state of the question of standards of length', *Proceedings of the American Academy of Arts and Sciences* 15, 1880, 273–312, and idem, 'Studies in metrology. First paper', *Proceedings of the American Academy of Arts and Sciences* 18, 1883, 287–398.

50 C.S. Peirce, 'Note on the progress of experiments for comparing a wave-length with a meter', *Nature* 20, 1879, 99, and *American Journal of Physics* 28, 1879, 51. See Victor F. Lenzen, 'The contributions of Charles S. Peirce to metrology', *Proceedings of the American Philosophical Society* 109, 1965, 29–46. Macé de Lepinay, 'Détermination de la valeur absolue de la longueur d'onde de la raie D_2', *Comptes Rendus* 102, 1886, 1153–5, and *Journal de physique théorique et appliquée* 5, 1886, 411–16. Louis Bell, 'On the absolute wave-length of light', *American Journal of Physics* 33, 1887, 167–82.

51 There were several exchanges relating to standards. See Darling, Brown & Sharpe to J.E. Hilgard, 29.6, 30.6 and 4.8.1875; Hilgard to Darling, Brown & Sharpe, 4.8.1875 in NARA RG 167 NC-76 Entry 7 Box 2. On the dividing machine see Darling, Brown & Sharpe to Hilgard, 24.9.1877, and for a (temporary) resolution their letter to Hilgard of 16.10.1877. See also their letters to Hilgard of 20 and 24.3, 7, 12, 17 and 24.4 and 13.5.1879 in Entry 7, Box 3.

52 G.M. Bond (ed.) *Standards of Length and their Practical Application*, Hartford, Conn.: Pratt & Whitney, 1887, p. 67.

53 See ibid., pp. 67–8, and Evans and Warner, 'Precision engineering', pp. 6–8.

54 See Rogers to F.A. Pratt, 12.8.1882 and Rogers to Hilgard, 12.9.1880 in NARA RG 167 NC-76, Entry 7, Box 3.

55 See Darling, Brown & Sharpe to Hilgard, 3 and 27.4.1883. (In the former they write: 'A good workman can do a good job with poor tools, but he can do a better job with good tools, and there are *some things* to be done that require the greatest skill, greatest experience, and the very best means, and the comparison of a measure of length is one of those things.') See also Darling to Hilgard, 17.7 and 8 and 15.8.1883, NARA RG 167 NC-76, Entry 7, Box 4.

56 See C. Evans, *Precision Engineering: An Evolutionary View*, Bedford: Cranfield Press, 1989, p. 79.

57 See Michelson and Morley, 'On a method'.

58 See Michelson and Morley, 'Relative motion', p. 339.

59 A.A. Michelson, 'Comparison of the international metre with the wave-length of the light of cadmium', *Astronomy and Astrophysics* 12, 1893, 556–60, p. 560.

60 Michelson and Morley, 'On a method', p. 430. Later George Hale described Michelson at the interferometer soon after his move to Chicago in 1893: 'I recall especially his uncanny precision in estimating the visibility of interference fringes, which he read off in quick sequences as though from a printed table while he rapidly shifted, step by step, the position of his interferometer mirror. In fact his skill as an observer matched his power of conceiving new methods and embodying them in instrumental form', Hale, 'Some of Michelson's researches', p. 177.

61 See Livingston, *Master*, pp. 136–7.

62 J.R. Benoît contacted Michelson and Morley through O.H. Tittman at the Coast Survey. Tittman to Benoît, 22.10.1890, as cited in Livingston, *Master*, pp. 137–8.

63 Michelson, 'Comparison', p. 557. The Michelson Collection contains a manuscript in Michelson's handwriting on the fine structure of spectrum emission lines (of sodium D) by the relative intensity of interference fringes, indicating the possibility that each of the two sodium lines is itself a double. It also reports on work done with the interferometer and vacuum tubes of cadmium and mercury. Michelson Collection X-16. Michelson's letters to Morley also chart his progress: Michelson to Morley, 29.3.1889, 3.12.1889. On 1.5.1891 he wrote of his observations of the structure revealed by the refractometer: 'every line examined thus far is *at least* double. Mercury looks like this [giving a diagram of the structure he had inferred] the final double lines are a thousandth the distance between the sodium lines!' (Michelson to Morley, 1 May 1891). All in the EWMP, Box 3.

64 A.A. Michelson, 'On the application of interference methods to spectroscopic measurements I', *Philosophical Magazine* 31, 1891, 338–45, and idem, 'On the application of interference methods to spectroscopic measurements II', *Philosophical Magazine* 34, 1892, 280–98. Occasionally one sees dramatic physical evidence of readership: the copy of these papers held in the University of Chicago Library is badly worn, torn in places, and held together by Sellotape, in marked contrast to the pages nearby. Michelson's discussion of the theoretical basis for the inference from visibility curves to the structure of the source (in II) drew an important contribution from Rayleigh, 'On the interference bands of approximately homogeneous light; in a letter to Prof. A. Michelson', *Philosophical Magazine* 34, 1892, 407–10.

65 On Warner & Swasey see E.J. Pershey, 'The early telescopes of Warner & Swasey', *Sky and Telescope* 67, 1984, 309–11, p. 311. On Brashear see John A. Brashear, *A Man Who Loved the Stars: The Autobiography of John A. Brashear*, Pittsburgh: University of Pittsburgh Press, 1988, and Bart Fried, 'The masterful techniques of John A. Brashear', *Sky and Telescope* 81, 1991, 432–8. Brashear was also responsible for the plates used in Rowland's diffraction gratings.

66 On the ether-drift apparatus see entries on 26.5 and 30.6.1887, Sales Book, Warner & Swasey Collection (hereafter W&S), Special Collections, Freiberger Library, Case Western Reserve University, Cleveland, Ohio. For the metre see 'Sketch of experimental apparatus for the determining of the length of a wave of sodium light, Profs. Morley & Michelson' (2 pp.), 23.1.1888, 'Experimental measuring machine for Prof. Michelson', 2.2.1888, 'Measuring machine for Prof. Michelson', 20.2.1888, Notebook 1886–1888, W&S; 31.10.1887 (this entry records the initial 593 hours spent making a 'metre sub-dividing' machine, including drawings, patterns, forging and machine work), 5.11.1887, 30.12.1887, Sales Book, W&S. See Stapleton, 'Context', pp. 17–18. The instrument was delivered ten months late in May 1888. For Morley's response see E.W. Morley to S.B. Morley, 10.5.1888, Morley Papers, Box 2.

67 J.A. Brashear to E.S. Holden, 5.4.1889, Mary Lea Shane Archives of the Lick Observatory, University of California – Santa Cruz.

68 J.A. Brashear to Lord Rayleigh, 2.8.1895, Rayleigh Archives [see n. 36].

69 Michelson, *Light Waves*, p. 41. Michelson's assistant F.L.O. Wadsworth described this process, with the hope that their practices would prove useful in industry. F.L.O. Wadsworth, 'On the manufacture of very accurate straight edges', *Journal of the Franklin Institute*, July (1894), 1–20.

70 B.A. Gould to G.S. Hall, 17.1.1892, Clark University Archives. Note that the maker was now the American Watch Tool Company, with the optics supplied by Brashear.

71 A.A. Michelson, 'Détermination de la valeur du mètre en longueurs d'onde lumineuses', *Travaux et Mémoires du Bureau International des poids et mesures* 11, 1895.

72 See A.A. Michelson, 'Radiation in a magnetic field', *Astrophysical Journal* 6, 1897, 48–54.

73 See D. Weaire and S. O'Connor, 'Unfulfilled renown: Thomas Preston (1860–1900) and the anomalous Zeeman effect', *Annals of Science* 44, 1987, 617–44.

74 T. Preston, 'Radiation phenomena in the magnetic field', *Nature* 53, 1899, 224–9.

75 T. Preston, 'Radiation in a magnetic field', *Nature* 59, 1899, 485 (letter to the editor). He was responding to Michelson, *Nature* 59, 1899, 440.
76 He also described a modification to improve the performance of the instrument: Rayleigh, 'The interferometer', *Nature* 59, 1899, 533 (letter to the editor).
77 Thomas Preston, 'The interferometer', *Nature* 59, 1899, 605 (letter to the editor).
78 Preston compared interferometric and grating methods of determining wavelength, expressing doubts about the former, in T. Preston, *The Theory of Light*, 2nd edn, London: Macmillan, 1895, see pp. 152–3 and 238–9.
79 For a summary of these debates which featured, among others, Jewell and Bell (of Johns Hopkins), Hartmann, Kayser and Fabry and Pérot, see H. Crew, 'Remarks on standard wave-lengths', *Astrophysical Journal* 20, 1904, 313–17, and H. Kayser, 'New standards of optical wave-length', *Philosophical Magazine* 8, 1904, 568–71.
80 C. Fabry and A. Pérot, 'Mesure de petites épaisseures en valeur absolue', *Comptes Rendus* 123 (1896), 802–5, and idem, 'Sur une nouvelle méthode de spectroscopie intéferentielle', *Comptes Rendus* 126 (1898), 34–6. On Fabry and Pérot see the entries in the *Dictionary of Scientific Biography*.
81 See for example C. Fabry and A. Pérot, 'Measures of absolute wave-lengths in the solar spectrum and in the spectrum of iron', *Astrophysical Journal* 15, 1902, 73–96, 261–73; L. Bell, 'On the discrepancy between grating and interference measurements', *Astrophysical Journal* 15, 1902, 157–71.
82 A.A. Michelson, 'The ruling of the diffraction grating', MS, no date (probably early 1915), Michelson Collection, X-8. From similarities in design, Evans and Warner infer that Michelson may have drawn on the ruling engines of Rogers in designing his own (Evans and Warner, 'Precision engineering', p. 5). See A.A. Michelson, 'Report of progress in ruling diffraction gratings', *American Physical Society* 20, 1905, 389–91, and idem, 'The ruling and performance of a ten-inch diffraction grating', *Proceedings of the American Philosophical Society* 54, 1915, 137–42.
83 See G.R. Harrison, 'The production of diffraction gratings: I. Development of the ruling art', *Journal of the Optical Society of America* 39, 1949, 413–26.

11 Travelling knowledge

Narratives, assemblage and encounters

David Turnbull

Knowledge, space, travel, and narratives have deep natural affinities with one another deriving from the way we locate and conceive events, actions and our conceptions of ourselves and the world as we move through it. These affinities are reflected in the embracing of the spatial by sociology of science, of travelling theory by critical social historians and of the narratological by the human sciences generally.[1] As yet there has been little in science studies that attempts a theoretical reorientation through linking these themes. My aim, in a volume that reflects what might be called the 'travelling turn' in the social history of science, is to explore the ways in which science can be conceived as being composed of 'travelling narratives'.

The task of linking travel and science, as I see it, consists in writing spatial history in the complex reflexive and slightly paradoxical sense Paul Carter meant when he described spatial history as 'history of the mapping of the world ahead'.[2] Why? Because we are all 'engineers of space' as Foucault put it in talking of modern scientists.[3] In constructing knowledge we simultaneously create space by travelling through it. Spatial history is then the reconstruction of the narratives of those journeys in which knowledge and space are brought into being.

The metalevel of spatial history is circular – the making of knowledge is the making of space, space is made in travelling, knowledge is travelling, and travelling like knowledge is a form of narrative. Some of these complex interweavings can be seen in a set of related root meanings; symbol – from the Greek to place together; metaphor – a conveyance, a device for being transported across space; theory – to contemplate, from *theorus* (in modern Greek one who travels to see things, an ambassador); travel – originally travail, to work; episteme – putting oneself in a good position. These terms which are central to knowledge and narrative indicate that all of them are based in the idea of active work, of moving through space, cognitively and physically.

From across a wide range of disciplinary perspectives it is starting to become apparent that narrative constitutes a 'fundamental way of knowing',[4] and further that a 'travelling narrative is always a narrative of space and difference'.[5] We construct ourselves, our relationships and our understandings of the world through narratives that forge meaningful links between the otherwise fractured components of our lives. Yet the processes of spatial translation inherent in narrative and knowledge construction are not readily apparent, partly because of the very power of those narratives. Our modernist predilections for displaying our knowledge as objective and universal denies human agency or movement in making knowledge. Our whiggish tendencies to retrospective linear analysis underline the need to recognise the key difference that Latour emphasises, between accounting for the construction of the 'black box' before it is closed – where agency and hybridity are exposed, and after – when it has been purified and networked into 'reality'.[6] Our vision is also obscured by the erasures and *méconnaissance* involved in misrepresenting to ourselves the role of others' knowledge, in the confusion of space and place and the allied elision from practice to theory.[7]

Lacan has drawn our attention to the misconstrual of knowledge in Plato's story of the elicitation of geometry from the slave. It is, I would argue, an example of *méconnaissance* because of the violence it does to the embodied practical knowledge of the artisan in the wrenching of it into abstract theoretical knowledge. A similar *méconnaissance* is committed in the slip from place to space. Space is not just an abstraction in which places are located. Space is linked places, and places are locations given shape and form by humans in their construction of knowledge.

The slave story is significant because, according to Lacan, it is a master's discourse about the appropriation of a servant's craft knowledge (connaissance), and its translation into abstract mathematised instrumental knowledge (science, savoir).[8] But as a narrative about an encounter with the other and the construction of objective universal knowledge from embodied performative knowledge with its attendant suppressions and denials, it raises the critical social question 'Can we use the master's knowledge tools to rebuild the master's house?'[9] To which the answer may be yes, provided that we follow Jane Jacobs's recommendation to adopt 'theory sensitised to movement and studies that are configured around diverse assemblages of people and places'.[10] It is in this light I want to recount a number of narratives that focus on movement and the assemblage of people and places.

Movement and assemblage have, I think, been central foci in recent science and cultural studies. The picture of science and technology that has emerged from these empirical and critical investigations of contemporary and historical makers and users of knowledge in Western and non-Western traditions is that all knowledge is constructed at specific sites through the embodied engagements of particular scientists with particular skills, materials, tools, theories and techniques. Such processes of knowledge production are revealed as thoroughly social and contingent, requiring judgements and negotiations by groups of scientists in specific socio-historical circumstances. Thus a fundamental characteristic of scientific knowledge is its localness and its inseparable link to practice.[11] In addition to being social, local and linked to practice it is also distributed.

> Scientific theory building is deeply heterogeneous: different viewpoints are constantly being adduced and reconciled . . . Each actor, site, or node of a

> scientific community has a viewpoint, a partial truth consisting of local beliefs, local practices, local constants, and resources, none of which are fully verifiable across all sites.[12]

If technoscientific knowledge is local, social, practical and distributed, how does it become universal? How does it move from its site of production and become accumulated at a centre? The answer is that knowledge travels.[13] Universality, objectivity, and accumulation are not characteristics of technoscientific knowledge itself, rather they are effects produced by the collective work of the technoscientific community. In order to be accumulated and rendered truthlike, technoscientific knowledge has to be standardised and homogenised. To move knowledge from the local site and moment of its production and application to other places and times in order to create assemblages, scientists deploy a variety of social strategies and technical devices for creating the equivalences and connections between otherwise heterogeneous and isolated knowledges.[14]

However, we need also to be aware that the standardised assemblages that network theory has brought to our attention should be set alongside other forms of travelling knowledge. As James Clifford remarks:

> Theory is no longer naturally 'at home' in the West – a powerful place of Knowledge, History or Science, a place to collect, sift, translate, and generalise. Or more cautiously, this privileged place is now increasingly contested, cut across, by other locations, claims, trajectories of knowledge articulating racial, gender, and cultural differences. But how is theory appropriated and resisted, located and displaced? How do theories travel among the unequal spaces of postcolonial confusion and contestation? What are their predicaments? How does theory travel and how do theories travel? Complex, unresolved questions.[15]

Assemblages and networks, discourses and narratives intersect, interact, jostle and vie with one another. There are dominant technoscientific assemblages and an increasing number of challenging ones, and something of the complexity of the interactions can be seen in their encounters as their knowledges travel.

Just as knowledge travels, so too is it spatial. By virtue of the fact that knowledge is local, it is also located; it has a place, and an assemblage is made up of linked sites, people and activities. Knowledge traditions create spaces within which defined kinds of subjects and objects can relate and certain kinds of actions are deemed appropriate.[16] Steven Shapin argued in his recent book, *A Social History of Truth*, that the basis of knowledge is not empirical verification, as we tend to think, but trust:

> Trust is, quite literally, the great civility. Mundane reason is the space across which trust plays. It provides a set of presuppositions about self, others, and the world which embed trust and which permit both consensus and civil dissensus to occur.[17]

Hence such spaces have social, practical, epistemological and moral dimensions. The cognitive and social order coproduce one anther and in so doing construct the kinds of mental and moral spaces we inhabit as knowledge producers. Knowledge and society do not merely interact or determine one another. They are constitutive of one another. Society consists in the relations between people and the natural world that produce what we take to be knowledge, and vice versa. But neither society nor our knowledge form anything like a coherent whole, both are complex and messy. Coherence results from our telling stories and making meaning. Seen in this light modern science is not, in essence, distinct from other so-called traditional knowledges. All knowledge traditions are in effect socially organised and sustained spaces in which trusted and authoritative knowledge is produced and transmitted. This permits an equitable basis for their analysis and comparison. It also brings into focus some of the hidden features and assumptions that must be recognised in encounters between knowledge traditions.

As Greenblatt remarks on such encounters: 'European contact with the New World natives is continually mediated by representations: indeed contact itself, at least where it does not consist entirely of acts of wounding and killing, is very often contact between representatives bearing representations.'[18] In order to give some flesh to the idea that such representations are in part narratives about what is to count as knowledge or truth and that they have largely invisible or concealed moral and spatial components that make contact between cultures problematic, I want to explore some instances of encounters between knowledge traditions through the examination of a number of travellers' tales or narratives of travelling knowledge.[19]

The King of Siam

The first of these travellers' tales concerns one of the minor characters in *A Social History of Truth*, though he has had an amazingly enduring bit part to play in the history of epistemology ever since Locke first created him – The King of Siam.[20]

In Locke's account of the rational evaluation of travellers' testimony he cites, in ironic reversal, an apocryphal encounter between the King of Siam and the Dutch ambassador. The king is initially prepared to accept the ambassador as a credible and trustworthy witness, but found he had to call him a liar because he didn't believe the ambassador's claim that water in Holland could become so hard as to bear the weight of an elephant.[21] According to Locke, a rational man should weigh the local implausibilities against the directness, knowledgeability and integrity of the source, and he should seek corroboration from other sources. Most versions of this story tend to display the King of Siam as one of the benighted 'other', and even when it is conceded that testimony and knowledge are based on trust and that he had, as it were, 'no option' but to trust his own local knowledge, the thrust is towards emphasising his error. Indeed, as Shapin points out, the King of Siam is recognisable as 'other' because he makes a mistake.[22]

Mongkut, the real nineteenth-century King of Siam (King Rama IV 1851–68), predicted an eclipse of the sun would occur in two years' time on 18 August 1868. From the point of view of his fellow astrologers, members of the royal court and his Siamese subjects, this was the equivalent of claiming elephants can stand on cold water and his prediction was 'greeted with widespread incredulity'.[23] By Siamese astrological standards a total solar eclipse was almost inconceivable and such long-term precision unlikely. Even though Mongkut was a Siamese astrologer of high repute, his judgement did not concur with that of the court astrologers.[24] Yet Mongkut had such strength of conviction in his ability to calculate with great exactness, he not only made the prediction two years in advance but he also set up a hugely risky 'optical event'.[25] Having determined the point in Siam where the eclipse would be of optimum visibility, he transported the entire court up the coast, including all his wives, all the visiting dignitaries in Bangkok, the Governor of Singapore Sir Harry Ord, the American medical missionary Dan Bradley, and ten French astronomers sent out for the event. Interestingly he did not include the chief astrologer or his fellow court astrologers; they remained in Bangkok. At Hua Wan (also called Waa Kor and Wako), near the present Malaysian frontier, he had an observatory and luxurious buildings for a thousand people especially constructed, providing not only an Italian *maître d'hôtel* and a French chef, but the ultimate ironic luxury – ice.[26]

> Come the day, the sky was clouded over, but just at the right moment the clouds parted to show the eclipse. After the totality had passed the king was so excited that he left his long telescope swinging on its axis, walked into his pavilion and addressed several of his wives saying 'Will you now believe the foreigners?' Meaning as we understood him, 'Will you henceforth believe what the foreigners say concerning the course and time of the eclipse?'[27]

His predictions were not only confirmed, his calculations proved to be better than those of the French by two seconds.[28]

The king not only differentiated himself from the court astrologers but also from the local villagers, who greeted the eclipse with drums and firecrackers. The villagers he dismissed with an amusing aside:

> The king remarked with a smile to his guests that they must not think these people were trying to frighten the demon [i.e. Rahu, who was believed to swallow the sun during an eclipse in Indian-derived folk tradition]: they were merely celebrating their sovereign's skill in having been able to calculate the moment of the eclipse more accurately than the European Astronomers.[29]

Sadly it turned out the villagers were right and it was indeed an evil portent; the king and many others got malaria, and he died shortly after his return to Bangkok.[30]

The court astrologers he dealt with much more severely. When Mongkut returned to Bangkok to find that the chief astrologer and other eminent astrologers had failed to observe and measure the eclipse in the scientific way he expected of them, he ordered them to polish stones for a day and then imprisoned them with chains of water-snail shells round their necks.[31] This was a classic instant at which the grounds for trust were reconfigured and the nature of the Siamese knowledge tradition was reshaped. The court astrologers may not have been inclined to accept Mongkut's prediction because they were incapable of following, or distrusted, the mathematical calculations that Mongkut used, since it seems likely that he augmented an unorthodox Siamese astrological tradition by adding Western astronomical and mathematical techniques.[32] Nonetheless the event clearly played a role in changing the grounds of trustworthy testimony in both Siam and the West, and in the complex and slightly contradictory way that is characteristic of such encounters Mongkut is championed by both sides. But the question remains how was Mongkut able to bridge the worlds of Siam and the West?

Shapin makes an interesting move in the direction I want to go when he describes the predicament of the traveller's tale:

> Whenever, and for whatever reasons, those who judge observation-claims cannot be at the place and time where the phenomena are on display, then judgement has to be made 'at a distance'. The trust relationship is, in that sense, inscribed in space . . .

This echoes the passage I cited earlier – 'Mundane reason is the *space* across which trust plays.'[33]

It is to the spatial and the narratological that we need to turn to answer my question about Mongkut, in order to consider the conditions under which self and other are created – the grounds upon which trust and its necessary corollary, distrust, are built. The geographer Winichakul Thongchai describes how the nation-state of modern Siam, now Thailand, was created through the displacement of indigenous maps and spatial conceptions by modern Western conceptions of space that are concomitant with modern mapping. He argues that 'Mapping produced a new Siam by transforming the meaning of boundary, sovereignty, and margin.'[34] The concept of what it is to be Siamese is co-produced with an understanding of what it is to be 'not Siamese' (to be other), which in turn depends on an understanding of what the state and its boundaries consist in – a totality Thongchai calls 'geobody', a complex interwoven narrative of space, self and knowledge.

Prior to the displacement of indigenous mapping, the boundary between Siam and its neighbours was an indistinct zone in which whole towns could have multiple overlords and allegiances. Consequently the concept of nationhood was variable and local. Engel has argued that for legal and political purposes Thai 'space' was defined in terms of 'hierarchical relationships between people and groups rather than in terms of bounded regions or territories'. But he also found that these spatialities overlapped with others defined in terms of control over rice fields or in terms of territorial spirits associated with natural objects, settings and markers. 'Who one was could not be separated from *where* one was.'[35] Hence the question of 'who is trustworthy and who is not' is in part spatial, and since this spatiality was itself transformed by the process of Western mapping it is also a question of evolving narratives about selfhood and difference.

Mongkut is now celebrated as the father of Thai science – 18 August, the day of his successful eclipse prediction, being National Science Day – and it was he who invited Britain and France to map Siam in his struggles with Burma and Vietnam.[36] The first bounded map of Siam appeared in the reign of Mongkut's son in 1893. Modern Thai scientific knowledge has thus been co-produced with a moral and political space, but it is not a spatiality that is identical with the Western one – it is very much a local Thai space.

While it is still a matter of debate among Thai astrologers today as to whether Mongkut's prediction was a betrayal or a triumph of Thai astrology, he is also claimed as the father of Thai science and Thai astrology.[37] Similar tensions and contradictions are revealed in this counter-narrative of an encounter between knowledge traditions and the development of a spatiality. Retelling the traveller's tale and the way knowledge is narrated in Thailand reveals a complex crumpling of space in the circulation and blending of knowledges.

Mpemba's physics

The story of 'Mpemba's physics' also shows that there is a multiplicity of knowledge circuits. It was Thomas Kuhn who laid the foundations for the 'network thesis' and the spatialisation of knowledge in his concept of the paradigm – a map that showed you the way around and provided a network within which all facts are connected and given existence. It was also Kuhn who first brought to contemporary attention another tale about the vicissitudes of travelling knowledge that could not be incorporated into a paradigm. Francis Bacon reported in 1620: 'Water slightly warm is more easily frozen than quite cold.'[38] Bacon's observation is cited by Kuhn as an example of the kind of fact gathered by natural histories prior to the emergence of a first paradigm 'that we are now quite unable to confirm'.[39] In network theory terms a claim is a not 'fact' until it is incorporated; it does not exist because it is outside of the network.

This phenomenon of warm water freezing faster than cold came up again in the sociology of science literature when Barry Barnes discussed it as the 'Mpemba effect' – an apparent experimental confirmation by Erasto Mpemba, a high school student in Tanzania. For Barnes the puzzle of the 'Mpemba effect' was its incompatibility with existing knowledge in the form of Newton's law of cooling, and the problem was how to render it compatible. In his view, it 'very effectively illustrates the *crucial role of existing knowledge* in assimilating and describing new findings'.[40]

But for this story to make sense it has to be put into the context in which Mpemba first encountered the effect. This process of telling a story, putting in context and 'making sense', both illustrates the network/assemblage thesis and the narratological nature of knowledge-making. When Mpemba was in the third form at Magmata Secondary School in Tanzania in 1963 he used to make ice cream, like other boys at the school, by boiling milk with sugar and freezing it. One day in the rush for space in the fridge he put his hot milk in without letting it cool first, and found that it froze before that of the other boys. He then found that this was a well-known practice among commercial ice-cream makers and vendors in Tanga, the local town, and went on to try some experiments of his own. He raised the question in class and was suspected by his teacher of being confused. When asked, 'But Mpemba did you understand your chapter on Newton's laws of cooling?' his response was 'Theory differs from practical.' The teacher concluded, 'Well all I can say is that this is Mpemba's physics and not the universal physics.'[41] The school was subsequently visited by a physicist, Dr D.G. Osborne from University College in Dar es Salaam, and Mpemba raised his question again, somewhat to the embarrassment of his schoolfellows. Osborne suppressed his initial scepticism and took the problem back to his own laboratory, where he conducted a series of experiments confirming that 'If you take two beakers with equal volumes of water, one at 35 °C and the other at 100 °C and put them into a refrigerator, the one that started at 100 °C freezes first.' Mpemba and Osborne published a joint article in the journal *Physics Education*, attributing the effect to faster cooling rates from the surface of hot liquids.[42]

The story then moved to London and was reported in the *New Scientist* later that year,[43] which started a series of letters to the editor, all reporting examples of the effect in a variety of local practices. For example, a Miss Blackwell wrote that in 1949 a member of the Birmingham and West Midland Bird Club 'stated that hot water put out to the birds froze faster than cold', and could also 'vaguely remember my late aunt's maids refraining from pouring hot water into a frozen drain, saying that the water would only freeze faster'. She finished by making the point that

> It seems that physicists are now discovering facts from the domestic front, just as physiologists occasionally publish 'discoveries' well known to the old gardener. This lack of communication between scientists and those 'in service' is not really surprising since the latter learned their art not from written manuals, but from long association with senior staff – by word and example. This fast-dying race will no doubt take their secrets with them. They are hardly likely in retirement to write text-books of instruction for the benefit of their aged one-time employers.[44]

Meanwhile, across the Atlantic, a Canadian chemist G.S. Kell published an account in the *American Journal of Physics* that subjected the effect to computer analysis and experimental test. He reported: 'It is widely believed, at least in Canada, that hot water will freeze more quickly than cold water.' Canadian folklore has it that cars should not be washed with hot water, skating rinks should be flooded with hot water, and that a bucket of hot water left out will freeze faster than cold.[45] Kell found that covered vessels follow Newton's law of cooling, but when evaporation can occur the hotter vessel may lose enough mass to gain an advantage. Counter to the finding of Mpemba and Osborne, which he does not mention, he argued that the situation is complicated by heat conduction in the fridge and the possibility of lowered thermal resistance by the formation of an ice layer under the hot vessel.

This point was also made by Adam Osborne, an American Shell employee who wrote to the *New Scientist* arguing that: 'This fact seems to be brought to light every year or so and the discussion is by no means new. The explanation is of course quite simple.' He went on to explain it in terms of the improved thermal contact established by the hot vessel, and asserted the effect would vanish if thermal insulation were used.[46] However, Sherwood, in another letter, pointed out that Mpemba and Osborne had taken both these points into account in their original experiment, and concluded that: 'Apparently it is not only in Tanzania that scientists are ready with a glib answer of which any medieval scholastic would be proud.'[47] The last letter in this series reported that the effect was well known in the frozen fish business. While another dismissed it as a 'cultural myth'.[48]

The story was then taken up by Ian Firth in the School of Physics at the University of St Andrews, as a project for second-year undergraduates who set out in classic style to reproduce Mpemba's results. Not surprisingly, 'The initial problem of reproducing the results proved more difficult than was expected.'[49] Firth felt that the whole episode showed the 'willingness of scientists to show their unscientific attitudes',[50] and noted that 'There is a wealth of experimental variation in the problem so that any laboratory undertaking such investigations is guaranteed different results from all others.'[51]

The Mpemba effect resurfaced again almost thirty years later in 1995 when David Auerbach of the Max Planck Institute for Fluid Dynamics in Göttingen claimed the phenomenon was due to supercooling. Auerbach, while on sabbatical at the Water Research Institute in Perth, Western Australia, put two sealed 100 ml beakers, one with water at 18 °C and the other at 90 °C, into a bath of cold ethanol monitored with a thermistor. After conducting 103 experiments he found the effect not to be a hard and fast rule, but that invariably the water did not start to freeze until well below 0 °C.[52] He concluded that the temperature at which supercooled liquid starts to freeze is unpredictable and that when the effect is produced it is because the sudden appearance of ice crystals from the supercooled liquid occurs at a higher temperature in water that was formerly hot. Cold water, however, is always the first to freeze completely. The following year Knight discussed the Mpemba effect in the same journal, also attributing it to supercooling, though suggesting a different causal mechanism by recalling the work of Ernest Dorsey in compiling all the known information on *The Properties of Ordinary Water Substance* for the US Bureau of Standards in 1940.[53]

One of the correspondents to the *New Scientist* also pointed out that it was not Bacon but Aristotle, in his *Meteorologica*, who was the first to record the effect:

> The fact that the water has previously been warmed contributes to its freezing quickly: for so it cools sooner. Hence many people when they want to

> cool water quickly begin by putting it in the sun. So the inhabitants of Pontus when they encamp on the ice to fish (they cut a hole in the ice and then fish) pour warm water round their reeds that it may freeze the quicker, for they use ice like lead to fix the reeds.[54]

Auerbach found that the effect had also been discussed by Descartes: 'one can see by experience that water that has been held on the fire for a long time freezes sooner than other water'.[55]

The 'Mpemba effect' is not a 'cultural myth', it arises in the context of human practice, out of endeavours to catch or freeze fish, deal with frozen drain pipes, make ice-cream, feed birds, maintain ice rinks, or perform experiments. This item of local knowledge circulates quite freely in a widely disparate set of places, social strata, times and cultures. It has small local specific networks that maintain it as fact, like those of servants in England or teenage Tanga town ice-cream vendors. It also seems to travel quite freely between those places and times, yet it appears to be a fact for which we have no accepted explanation and which seems to stand in contradiction with an established and uncontested scientific law.

It is tempting to suggest that problems of reproducibility, interpretative flexibility and incompatibility with existing knowledge serve to explain why the Mpemba effect has not yet been integrated into any universal scientific network, despite its materiality as an observed constraint on human practice where freezing water is of consequence; and further, that once the complex variables inherent in the phenomenon have been controlled and measured with sufficient precision such an integration will occur. However, I think such an approach overemphasises standardised networks and underemphasises ways in which knowledge is moved and assembled. The effect has moved quite readily, not only in local practices but also in the work of pre-moderns like Aristotle, Bacon and Descartes, in the work of sociologists of scientific knowledge like Kuhn and Bloor, as well as in the scientific and educational literature of the last thirty years. Far from being a 'fact that we are now quite unable to confirm', or simply being problematic because it is 'incompatible with existing knowledge', as Kuhn and Barnes would have it, despite its fuzziness and incoherence, the effect circulates in all kinds of narratives and practices. Indeed Kuhn and Barnes both treat it as a non-problematic 'social fact' in their narratives, thereby attributing to it a form of existence and mobility.

While the invocation of social processes in the assemblage of local knowledge has been immensely fruitful, it has been overly representationalist. A more performative account allows recognition to the element of praxis where our theoretical constructs are constrained by the material world.[56] On the other hand the accounts in terms of a domain, field, network, paradigm, form of life, or pre-existing body of knowledge have tended to deploy a rather restricted sense of the social, limiting it largely to judgement and negotiation. Following the Mpemba effect around shows, I think, the advantages of an account that is simultaneously more performative and more social, and that the travelling problem is not just one of mobility. The master narrative of network theory needs to be offset by small narratives of incomplete theories and disparate practices.[57]

Australian Aboriginal maps

The following story is cartographical and has been told elsewhere in a different version, largely in the context of reframing indigenous knowledge.[58] I want to retell it here because maps are a pre-eminent form of space-making narrative in which knowledge is moved and assembled. The story concerns ways in which differing spatialities emerge in encounters between disparate ways of travelling and narrating the world.

Spatial crumplings, political tensions and epistemological contradictions are ever present in the cartographic encounter and can be clearly seen in contemporary Australia where Aboriginal maps are at the centre of the struggle over who is to be trusted, what is to accepted as authoritative knowledge, and the nature of spatiality. One of the commonest forms of representation in Australian Aboriginal culture is the map. Bark paintings are often maps, as are sand sculptures, body painting and rock art. Aboriginal maps may be landscape maps depicting known places in the geographical environment; they may be territorial maps showing ownership of a story or region. They may show events from the Dreaming, vegetation types, or land forms, and the knowledge displayed in an Aboriginal map can be structured by kinship, territorial relations, topography or mythology.[59]

The way the Yolgnu of eastern Arnhem Land structure their knowledge system is typical of the kinds of ways in which it is possible for Aboriginal groups to have a detailed understanding of their environment. The Yolgnu system is dependent on the joint articulation of two modes of patterning. One is genealogical – *gurrutu*, the systematic structure of kinship relations; the other is spatial – *djalkiri*, the footsteps of the Ancestors or the Dreaming tracks.[60]

The genealogical structure provides an unlimited process of recursion that enables all things to be named and related and thus imposes an order on the social and natural world that gives it coherence and value. It provides the framework within which social obligations with regard to life, death, marriage and land can be negotiated. The other mode of patterning is provided by the stories, myths or dreamings that relate the travels and activities of the Ancestors in creating the landscape in the form of tracks or songlines that traverse the whole country. The kinship system and the songlines together form a knowledge network that allows for everything to be connected. The concept of connectedness is an extremely powerful one in Aboriginal and Yolgnu culture and is exemplified by the Yolgnu term *likan*, which in the mundane sphere means 'elbow' (the connection of the upper and lower arm), and in the spiritual sphere connotes the connections among ancestors, persons, places and ceremonies.[61] A wide variety of Australian Aboriginal paintings can be interpreted as being simultaneously social and geographic, representing both the tracks of the Ancestors and detailed maps of places. Hence bark paintings are encoded knowledge of connections.

However, while the land may have boundaries which can be known with precision, it is not typically good custom to display them, because they are permeable rather than fixed entities with rights of access being required and most frequently granted. Areas can be owned by more than one group and routes can be common property. Boundaries are more properly the subject of negotiation and exchange in ceremony and ritual. Moreover, Yolgnu conceptions of land or territory do not correspond to Western legal notions of enclosure but are typically of open and extendable 'strings' of connectedness.[62]

The tensions, contradictions and classificatory difficulties re-emerge when you consider the counter arguments to this position. Peter Sutton, an anthropologist who is one of the most informed non-indigenous writers on Aboriginal mapping, argues that bark paintings are not maps but icons. They are maplike because of the intimate role of the land in Aboriginal spiritual cosmology, but in his view a drawing is only properly a map if it is made to explain geographic landscapes and geopolitical distributions to non-Aboriginals.[63] He insists on this fairly severe distinction, because of his antipathy to appropriation by non-Aboriginals of Aboriginal designs. By categorising and classifying them in our terms, we erase their Aboriginal context and meaning in order to commodify and own them. Sutton's position is, of course, commendable but requires additional dimensions when it comes to the contemporary political context in which struggle for control of history, territory and resources is at the heart of the native title debate.

Australian Aboriginal groups do constantly map their land, though in a very different process from that of the dominant white society. But being mapped in the white manner may have advantages – for example, in making land claims. In fact it is standard procedure for anthropologists to record Dreaming tracks on Western topographical maps as evidence that this knowledge, and hence the land, is the property of the claimants.

However, this melding or working of two different knowledge traditions together, and in effect creating a third space, is a very different process from a recent controversial project of mapping the 'precise boundaries of all Aboriginal groups in Australia'.[64] As we have seen, Aboriginal conceptions of identity with the land do not equate with the notions of boundary precision, exclusion and individual property rights and the linkages to the state implicit in Western maps. Aboriginal maps celebrate that identity by providing connectedness, leaving their permeability, imprecision and inclusion open to negotiation in ceremony and ritual. Western modes of mapping by contrast make land, its use and resources, tradeable commodities, and, as a consequence in contemporary Australia, highly contested sites of struggle for control and ownership.

Further evidence of the complex ways in which spaces fold in on one another, and in which tensions and contradictions are generated, is to be found in the use by Central Desert Aboriginals of GIS/GPS to map their territory, in the incorporation of Aboriginal knowledge in mapping desert water resources, and in the overtly spatial and cartographical character of contemporary Aboriginal art. Art which is painted in a secular, even commercial, context, though using purely Aboriginal iconography and partly traditional techniques, is thoroughly postmodern, demanding to be read across the boundaries. Simultaneously cartographic and iconic, it also insists that the cultural processes that informed it are as much part of the work as the representation itself.

In June 1997 a group of sixty people got together and painted the largest ever (10 m by 8 m) collective Aboriginal canvas as evidence of their Sandy Desert land claim. Half the group were Kimberley artists from four different language groups (including Walmatjarri, Wangkajungka, Mangala and Juwaliny), and the other half were advisers chosen for their local knowledge of the land.

Ngarralja Tommy May, chairperson of the co-ordinating body Mangkaja Arts, explained the painting's purpose:

> Native Title is properly blackfella story, blackfella law I reckon. That painting is only for proof. When you go to that High Court and tell your story, listening really carefully first before we open our mouths, 'we don't believe' might be what *kartiya* (white fellas) say. They might say 'we don't believe, might be you're lying.' There's plenty like that I reckon: some might be all right, some might believe, some don't.
>
> That story for Native Title, sometimes we can paint, for proof; to let people understand, so that they can know.[65]

There is a double irony in the complex folding of this painting. Jimmy Pike, a renowned Aboriginal artist responsible for getting the project going and on whose place the painting was done, started it off with a wide stripe of rust-coloured acrylic across the canvas representing the Canning stock route which bisects the claim area and provides a reference point for each artist.[66] Not only does a European trading route form the axis of the painting but the map is also a classic example of the modernist archive project to assemble heterogeneous local knowledge in one space and time.

It is also subject to the processes of struggle, negotiation and concern with 'accuracy' which are concealed in Western maps. An earlier version of the big map was rejected on the grounds of inaccuracy with respect to the location of certain areas, the absence of certain owners and concerns over the participation of artists who were not parties to the land claim.[67]

This painting is now part of the evidence before the court hearing the land claim and the outcome is as yet unknown. It avoids, I think, some of the appropriation problems that concern Sutton and many of the commentators on counter-mapping or resistance mapping[68] by being so firmly Aboriginal in its iconography and performance, while at the same time being 'readable' in a non-Aboriginal court setting. However, that readability is achieved through a transformation of place, space and trusted knowledge in both Aboriginal and Western traditions. That transformation has been going on for some time, as can be seen in the establishment and acceptance of the equivalences whereby the presentation in court of anthropologists' maps of knowledge of dreaming tracks is taken as evidence of landownership in the same way as cartographically co-ordinated title deeds. Similar transformations and hybridisations can be seen in the so-called bark petitions from the area the crocodile dreaming depicts in north-eastern Arnhem Land.

Figure 11.1 The Barunga Statement calling for a *Makarrata* or treaty, displayed at the Barunga Festival in the Northern Territory, Australia, just before its presentation to Prime Minister Bob Hawke on the Queen's birthday in the year of the bicentenary of the arrival of the first fleet, June 1988. Photograph: Christine Colton.

Galarrawuy Yunupingu, a Yolgnu elder and chairman of the Northern Land Council, has described what happened when on 28 August 1963 the Yolgnu petitioned Parliament to stop compulsory acquisition for mining:

> It was not an ordinary petition: it was presented as a bark painting and showed the clan designs of all the areas that were threatened by the mining. It showed, in ways a coloured piece of bunting could never do, the ancient rights and responsibilities we have to our country. It showed that we are not people who could be painted out of the picture, or left at the edge of history.

In 1970 the judge refused to recognise the bark painting as title to their land and on 12 June 1988 another bark was presented to Bob Hawke, the then Labour Prime Minister, calling for a *Makarrata* or treaty. The painting hangs in Parliament today, but there is still no treaty. Yunupingu described the jointly political and narratological nature of the process:

> I and other northern Australian Aboriginal artists did the paintings on the Barunga statement. The dot-style paintings of Central Australia and the cross hatching painting of Northeast Arnhemland show that Aboriginal people of different countries, speaking different languages, can unite in the

> same struggle. The painting surrounds the words of the statement, showing that our painting for the land and the words that express it in English speak equally. They cannot be differentiated. One form of writing, writing for land, is in the form of a painting; whereas in the modern sense, in English terms, its [*sic*] words on a piece of paper. The painting is simply a title to the land.
>
> When we paint – whether its [*sic*] on our bodies for ceremony or on bark canvas for the market – we are not just painting for fun or profit. We are painting as we have always done, to demonstrate our continuing link with our country and the rights and responsibilities we have to it. We paint to show the rest of the world that we own the land and that the land owns us. Our painting is a political act.[69]

Yunupingu succinctly captures the underlying commonalities; writing and painting are both ways of telling stories about the land and are political acts in that they establish spaces for trusted and authoritative knowledge. Spaces which are themselves transformed as differing narratives are performed in them.

Conclusion

But the three stories I have told you about Mongkut, Mpemba, Aboriginal artists from Kimberley, while all being encounters between knowledge traditions, may appear to have little in common. What unites them and makes a narrative of their telling is the recognition that all knowledge traditions are heterogeneous with conflicting and contradictory variants constantly in circulation. Hiatuses, discontinuities and indeterminacies separate theory and practice, place and space, distributed and centralised knowledge, the local and the global. Though a plethora of social strategies, technical devices, spatial and literary practices have been developed in every culture enabling knowledge to travel, all knowledge traditions remain messy and complex, replete with unbridged gaps and overlapping spatialities. Every tradition has to cope with a variety of 'travelling problems', which include finding ways to move and assemble disparate elements that have sufficient propinquity to permit commensuration and precision, thereby forming knowledge spaces; strategies for moving between and linking elements in such spaces; developing ways in which to make visible the varieties of space-making and their complex interactions. Narratives, I suggest, enable us to bridge the hiatuses and create spatial coherence, linking people, places and practices. They also enable us to move through knowledge spaces by making connections. Spatial histories of movement and assemblage are themselves narratives that bring into conjunction otherwise disparate spatialities, creating new ones.

All knowledge, including science, consists of travellers' tales, accounts of events that have happened at a distance. The problem for all knowledge traditions is to establish ways of telling which tales and narrators to trust, how to integrate the local and specific into the general, which in turn means finding commonalities between the stories so that in their retelling they appear cohesive. Knowledge is the assemblage of a messy multiplicity of accounts, people, practices, places, objects and instruments in a linguistic and classificatory structure. Knowledge is then a spatialised narrative of human actions.

What we now count as a specially authoritative form of knowledge – Western or modern science – is a tradition which has devised social strategies, narrative forms and instrumental practices that enable local knowledge to travel, to be assembled at a centre of calculation and then to be put into use or transmitted as a unified body to other centres. This process of assemblage and unification is accompanied by the erasure of the local, the heterogeneous and the narratological, but closure is never complete, heterogeneity always re-emerges, different voices, different practices, different accounts vie with one another for dominance as their narrators and practitioners struggle for authority.[70]

This chapter is primarily about the ways in which differing narratives travel, collide and mix with one another. Somewhat serendipitously it also turns out to be a set of narratives that coalesce, interact and contradict the narratives in the rest of the collection. The other chapters are explicitly focused on the ways in which scientific instruments and their measurements both travel and establish the conditions for travelling and are implicitly narratological. The entwined theme that runs through all the chapters is that of co-ordinating commensurability through the creation of the conditions for the possibility of cumulativity, communicability, credibility and civility; conditions that are material, linguistic, practical and, ultimately, economic and political.

What is revealed in each of the chapters is that as historical accounts not only are they inescapably narratological themselves but, in their revelation of the widely differing components and strategies imbricated in the co-ordination of commensurability across the spectra of the sciences, they are also narratives of subversion and conversion.

The various ways in which a knowledge tradition has been conceived as an episteme, a discourse, a network, a form of life, an assemblage, or a knowledge space and the hegemonic have a great deal in common. A communality which Jean and John Comaroff capture in their definition of the hegemonic as 'the order of signs, practices, relations, distinctions, images and epistemologies – drawn from an historically situated field – that come to be taken-for-granted as the natural and received shape of the world and everything that inhabits it'.[71] From this perspective what counts as scientific knowledge is a naturalised ordering of the world that we have come to take as self-evident. In their insightful discussion of narrative and the law Patricia Ewick and Susan Silbey show that the hegemonic can be both sustained and subverted by narratives. They suggest that

> when narratives emphasise particularity, and when they efface the connection between the particular and the general, they help sustain hegemony. Conversely when narrativity helps bridge particularities and makes connections across individual experiences and subjectivities it can function as a subversive social practice.[72]

What each of the narratives does in this collection is subvert the hegemonic. They reverse the naturalisation of the arbitrary and show the historically contingent process of knowledge movement and assemblage by displaying the connections through narrative of the local and universal that are central to, and have been erased in, the hegemonic account.

Richard Harvey Brown in *Toward a Democratic Science*, perhaps more effectively than anyone else to date, has articulated and extended the narratological approach to science. For Brown stories of discovery and truth are narratives of conversion – 'those that convey the reader into worlds that at first seem wholly other than anything seen before'.[73] Such stories are also narratives of conversion in that they are concerned with the metaphorical conversion of the unfamiliar into the familiar, with the establishment of judgements and relations of similarity. All the chapters in this collection are thus narratives of subversion and conversion, because they are all concerned with ways of moving the distant and the unfamiliar into the sphere of the 'here', and the familiar. Their common concern is with the ways we make objects, processes and events conceivable and measurable by telling stories that enable theories and instruments to travel.

Just as assembling and moving knowledge is common to all Western knowledge traditions, though its modes and forms vary with historical context, so too does it have a multiplicity of instantiations in other knowledge traditions. Many of the tales of travelling knowledge told here are devoted to showing how, for example, knowledge like Mpemba's physics can move in quite different circuits from those of more orthodox knowledge.

It is in the case of Australian Aboriginal maps that all the various threads of narrative, space, commensurability, travel and civility come together. In the making and performing of the Kimberley land claim Australian Aboriginals have created alternative ways of both assembling their own knowledge and stories and of linking those narratives to non-indigenous accounts. By so doing they have given explicit recognition to the political dimensions of knowledge assemblage. The creation of a knowledge space is simultaneously the creation of a civility, a common discourse of approved behaviour and trusted authority. The civility that their work proposes is multivocal, one in which many voices speak, many rationalities, spatialities and narratives are acknowledged and encouraged to rub up against one another.

In the end the intention of the kind of narrativised approach advocated here is not, as the like of Sokal and Bricmont would have it, to show that science is all just pointless fabulation. It is instead to celebrate the messy multiplicity of the stories that are part and parcel of all knowledge and thereby allow both a stronger science and a stronger democracy. The strength of this narratological position should be assessed in terms of its ability to face Meera Nanda's criticism of the constructivist approach: that unless we have the capacity to evaluate claims to truth, superstition and oppression will be unopposable.[74] Narratives that conceal the ways the local and the universal are joined, narratives that do not expose such conditions as those that make assemblage and travel possible serve to make the arbitrary seem natural.[75] What a narratological approach can allow is a co-ordinated civility in which many bodies and voices can travel and narrate, and in which the conflict of differing accounts can ensure criticism and change.

Notes

1 On the spatiality of knowledge see D. Turnbull, 'Constructing knowledge spaces and locating sites of resistance in the early modern cartographic transformation', in R. Paulston (ed.) *Social Cartography: Mapping Ways of Seeing Social and Educational Change*, New York: Garland Publishing Inc., 1996, pp. 53–79; see also J. Jacobs, *Edge of Empire: Postcolonialism and the City*, London: Routledge, 1996; D. Livingstone, 'The spaces of knowledge: contributions towards a historical geography of science', *Environment and Planning D: Space and Society* 13, 1995: 5–34; S. Shapin, 'Placing the view from nowhere: historical and sociological problems in the location of science', *Transactions of the Institute of British Geographers* 23, 1998: 5–12. On travelling theory see J. Clifford, 'Traveling cultures', in L. Grossberg, C. Nelson and P. Treichler (eds) *Cultural Studies*, New York: Routledge, 1992, pp. 96–116; idem, 'Notes on theory and travel', in J. Clifford and V. Dhareshwar (eds) *Traveling Theory Traveling Theorists*, Santa Cruz: Center for Cultural Studies, 1989, pp. 177–88; M.L. Pratt, *Imperial Eyes: Travel Writing and Transculturation*, London: Routledge, 1992. On the narratological in the human sciences see R.H. Brown, *Toward a Democratic Science: Scientific Narration and Civic Communication*, New Haven: Yale University Press, 1998; D.E. Polkinghorne, *Narrative Knowing in the Human Sciences*, Albany: State University of New York Press, 1991; J. Bruner, *Acts of Meaning*, Cambridge, Mass.: Harvard University Press, 1990.

2 P. Carter and D. Malouf, 'Spatial history', *Textual Practice* 3, 2, 1989: 173–83, p. 175.

3 Cited in A. Barry, 'The history of measurement and the engineers of space', *BJHS* 26, 1993: 459–68, p. 466.

4 C. Mattingly, 'Time, narrative and cultural action', *American Anthropologist* 100, 1, 1998: 184–6, reviewing C. Briggs (ed.) *Disorderly Discourses: Narrative, Conflict and Inequality*, New York: Oxford University Press, 1997, and L. Hinchman and S. Hinchman (eds) *Memory, Identity, Community: The Idea of Narrative in the Human Sciences*, Albany: State University of New York Press, 1997, p. 185.

5 G. Robertson, M. Mash *et al.* (eds) *Travellers' Tales: Narratives of Home and Displacement*, London: Routledge, 1994, p. 2.

6 B. Latour, *Science in Action*, Milton Keynes: Open University Press, 1987; idem, *We Have Never Been Modern*, Cambridge, Mass.: Harvard University Press, 1993.

7 M. Curry, *The Work in the World: Geographical Practice and the Written World*, Minneapolis: University of Minnesota Press, 1996.

8 J. Lacan, *Le Seminar: Livre 17 1969–70. L'envers de Psychanalyse*, Paris: Editions de Cie, 1991.

9 For discussion see M. Nanda, 'The science question in postcolonial feminism', in P. Gross, N. Levitt and M. Lewis (eds) *The Flight from Science and Reason*, New York: New York Academy of Sciences, 1997, pp. 420–36.

10 Jacobs, *Edge of Empire*, p. 7.

11 D. Turnbull, 'Local knowledge and comparative scientific traditions', *Knowledge and Policy* 6, 1993: 29–54.

12 S.L. Star, 'The structure of ill-structured solutions: boundary objects and heterogeneous distributed problem solving', in L. Gasser and N. Huhns (eds) *Distributed Artificial Intelligence*, New York: Morgan Kauffman Publications, 1989, pp. 37–54, p. 46.

13 S. Shapin, 'Here and everywhere: sociology of scientific knowledge', in J. Hagan and K. Cook (eds) *Annual Review of Sociology*, Palo Alto, Calif.: Annual Reviews, 1995,

pp. 289–321. N. Wise, 'Precision: agent of unity and product of agreement. Part 1: Travelling', in N. Wise (ed.) *The Values of Precision*, Princeton, N.J.: Princeton University Press, 1995, pp. 92–100, p. 98.

14 D. Turnbull, 'The ad hoc collective work of building gothic cathedrals with templates, string, and geometry', *Science Technology and Human Values* 18, 1993: 315–40.

15 Clifford, 'Notes on theory and travel', p. 179.

16 Kathy Addelson points out that the kind of world prescribed by the supposition of a judging scientific observer makes assumptions about subjects, objects, and their relationships in the world. K.P. Addelson, *Moral Passages: Toward a Collectivist Moral Theory*, New York: Routledge, 1994.

17 S. Shapin, *A Social History of Truth: Civility and Science in 17th Century England*, Chicago: University of Chicago Press, 1994, p. 36.

18 S. Greenblatt, *Marvellous Possessions: The Wonder of the New World*, Chicago: University of Chicago Press, 1991, p. 119.

19 U. Bitterli, *Cultures in Conflict: Encounters Between European and Non-European Cultures, 1492–1800*, Cambridge: Polity Press, 1989. Bitterli points out that encounters with the other are intensely difficult epistemological and moral denials.

20 Mongkut also had roles in travel literature, film and post-colonial critique of the oriental other and gender relations, being portrayed as the arbitrary and exotic keeper of a harem by the Western nanny he employed – Anna Leonowen. See C. Kaplan, '"Getting to know you": travel, gender and the politics of representation in *Anna and the King of Siam* and *The King and I*', in R. de la Campa, A. Kaplan and M. Sprinker (eds) *Late Imperial Culture*, London: Verso, 1995, pp. 33–52.

21 Shapin, *A Social History of Truth*, p. 229. J. Locke, *Essay concerning Human Understanding*, New York: Dover, 1959 [1690], Bk. IV, ch. 15, sect. 5. The King of Siam example is also discussed by Leibniz and Hume and by many commentators since – for example, L. Daston, 'The moral economy of science', *Osiris* 10, 1995: 1–26.

22 Shapin, *A Social History of Truth*, p. 244.

23 N. Cook, 'A tale of two city pillars: Mongkut and Thai astrology on the eve of modernization', in G. Wijeyewardene and E.C. Chapman (eds) *Patterns and Illusions: Thai History and Thought*, Canberra: The Richard Davies Fund and Department of Anthropology, ANU, 1993, pp. 276–304, p. 293.

24 A. Moffat, *Mongkut, The King of Siam*, Ithaca, N.Y.: Cornell University Press, 1961, p. 171.

25 On staging optical events see B. Latour, 'Visualisation and cognition: thinking with eyes and hands', *Knowledge and Society* 6, 1986: 1–40.

26 J. Blofeld, *King Maha Mongkut of Siam*, Bangkok: The Siam Society, 1987 [1st edn 1972], p. 85.

27 G. Feltus (ed.) *Abstract of the Journal of Rev. Dan Beach Bradley, M.D. Medical Missionary in Siam 1835–1873*, Cleveland, OH.: Dan Bradley, Pilgrim Church, 1936, p. 278.

28 Ibid.

29 A.B. Griswold, *King Mongkut of Siam*, New York: Asia Society, 1961, p. 51; editorial note in Cook, 'A tale of two city pillars', p. 296.

30 Hundreds of labourers employed in setting up the observation encampment in the jungle also died, showing that moving knowledge is costly and difficult in all cultures.

31 Cook, 'A tale of two city pillars', p. 297.

32 According to Winichakul Thongchai, *Siam Mapped: A History of the Geo-Body of a Nation*, Honolulu: University of Hawaii Press, 1994, p. 45, Mongkut preferred the unorthodox *Saram* astrological tradition. That the king succeeded in making an accurate prediction is evident in the events of 18 August 1868, but the difficulties of knowing what exactly was involved in his calculations and the problems of translation in an encounter between knowledge traditions are evident in the letter he wrote to the French astronomer Stephan beforehand (included here in full and as it appears in Stephan's report):

The astronomical statement of the king of Siam. I beg to state truely that my knowledge of astronomical science was very less, almost inconsiderable. I have studied this science firstly in Siamese and Peguan astronomy, which had been adopted and somewhat translated from the ancient book of Hindu, in titles Suruyasuddhant, Varoha-mihirat, Kaju-multant, etc – I have afterwards only tested certain european books of astronomy and astronomical navigation, geometry, so I have understood better manner of calculation in use of logarithm, of secant, cosecant, sine, cosine, tangent, cotangent, logarithm of number etc. Became acquainted with various astronomical terms in Latin and english on certain way. But my knowledge of algebra, etc, is not sufficient for accurate calculation. I have compared with some knowledge of geographical observation and apprehended the place of the central eclipse, which we may able to stand and see the present solar total eclipse in duration as great as obtainable on land; but I observed the more duration than that which will be here, will be fallen at about the middle of the gulf of Siam towards east most inclined towards southeast where there is no land, to be standing steadily and see, while duration of the total darkness may be more than here about only 2 or 3 seconds of minute. But to point directly the place of most durable point on land, my knowledge is not sufficient. I do not understand of using various instruments which were not in my possession. I have only a few telescopes large and small: they in comparison with those newly invented and improved must be considered as very common. The knowledge of Siamese and Peguan astronomers are thus.

1. The node, either ascending or descending, comes near to the sun's apogy, either before or behind, at a distance at least 60 degrees on the ecliptic course.
2. The moon's peregy comes near to the place of syzygy, or strait between the centres of the sun and moon and the earth, or near to said nodal position.
3. The sun comes as near to the said nodal position as 720 miles, or 12 degrees in ecliptic.

They said also that whenever the interior planets, Mercury or Venus, and moon, come below the sun directly at any latitude or place on the surface of the earth, the extractive powers of those said bodies attracts the surface of the earth and produces greater wind and clouds more than usual on that point for several days. It is evident that the like or similar occurence can be comparated with attraction of the sun and the moon, and produce high water on the surface of the earth in the days of the new moon and full moon always, and lowest water appears in first and last quarter of the lunar month, for the sun and the moon are in a very different position toward the earth.

But the wider knowledge was not to me more than is before said indea. S.P.P.M Mong Kut, K.S.

From E. Stephan, *Rapport sur l'observation de l'éclipse de Soleil du Août 1868*, Paris: Imprimerie Impériale, 1869, pp. 15–16.

33 Shapin, *A Social History of Truth*, pp. 245 and 36.

34 Thongchai, *Siam Mapped*, pp. 129–30.

35 D. Engel, 'Litigation across space and time: courts conflict and social change', *Law and Society Review* 24, 2, 1990: 333–44, pp. 338–9.

36 Thongchai, *Siam Mapped*, p. 128.

37 Cook, 'A tale of two city pillars', p. 299.

38 F. Bacon, *Novum Organum*, vol. III of *The Works of Francis Bacon*, eds J. Spedding, R.L. Ellis and D.D. Heath, New York, 1869, pp. 235, 337.

39 T. Kuhn, *Structure of Scientific Revolutions*, Chicago: University of Chicago Press, 1970, p. 16.

40 B. Barnes, *About Science*, Oxford: Blackwell, 1985, p. 61.
41 E.B. Mpemba and D.G. Osborne, 'Cool?', *Physics Education* 4, 1969: 172–5, p. 173.
42 Ibid.
43 'Mpemba's ice cream froze quicker when hot', *New Scientist* 42, 1969: 515.
44 F.L.C. Blackwell, *New Scientist* 43, 1969: 89.
45 G.S. Kell, 'The freezing of cold and hot water', *American Journal of Physics* 37, 1969: 564–5.
46 *New Scientist* 43, 1969: 662.
47 *New Scientist* 44, 1969: 205.
48 M.B.F. Ranken, *New Scientist* 45, 1970: 25.
49 I. Firth, 'Cooler?', *Physics Education* 6, 1971: 32–40. See subsequent articles in the same journal: E. Deesu, 'Cooler lower down', *Physics Education* 6, 1971: 42; M. Freeman, 'Cooler still', *Physics Education* 14, 1979: 417; D.G. Osborne, 'Mind on ice', *Physics Education* 14, 1979: 414; and also J. Walker, 'Hot water freezes faster than cold water: why does it do so?', *Scientific American*, September 1977: 246–54.
50 Firth, 'Cooler?', p. 33.
51 Ibid., p. 40.
52 D. Auerbach, 'Supercooling and the Mpemba effect: when hot water freezes quicker than cold', *American Journal of Physics* 63, 1995: 882–5, p. 882; M. Chown, 'Supercool theory solves hot ice cream puzzle', *New Scientist* 148, 1995: 22.
53 C. Knight, 'The Mpemba effect: the freezing times of hot and cold water', *American Journal of Physics* 64; 1996, 524 N.E. Dorsey, *The Properties of Ordinary Water Substance in all its Phases: Water Vapour and all the Ices*, New York: Reinhold, 1940.
54 *The Works of Aristotle*, ed. W.D. Ross, Oxford, 1931: *Meteorologica* Bk. I.12, 348b.
55 C. Adam and P. Tannery (eds) *Oeuvres de Descartes: Vol. 6*, Paris: Librairie Philosophique J. Vrin, 1982, p. 238 (my translation).
56 A. Pickering, *The Mangle of Practice: Time, Agency, and Science*, Chicago: University of Chicago Press, 1995.
57 B. Szerszynski and B. Wynne (eds) *Risk, Environment and Modernity: Towards a New Ecology*, London: Sage, 1996, pp. 44–83.
58 D. Turnbull, 'Mapping encounters and (en)countering maps: a critical examination of cartographic resistance', in Shirey Gorenstein (ed.) *Research in Science and Technology Studies: Knowledge Systems. Knowledge and Society*, Stanford, Conn.: JAI Press, 1998, pp. 15–44; idem, 'Cook and Tupaia, a tale of cartographic méconnaissance?', in Margaret Lincoln (ed.) *Science and Exploration: European Voyages to the Southern Oceans in the Eighteenth Century*, London: Boydell & Brewer, 1998: idem, *Masons, Tricksters and Cartographers: Comparative Studies in the Sociology of Science and Indigenous Knowledge*, Reading: Harwood Academic Publishers, 2000.
59 From D. Turnbull, 'Maps and map-making of the Australian Aboriginal people', in H. Selin (ed.) *Encyclopedia of the History of Science, Technology and Medicine in Non-western Cultures*, Dordrecht: Kluwer Academic Publishers, 1997, pp. 560–2.
60 This section is based on the work of Helen Watson in D. Turnbull, *Maps Are Territories; Science is an Atlas*, Chicago: University of Chicago Press, 1993, pp. 28–36, and H. Watson (with the Yolgnu community at Yirrkala) and D.W. Chambers, *Singing The Land, Signing The Land*, Geelong: Deakin University Press, 1989, pp. 36–7.
61 H. Morphy, *Ancestral Connections: Art and an Aboriginal System of Knowledge*, Chicago: University of Chicago Press, 1991, pp. 187–8; I. Keen, 'Metaphor and the metalanguage: "Groups" in northeast Arnhemland', *American Ethnologist* 18 (1994); N.M. Williams, *The Yolgnu and Their Land: A System of Land Tenure and the Fight for Its Recognition*, Canberra: Australian Institute of Aboriginal Studies, 1986, p. 44; L. Taylor, *Seeing The Inside: Bark Painting in Western Arnhem Land*, Oxford: Clarendon Press, 1996.
62 Keen, 'Metaphor', p. 2.
63 P. Sutton, 'Icons of country: topographic representations in classical aboriginal traditions', in D. Woodward and G.M. Lewis (eds) *The History of Cartography, Vol. 2, Book 3:*

Cartography in the Traditional African, American, Arctic, Australian, and Pacific Societies, Chicago: University of Chicago Press, 1998, pp. 353–86, pp. 362–4.

64 S.L. Davis and J.R.V. Prescott, *Aboriginal Frontiers and Boundaries in Australia*, Melbourne: Melbourne University Press, 1992. The map is *Australia's Extant and Imputed Aboriginal Territories*, Melbourne: Melbourne University Press, 1994. For critique see P. Sutton (ed.) *Country: Aboriginal Boundaries and Land Ownership in Australia. Aboriginal History Monograph 3*, Canberra: Aboriginal History Inc., 1995.

65 H. Fink and H. Perkins, 'Writing for land', *Art and Australia* 35, 1, 1997: 60–3. Artists' statement April 1997, p. 63.

66 E. Wiesmann and D. Jagger, 'Desert hearts on canvas', *Australian Weekend Review*, 1997: 12.

67 Interview, John Kean, exhibition producer Victoria Museum, 29 July 1997.

68 See discussion in Turnbull, 'Mapping encounters'.

69 Galarrawuy Yunupingu, 'Indigenous art in the Olympic age', *Art and Australia* 35, 1, 1997: 65–7.

70 For discussion of the ways narratives are incomplete and incorporate conflict, see C. Briggs (ed.) *Disorderly Discourse: Narrative, Conflict and Inequality*, New York: Oxford University Press, 1996.

71 J. Comaroff and J. Comaroff, *Of Revelation and Revolution*, Chicago: University of Chicago Press, 1991, p. 23

72 P. Ewick and S. Silbey, 'Subversive stories and hegemonic tales: towards a sociology of narrative', *Law and Society Review* 29, 2, 1995: 197–226, p. 200.

73 R.H. Brown, *Toward a Democratic Science: Scientific Narration and Civic Communication*, New Haven, Conn.: Yale University Press, 1998, p. 65.

74 M. Nanda, 'The science question in postcolonial feminism', in P. Gross, N. Levitt and M. Lewis (eds) *The Flight from Science and Reason*, New York: New York Academy of Sciences, 1997, pp. 420–36; idem, 'The epistemic charity of the constructivist critics of science and why the Third World should refuse the offer', in N. Koertge (ed.) *A House Built on Sand: Exposing Postmodernist Myths about Science*, New York: Oxford University Press, 1998.

75 Brown, *Toward a Democratic Science.*

Index